Oxidative Stress and Exercise

Oxidative Stress and Exercise

Editors

Gareth Davison
Conor McClean

MDPI • Basel • Beijing • Wuhan • Barcelona • Belgrade • Manchester • Tokyo • Cluj • Tianjin

Editors
Gareth Davison
Faculty of Life and Health
Sciences, Ulster University
UK

Conor McClean
Faculty of Life and Health
Sciences, Ulster University
UK

Editorial Office
MDPI
St. Alban-Anlage 66
4052 Basel, Switzerland

This is a reprint of articles from the Special Issue published online in the open access journal *Antioxidants* (ISSN 2076-3921) (available at: https://www.mdpi.com/journal/antioxidants/special_issues/Oxidative_Stress_Exercise).

For citation purposes, cite each article independently as indicated on the article page online and as indicated below:

LastName, A.A.; LastName, B.B.; LastName, C.C. Article Title. *Journal Name* **Year**, *Volume Number*, Page Range.

ISBN 978-3-0365-4201-0 (Hbk)
ISBN 978-3-0365-4202-7 (PDF)

Contents

About the Editors

Gareth Davison

Gareth Davison is a Professor of Exercise Biochemistry and Physiology and Director of Research at the Sport and Exercise Sciences Research Institute at Ulster University in the UK. He holds a BA, MSc, and MSt in Genomic Medicine from the University of Cambridge and was awarded his PhD in Biochemistry and Physiology in 2002. Professor Davison is a Fellow of the American College of Sports Medicine and currently serves on several editorial boards, holding Editor roles with the *Journal of Sports Sciences*, *Frontiers in Physiology* (Redox Physiology Section), and *Antioxidants*. His research interests are aligned to exercise, DNA damage, and antioxidant function. Recently, his laboratory has focused on bridging the gap between intracellular redox metabolism and DNA methylation in health and disease.

Conor McClean

Conor McClean is currently a Lecturer in Exercise Physiology at Ulster University. He completed his undergraduate degree in Biomedical Science at Queen's University Belfast prior to completing his PhD in Exercise Physiology at Ulster University. McClean's research involves the role of acute and chronic exercise interventions in conditions of metabolic disturbance and the corresponding relationships with peripheral vascular function and indices of oxidative stress. His current PhD students are investigating the role of exercise and oxidative stress in circadian biology.

 *antioxidants*

MDPI

Editorial

Oxidative Stress and Exercise

Gareth W. Davison * and Conor McClean

Sport and Exercise Sciences Research Institute, Ulster University, Newtownabbey BT37 0QB, UK;
cm.mcclean@ulster.ac.uk
* Correspondence: gw.davison@ulster.ac.uk

Citation: Davison, G.W.; McClean, C. Oxidative Stress and Exercise. *Antioxidants* **2022**, *11*, 840. https://doi.org/10.3390/antiox11050840

Received: 15 April 2022
Accepted: 21 April 2022
Published: 26 April 2022

Publisher's Note: MDPI stays neutral with regard to jurisdictional claims in published maps and institutional affiliations.

It is now well-established that regular moderate-intensity exercise training can activate salient cell adaptive properties, leading to a state of oxidative eustress. At the other end of the continuum, both sporadic and high-intensity bouts of exercise can induce an intracellular state of oxidative stress, due primarily to an augmented production of reactive metabolites of oxygen such as the superoxide anion and hydroxyl radical species. Exercise-induced free-radical formation impairs cell function by oxidatively modifying nucleic acids, where DNA damage and insufficient repair may lead to genomic instability. Likewise, lipid and protein damage are significant cellular events that can elicit potentially toxic perturbations in cellular homeostasis.

This Special Issue on *Oxidative Stress and Exercise* has attracted a broad range of 10 articles (three narrative reviews and seven original research manuscripts) with over 11,000 views to date. The original research examined oxidative stress and exercise in the context of ageing, probiotics, metabolic acidosis, antioxidant enzyme capacity, obesity, apoptosis, and genotyping, respectively. Considering ageing, Guerrero and colleagues [1] elucidated the impact of ultramarathon exercise (107.4 km) on oxidative stress (MDA, protein carbonyls) with a particular emphasis on outcomes related to sex and age. Their intriguing data showed that sex and ageing per se are important variables for consideration in ultra-endurance exercise-induced macromolecular damage, with more senior runners (45–53 years) having a much higher level of oxidative stress compared to their junior counterparts (31–37 years). The authors suggest that the difference is perhaps due to a higher percentage of body fat and increased loss of muscle mass in the older athletes. The sex differences observed in this investigation are also notable but require comparative data in similar ultra-endurance events.

The antioxidant capacity of erythrocytes is known to protect against severe oxidative stress, however, despite improved haemodynamic efficiency, the effect of eccentric exercise on erythrocyte antioxidant capacity remains unclear. Using a sedentary human model, Huang et al. [2] utilised cycling exercise to ascertain the difference between concentric versus eccentric exercise on a plethora of biochemical molecules, including O_2 release capacity, glutathione, glutathione disulfide, glucose and lactate production. Both exercise regimes were shown to increase the metabolic state of glycolysis through glucose and lactate, leading to intracellular acidosis and enhanced O_2 release from erythrocytes. Moreover, concentric and eccentric exercise increased antioxidant capacity, protecting against severe exercise-evoked circulatory oxidative stress. In keeping with the endogenous antioxidant capacity theme, Martinez-Noguera et al. [3] quantified both oxidised (GSSG) and reduced (GSH) glutathione alongside the well-known enzymatic antioxidant's catalase and superoxide dismutase in different categories of cyclists, concluding that professional cyclists have a higher catalase and oxidised glutathione capacity, respectively, compared to amateur cyclists. They also provided tentative evidence to suggest that antioxidant status (based on their GSSG/GSH data) may be related to power output.

Efficient electron-donor antioxidants are known to minimise exercise-induced oxidative damage to susceptible macromolecules. In a novel study, Sánchez-Macarro et al., used a probiotic product (*Bifidobacterium longum, Lactobacillus casei, Lactobacillus rhamnosus*) as an antioxidant activator to examine high-intensity exercise [4]. Using an impressive array of biomarkers (urinary isoprostane, malondialdehyde, oxidized low-density lipoprotein, urinary 8-hydroxy-2′-deoxiguanosine, protein carbonyls, glutathione peroxidase, superoxide dismutase), the authors determined that the consumption of the three-strain probiotic product for 6 weeks in male cyclists undergoing high-intensity exercise was associated with a reduction in lipid-related oxidative stress biomarkers only, with no change in antioxidant enzyme capacity. They conclude that their findings suggest an antioxidant effect of the probiotic product on underlying oxidative stress mechanisms and their modulation in healthy subjects [4]. Betaine is a natural compound, commercially obtained from sugar beet, and is known to protect mitochondrial and wider cell function. To this end, Yang et al. hypothesized that 2 weeks of betaine supplementation can attenuate apoptosis and exercise-induced oxidative stress [5]. Their study demonstrated that placebo ingestion increases lymphocyte apoptosis immediately following, and 3 h post exhaustive exercise, however, betaine supplementation exhibited no change in either apoptosis or oxidative stress.

Two of the published studies opted to use an animal model to probe the relationship between oxidative stress and exercise in a clinical context. Obesity is known to induce insulin resistance, hyperinsulinemia and trigger oxidative stress. In obese male Wister rats, Gómez-Barroso and colleagues [6] tested the effects of diazoxide (*a K_{ATP} channel activator*) and exercise training, both independently and combined, on muscle contraction and subsequent oxidative stress (ROS were determined by oxidation of the $2', 7'$-dichlorodihydrofluorescein diacetate fluorescent probe). Outcomes showed that exercise and diazoxide both reduce ROS production and lipid peroxidation and improve glutathione redox state. Diazoxide treatment and exercise, independently and combined, also leads to a greater increase in maximum and total tension in the soleus muscle of obese rats, suggesting that the K_{ATP} channels may play a prominent role in muscle function in obesity. Chaudhari et al. [7] determined whether dietary vitamin E and C are viable alternatives to exercise in improving cognitive and motor performance in a mouse model of Alzheimer's disease risk. Mice expressing human Apolipoprotein E3 (GFAP-ApoE3) or E4 (GFAP-ApoE4) fed an 8-week control or antioxidant-rich diet were subjected to various tests to quantify reflex and motor, cognitive and affective function. The antioxidant diet (and exercise) improved balance, learning, and cognitive flexibility in the GFAP-ApoE3 only group, with the authors concluding that genotyping needs to be considered in interventions designed to improve brain function during ageing, and particularly in neurodegenerative disease.

Three narrative reviews were published in this *Special Issue*. In their review, Zeng et al. [8] summarised the current evidence available for the use of whole dietary strategies to reduce exercise-induced oxidative stress in humans. The review included an overview of 28 studies, with the majority outlining the importance of dietary antioxidant consumption as a scavenging technique to curb exercise-related oxidative stress. Yet, the investigative protocols are heterogeneous in nature, and further enquiry is needed to strengthen the evidence base. Other dietary factors, such as excessive consumption of nutrients, are tightly linked to obesity, where oxidative stress plays a prominent role. For example, nutrients such as glycerol and fatty acids can accumulate in white adipose tissue, while subsequently secreted molecules such as adipocytokines (IL-6, TNF-α) and reactive species can lead to inflammation and oxidative stress. Taherkhani and colleagues propose in their review [9] that antioxidant supplementation can therapeutically help to neutralise oxidative stress by removing reactive oxygen metabolites in animal adipose tissue. The authors acknowledge that data in this domain are currently inconsistent due to differences in study design relating to experimental duration and diversity (strain, age sex) in animal models used [9]. As outlined by McClean and Davison [10], compelling research now documents how the circadian system is essential for the maintenance of homeostasis and glucose metabolism, and disruptions to circadian rhythms may well be instrumental in the development of obesity and other conditions. Although a direct relationship has been established between circadian rhythms and oxidative stress, it appears that the emerging interface between circadian rhythmicity and oxidative stress has not been explored in relation to exercise. In their novel review [10], a summary of the evidence supporting the conceptual link between the circadian clock, oxidative stress/redox homeostasis, and exercise stimuli is offered, while the latter section of the review outlines that carefully designed investigations of this nexus are required to further progress this important area of study.

The guest editors would like to acknowledge all author contributions to the Special Issue *Oxidative Stress in Exercise*. While we commend this Special Issue to the readers of *Antioxidants*, we trust that the work contained within will be viewed as a starting point for further exploration into the complex relationship between oxidative stress and exercise.

Funding: This research received no external funding.

Conflicts of Interest: The authors declare no conflict of interest.

References

1. Guerrero, C.; Collado-Boira, E.; Martinez-Navarro, I.; Hernando, B.; Hernando, C.; Balino, P.; Muriach, M. Impact of Plasma Oxidative Stress Markers on Post-race Recovery in Ultramarathon Runners: A Sex and Age Perspective Overview. *Antioxidants* **2021**, *10*, 355. [CrossRef] [PubMed]
2. Huang, Y.-C.; Cheng, M.-L.; Tang, H.-Y.; Huang, C.-Y.; Chen, K.-M.; Wang, J.-S. Eccentric Cycling Training Improves Erythrocyte Antioxidant and Oxygen Releasing Capacity Associated with Enhanced Anaerobic Glycolysis and Intracellular Acidosis. *Antioxidants* **2021**, *10*, 285. [CrossRef] [PubMed]
3. Martínez-Noguera, F.J.; Alcaraz, P.E.; Ortolano-Ríos, R.; Dufour, S.P.; Marín-Pagán, C. Differences between Professional and Amateur Cyclists in Endogenous Antioxidant System Profile. *Antioxidants* **2021**, *10*, 282. [CrossRef] [PubMed]
4. Sánchez Macarro, M.; Ávila-Gandía, V.; Pérez-Piñero, S.; Cánovas, F.; García-Muñoz, A.M.; Abellán-Ruiz, M.S.; Victoria-Montesinos, D.; Luque-Rubia, A.J.; Climent, E.; Genovés, S.; et al. Antioxidant Effect of a Probiotic Product on a Model of Oxidative Stress Induced by High-Intensity and Duration Physical Exercise. *Antioxidants* **2021**, *10*, 323. [CrossRef] [PubMed]
5. Yang, M.-T.; Lee, X.-X.; Huang, B.-H.; Chien, L.-H.; Wang, C.-C.; Chan, K.-H. Effects of Two-Week Betaine Supplementation on Apoptosis, Oxidative Stress, and Aerobic Capacity after Exhaustive Endurance Exercise. *Antioxidants* **2020**, *9*, 1189. [CrossRef] [PubMed]
6. Gómez-Barroso, M.; Moreno-Calderón, K.M.; Sánchez-Duarte, E.; Cortés-Rojo, C.; Saavedra-Molina, A.; Rodríguez-Orozco, A.R.; Montoya-Pérez, R. Diazoxide and Exercise Enhance Muscle Contraction during Obesity by Decreasing ROS Levels, Lipid Peroxidation, and Improving Glutathione Redox Status. *Antioxidants* **2020**, *9*, 1232. [CrossRef] [PubMed]
7. Chaudhari, K.; Wong, J.M.; Vann, P.H.; Como, T.; O'Bryant, S.E.; Sumien, N. ApoE Genotype-Dependent Response to Antioxidant and Exercise Interventions on Brain Function. *Antioxidants* **2020**, *9*, 553. [CrossRef] [PubMed]
8. Zeng, Z.; Centner, C.; Gollhofer, A.; König, D. Effects of Dietary Strategies on Exercise-Induced Oxidative Stress: A Narrative Review of Human Studies. *Antioxidants* **2021**, *10*, 542. [CrossRef] [PubMed]
9. Taherkhani, S.; Suzuki, K.; Ruhee, R.T. A Brief Overview of Oxidative Stress in Adipose Tissue with a Therapeutic Approach to Taking Antioxidant Supplements. *Antioxidants* **2021**, *10*, 594. [CrossRef] [PubMed]
10. McClean, C.; Davison, G.W. Circadian Clocks, Redox Homeostasis, and Exercise: Time to Connect the Dots? *Antioxidants* **2022**, *11*, 256. [CrossRef] [PubMed]

 antioxidants

Article

Impact of Plasma Oxidative Stress Markers on Post-race Recovery in Ultramarathon Runners: A Sex and Age Perspective Overview

Carlos Guerrero [1,†], Eladio Collado-Boira [2,†], Ignacio Martinez-Navarro [3,4], Barbara Hernando [1], Carlos Hernando [5,6], Pablo Balino [1] and María Muriach [1,*]

[1] Department of Medicine, Jaume I University, 12001 Castellon, Spain; cguerrer@uji.es (C.G.); hernandb@uji.es (B.H.); balino@uji.es (P.B.)
[2] Faculty of Health Sciences, Jaume I University, 12001 Castellon, Spain; colladoe@uji.es
[3] Department of Physical Education and Sport, University of Valencia, 46010 Valencia, Spain; ignacio.martinez-navarro@uv.es
[4] Sports Health Unit, Vithas-Nisa 9 de Octubre Hospital, 46001 Valencia, Spain
[5] Sport Service, Jaume I University, 12001 Castellon, Spain; hernando@uji.es
[6] Department of Education and Specific Didactics, Jaume I University, 12001 Castellon, Spain
* Correspondence: muriach@uji.es
† Authors contributed equally to this work.

Citation: Guerrero, C.; Collado-Boira, E.; Martinez-Navarro, I.; Hernando, B.; Hernando, C.; Balino, P.; Muriach, M. Impact of Plasma Oxidative Stress Markers on Post-race Recovery in Ultramarathon Runners: A Sex and Age Perspective Overview. *Antioxidants* **2021**, *10*, 355. https://doi.org/10.3390/antiox10030355

Academic Editors: Gareth Davison and Conor McClean

Received: 21 January 2021
Accepted: 23 February 2021
Published: 27 February 2021

Publisher's Note: MDPI stays neutral with regard to jurisdictional claims in published maps and institutional affiliations.

Abstract: Oxidative stress has been widely studied in association to ultra-endurance sports. Although it is clearly demonstrated the increase in reactive oxygen species and free radicals after these extreme endurance exercises, the effects on the antioxidant defenses and the oxidative damage to macromolecules, remain to be fully clarified. Therefore, the aim of this study was to elucidate the impact of an ultramarathon race on the plasma markers of oxidative stress of 32 runners and their post-race recovery, with especial focused on sex and age effect. For this purpose, the antioxidant enzymes glutathione peroxidase (GPx) and glutathione reductase (GR) activity, as well as the lipid peroxidation product malondialdehyde (MDA) and the carbonyl groups (CG) content were measured before the race, in the finish line and 24 and 48 h after the race. We have reported an increase of the oxidative damage to lipids and proteins (MDA and CG) after the race and 48 h later. Moreover, there was an increase of the GR activity after the race. No changes were observed in runners' plasma GPx activity throughout the study. Finally, we have observed sex and age differences regarding damage to macromolecules, but no differences were found regarding the antioxidant enzymes measured. Our results suggest that several basal plasma markers of oxidative stress might be related to the extent of muscle damage after an ultraendurance race and also might affect the muscle strength evolution.

Keywords: ultraendurance exercise; oxidative stress; antioxidants; muscle injury

1. Introduction

Ultramarathon races are defined as sport events that involve running and/or walking distances greater than the 42,195 km of a marathon. In the recent years, these short of competitive events have gained a lot of popularity. These extremely long races defiance our physiological systems inducing muscle injuries (muscle membrane disruption), respiratory fatigue, cardiac and renal damage, representing an outstanding model to evaluate the ultra-endurance exercises/sports on human body physiology [1–4].

Oxidative stress is defined as the imbalance between the body oxidants and antioxidants in favor of the former [5] and has been widely studied in association to ultra-endurance sports. However, although it is clearly demonstrated the increase in reactive oxygen species (ROS) and free radicals, the effects of these extreme endurance exercises on the cellular antioxidant defense system, and the oxidative damage to macromolecules, remain to be fully clarified [6]. In this regard, an increase in different plasmatic antioxidant compounds such as glutathione (GSH), thioredoxin or paraoxonse, as well as in the total antioxidant capacity [7–10], supports the hypothesis of a compensatory mechanism to counteract the increased oxidative stress elicited by the intense exercise in these runners. Conversely, other researchers also showed contradictory and inconclusive results regarding the effects of intense activity on the antioxidant system. At this respect, several studies showed that the antioxidant system remained unchanged or even a decrease in several antioxidant molecules such as glutathione, catalase, or superoxide dismutase [11–14].

In this line of evidence, similar inconclusive results have been found regarding ultra-endurance activity effects on oxidative damage to macromolecules such as proteins, deoxyribonucleic acid (DNA) or lipids [15,16]. It has been demonstrated that the optimal muscle contractile function depends on the cellular redox state. However, the effects of ROS as well as several antioxidant compounds on contractile function during fatigue and recovery are still being debated [17,18].

Given this, the purpose of this study was to elucidate the impact of ultratrail endurance exercise on several plasma oxidative stress markers with especial focus on the post-race recovery. The objectives of this study are:

i.   To determine four plasma oxidative stress markers of long-distance amateur runners throughout the study. Two of them are macromolecule oxidative damage markers (MDA and CG) and the other two are antioxidant enzymes (GR and GPx).
ii.   To evaluate the influence of physical variables (sex and age) on these plasma oxidative stress markers after the race, 24 and 48 h afterwards.
iii. To study the possible correlation between baseline values of these plasma oxidative stress markers and the post-race degree of systemic inflammatory processes, loss of skeletal muscle strength and muscle membrane disruption.

2. Materials and Methods

2.1. Participants

Forty-seven recreational ultra-endurance athletes (29 males and 18 females) were recruited to participate in the study that was developed at the Penyagolosa Trails CSP race in 2019. The track consisted of 107.4 km, starting at an altitude of 40 m and finishing at 1280 m above the sea level, with a total positive and negative elevation of 5604 and 4356 m respectively. All volunteers were fully informed of the procedure and gave their written consent to participate. They were also allowed to withdraw from the study at will. A questionnaire was used to collect demographic information, the history of training and competition and the consumption of antioxidant supplements in the preparation of the race and in the development of the same. The investigation was conducted according to the Declaration of Helsinki and approval for the project was obtained from the research Ethics Committee of the University Jaume I of Castellon (Expedient Number CD/007/2019). This study is enrolled in the ClinicalTrails.gov database, with the code number NCT03990259.

In order to have homogeneous groups, the participants were grouped by age as follows; under 38, between 38 and 45, and over 45 years.

2.2. Blood Sampling and Analysis

Blood samples were collected from an antecubital vein by venipuncture at the time of race number collection which was 8 to 6 h before the start, after crossing the finishing line, 24 and 48 h post-race using BD Vacutainer PST II tubes. Samples were centrifuged at 3500 rpm for ten minutes and kept at 4 °C during transport to Vithas Rey Don Jaime Hospital (Castellon), where they were processed using the modular platform Roche/Hitachi clinical chemistry analyzer Cobas c311 (Roche Diagnostics, Penzberg, Germany), as previously published [1,19]. Lactate dehydrogenase (LDH) and creatin kinase (CK) were used to assess muscle membrane disruption, as a surrogate for muscle damage. C-reactive protein (CRP) as an indicator of acute inflammatory reaction [3] (supplementary Table S1).

The oxidative stress biomarkers used in the present investigation were GR, GPx, MDA and CG, which were analyzed as follows:

GPx activity, which catalyzes the oxidation by H_2O_2 of glutathione (GSH)to its disulfide (GSSG), was assayed spectrophotometrically as reported by Lawrence et al. [20] toward hydrogen peroxide, by monitoring the oxidation of nicotinamide adenine dinucleotide phosphate (NADPH) at 340 nm. The reaction mixture consisted of 240 mU/mL of GSH disulfide reductase, 1 mM GSH, 0.15 mM (NADPH) in 0.1 M potassium phosphate buffer, pH 7.0, containing 1 mM ethylethylenediaminetetraacetic acid (EDTA) and 1m sodiumazide; a 50 μL sample was added to this mixture and allowed to equilibrate at 37 °C for 3 min. Reaction was started by the addition of hydrogen peroxide to adjust the final volume of the assay mixture to 1 mL.

GR activity was determined spectrophotometrically using Smith proposed method [21]. Briefly, when the GR catalyzed reduction of GSSG to GSH is produced in presence of 5,5′-dithiobis (2-nitrobenzoic acid) (DTNB), 2-nitrobenzoic acid is formed as a subproduct, which formation is monitored at 412 nm. The GSSG reduction was started by adding 25 μL of brain sample to a solution containing DTNB 3 mM prepared in 10 mM phosphate buffer, 2 mM NADPH, 10 mM MEDTA in 0.2 M pH 7.5 phosphate buffer.

MDA concentration was measured by liquid chromatography according to a modification of the method of Richard and coworkers [22], as previously reported [23]. Briefly, 0.1 mL of sample (or standard solutions prepared daily from 1,1,3,3-tetramethoxypropane) and 0.75 mL of working solution (thiobarbituric acid 0.37% and perchloric acid 6.4%; 2:1, *v/v*) were mixed and heated to 95 °C for 1 h. After cooling (10 min in ice water bath), the flocculent precipitate was removed by centrifugation at $3200\times g$ for 10 min. The supernatant was neutralized and filtered (0.22 μm) prior to injection on an ODS 5 μm column (250 × 4.6 mm). Mobile phase consisted in 50 mM phosphate buffer (pH 6.0): methanol (58:42, *v/v*). Isocratic separation was performed with 1.0 mL/min flow and detection at 532 nm.

CG were determined to evaluate protein oxidation in milk samples. The CGs released during incubation with 2,4-dinitrophenylhydrazine were measured using the method reported by Levine et al. (1990) [24] with some modifications introduced by Tiana et al. (1998) [25]. Briefly, the samples were centrifuged at $13,000\times g$ for 10 min. Then, 20 mL of brain homogenate was placed in a 1.5 mL Eppendorf tube, and 400 mL of 10 mM 2,4 dinitrophenylhydrazine/2.5 M hydrochloric acid (HCl) and 400 mL of 2.5 M HCl were added. This mixture was incubated for 1 h at room temperature. Protein precipitation was performed using 1 mL of 100% of TCA, washed twice with ethanol/ethyl acetate (1/1, *v/v*) and centrifuged at $12,600\times g$ for 3 min. Finally, 1.5 mL of 6 N guanidine, pH 2.3, was added, and the samples were incubated in a 37 °C water bath for 30 min and were centrifuged at $12,600\times g$ for 3 min. The carbonyl content was calculated from peak absorption (373 nm) using an absorption coefficient of $22,000\ M^{-1}cm^{-1}$ and was expressed as nmol/mg protein.

Biochemical results obtained immediately post-race were adjusted by employing the Dill and Costill method [26], using hematocrit and hemoglobin to determine the magnitude of plasma volume changes after the race in each participant.

2.3. Muscle Strength. Squat Jump (SJ) and Handgrip (HG) Strength Assessment

The SJ is a validated research test based on three parameters (body mass, jump height and push distance), which allows to accurately assess the strength, speed and power developed by the extensor muscles of the lower extremities during squat jumps [27].

Grip strength, short-term maximal voluntary force of the forearm muscles, measured by dynamometry, is well established as an indicator of muscle status [28]. Grip strength provides a direct measure of the hand skeletal muscle strength. It has been described as an strength index, endurance and general muscular status because its association between peripheral strength and exercise capacity [29].

Previous studies have also suggested that the strength decline index (SDI), calculated as the decline in strength as a proportion of baseline values, measured through tests such as the HG and SJ, is a useful assessment of muscle fatigue [30].

Volunteers were familiarized with procedures concerning strength assessment during an informative session prior to the investigation. HG and SJ tests were performed before the race and within 15 min after the race. In the HG assessment, volunteers remained in standing position, arm by their side with full elbow extension, holding the grip dynamometer (T.K.K. 5401 GRIP-D, Takei Scientific Instruments Co., Tokyo, Japan) in their dominant hand. They were asked to squeeze the dynamometer for 5 s and the test was performed twice, with 30 s of rest in between attempts. Each individual's peak value was retained for statistical analysis. Following previous studies [31,32], pre to post-race change in HG, given that upper-limb muscles could be considered as being hardly no-exercising muscles during the race. In the SJ assessment, participants were asked to jump as high as possible from a starting position with hips and knees flexed 80 degrees and hands stabilized on hips to avoid arm-swing. Jump height was estimated by the flight time measured with a contact platform (Chronojump, Barcelona, Spain). The test was performed twice, with 90 s of rest in between attempts. Each individual's best performance was retained for statistical analysis [3] (Supplementary Table S1).

2.4. Basal Metabolic Rate (BMR), Body Mass Index (BMI), and Body Composition Assesment

Volunteers height and weight were measured before the start the day of the race. Participants were also subjected to a body composition evaluation test (Tanita BC-780MA, Tanita Corp., Tokyo, Japan). The BMR is defined as the daily rate of energy metabolism an individual needs to sustain in order to preserve the integrity of vital functions. The BMR formula (BMR = Kg × 1 Kcal/h) was calculated based on previous studies [33].

2.5. Statistical Analysis

Statistical analyses were carried out using the Statistical Package for the Social Sciences software (IBM SPSS Statistics for Windows, version 25.0, IBM Corp., Armonk, NY). Normal distribution of the variables was verified through the Shapiro-Wilk test ($p > 0.05$) [34].

A Pearson or Rho Spearman correlation analysis was used to assess whether the concentration of oxidative stress biomarkers (GR, GPX, MDA and CG) was interrelated or related to the loss of upper (HG) and lower limb (SJ) strength, hematologic variables of systemic inflammation (CRP) and muscle damage (CK, LDH), as well as cardiopulmonary exercise test results. Post-race and at 24 and 48 h values for these variables (GR, GPX, MDA, CG, HG, SJ, CRP, CK and LDH) for each participant were related to the individual pre-race level to define the delta scores (Δ): Δ (fold increase) = (post-race value − Pre-race value)/Pre-race value [3].

On the other hand, the quantitative variables of oxidative stress were compared using the Student method Tests or U Mann Whitney in each of the sectors where measurements were taken (pre-race, finish line, 24 and 48 h after the race) when they existed two categories and ANOVA test or Kruskall Wallis when there were more categories. Post-hoc comparisons were performed using Bonferroni adjustment for multiple comparisons.

The meaningfulness of the outcomes was estimated through the partial estimated effect size (η2 partial) for ANOVA and Cohen's d effect size for pair wise comparisons. In the latter case, a Cohen's d < 0.5 was considered small; between 0.5–0.8, moderate; and greater than 0.8, large [35]. Likewise, correlations > 0.5 were considered strong, 0.3–0.5, moderate and <0.3, small. The significance level was set at p-value < 0.05 and data are presented as means and standard error of the means (±SEM).

Finally, the multiple regression analysis was performed using the forward stepwise method. Only normally distributed variables were used as dependent variables. Among the different models obtained, the parsimony principle was applied [36]. Given our limited sample size and the non-normal distribution of independent variables, residual errors from the resulting models were inspected to ensure their normal distribution and thus the reliability of our regression models [37]. To identify the predictive value of the model, the Cohen criterion [38] was applied to one-way ANOVA models. This criterion indicates that R^2 values less than 0.10 do not present a relevant explanatory value; an R^2 between 0.10

and 0.25 indicates a dependency of the analyzed variables variance explanation for the identified factors; and R^2 values above 0.25 is possible to affirm that the explanatory model clinically relevant.

3. Results

3.1. Demographic Characteristics of The Participants

Thirty-two runners reached the finish line. Nineteen were male and thirteen females, with an average finish time of 21 h 21 min $\pm$ 3 h 28 min. All levels of performance were represented in our sample, as shown by their rank, ranging from 7th to 32nd. The main characteristics of these runners are described in Table 1, including sex and age differences. As expected, males showed significant higher pre-race values in weight, BMR, BMI and percentage of muscular mass when compared with female runners. No differences were found in training characteristics or experience between male and female runners. The runner's age did not affect any of the parameters measured except for the weekly running volume that was smaller in the senior runner group.

Table 1. Baseline characteristics of the runners which completed the race by sex and age (Average $\pm$ SE).

	Total (n = 32)	Males (n = 19)	Females (n = 13)	Young (n = 10)	Medium (n = 14)	Senior (n = 8)
Age (years)	40.9 $\pm$ 1.0	40.1 $\pm$ 1.2	42.2 $\pm$ 1.7	34.6 $\pm$ 0.6	41.4 $\pm$ 0.5	48 $\pm$ 1.1
Weight Pre-race (Kg)	66.1 $\pm$ 1.9	73.2 $\pm$ 1.5	55.7 $\pm$ 1.4 *	68.4 $\pm$ 2.9	63.1 $\pm$ 2.3	68.4 $\pm$ 5.1
BMR Pre-race (Kcal)	1619 $\pm$ 52	1835 $\pm$ 30	1302 $\pm$ 32 *	1708 $\pm$ 84	1563 $\pm$ 69	1604 $\pm$ 135
BMI Pre-race (kg/m^2)	22.9 $\pm$ 0.4	23.8 $\pm$ 0.4	21.7 $\pm$ 0.6 *	23.2 $\pm$ 0.4	22.25 $\pm$ 0.5	23.8 $\pm$ 1.1
% Body Fatty Pre-race (%)	15.9 $\pm$ 1.1	12.4 $\pm$ 0.8	20.9 $\pm$ 1.2 *	14.3 $\pm$ 1.8	14.9 $\pm$ 1.2	19.5 $\pm$ 2.7
% Muscular Mass Pre-race (%)	84.1 $\pm$ 1.1	87.6 $\pm$ 0.8	79.1 $\pm$ 1.3 *	85.7 $\pm$ 1.8	85.1 $\pm$ 1.2	80.5 $\pm$ 2.7
Number of years running	8.0 $\pm$ 0.5	8.0 $\pm$ 0.6	8.1 $\pm$ 0.9	7.7 $\pm$ 1.0	7.7 $\pm$ 0.9	9.0 $\pm$ 1.0
Number of races > 100 km	2.5 $\pm$ 3.3	3.0 $\pm$ 0.6	2.0 $\pm$ 1.1	2 $\pm$ 1.1	2.4 $\pm$ 1.0	3.1 $\pm$ 0.5
Weekly training days	4.8 $\pm$ 1.2	4.7 $\pm$ 0.3	4.8 $\pm$ 0.3	4.6 $\pm$ 0.3	5.2 $\pm$ 0.3	4.3 $\pm$ 1.0
Weekly running volume (km)	70 $\pm$ 22	71 $\pm$ 5.8	74 $\pm$ 3.7	79.3 $\pm$ 8.4	73.4 $\pm$ 5.1	53.3 $\pm$ 4.5 [#]
Weekly positive elevation (m)	1771 $\pm$ 691	1869 $\pm$ 175	1631 $\pm$ 157	1600 $\pm$ 175	2057 $\pm$ 154	1488 $\pm$ 318
Weekly training hours	9.6 $\pm$ 4.2	10 $\pm$ 0.9	9 $\pm$ 1.3	10 $\pm$ 1.2	9.9 $\pm$ 1.2	8.6 $\pm$ 1.6

Data partially published previously by our group (Martinez-Navarro et al., 2020) [3]. Abbreviations: BMR: Basal Metabolic Rate; BMI: Body Mass Index. * $p < 0.05$ vs. Males; [#] $p < 0.05$ vs. Young and Medium.

3.2. Analysis of Plasma Markers of Oxidative Stress

Descriptive data of oxidative stress biomarkers pre-race (baseline), finish line and after 24 and 48 h post-race are depicted in Table 2. Regarding the antioxidant defenses, no significant changes were observed in GPx activity. The GR activity was significantly enhanced in the finish line. The GR enzymatic activity reached the highest value 24 h post-race and returned to normal values after 48 h. Lipid peroxidation (MDA concentration) was also increased in the finish line, declined 24 h post-race, and was significantly increased after 48 h. Oxidative damage to proteins (CG content) also increased immediately after the race and remained elevated 48 h later.

Table 2. Changes in plasma markers of oxidative stress throughout the study period (Average $\pm$ SE).

	Baseline	Finish Line	24 H Post-Race	48 H Post-Race
GPx (μmol/L $\times$ min)	87 $\pm$ 7	97 $\pm$ 5	98 $\pm$ 6	96 $\pm$ 3
GR (UI/mL)	2.6 $\pm$ 0.1	3.3 $\pm$ 0.2 [#]	7.3 $\pm$ 0.5 *[#]	2.6 $\pm$ 0.1 *
MDA (μM)	0.8 $\pm$ 0.1	1.3 $\pm$ 0.1 [#]	0.9 $\pm$ 0.1 *	1.4 $\pm$ 0.1 *[#]
CG (nmol/mL)	1.4 $\pm$ 0.1	2.2 $\pm$ 0.2 [#]	1.9 $\pm$ 0.2 [#]	2.1 $\pm$ 0.3 [#]

* $p < 0.05$ vs. preceding time point; # $p < 0.05$ vs. baseline value.

3.3. Influence of Sex and Age in the Plasma Markers of Oxidative Stress Evolution of the Runners

Data depicted in Figure 1, demonstrate that female runners have a significantly higher CG content when compared to males at the end of the race and 48 h later. To the contrary, MDA concentration 48 h post-race is higher in male compared to female runners. The antioxidant defenses, measured as GR and GPx enzymatic activity, are not conditioned by runner's sex.

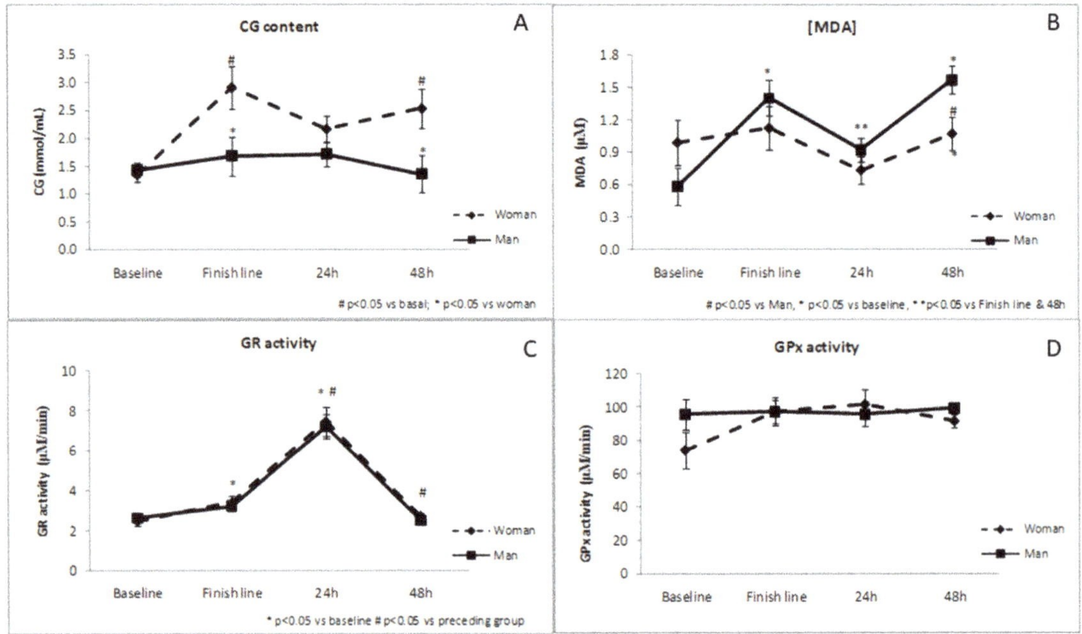

Figure 1. Changes in plasma markers of oxidative status throughout the study according to runners' gender. CG (carbonyl groups) content (**A**), MDA (malondialdehyde) (**B**), GR (glutathione reductase) activity (**C**) and GPx (glutathione peroxidase) activity (**D**).

Data from Figure 2 shows the effect of the runner's age. A tendency can be observed in the plasma oxidative status of the senior runners which appears to be globally worse compared to the in the younger runner's values. Briefly, less antioxidant defenses and higher levels of oxidative damage to lipids and proteins. However, only the lipid peroxidation (MDA concentration) was significantly affected by the age. Thus, the MDA concentration was significantly higher in the senior runners when compared to the middle age and young competitors at the finish line and 48 h post-race.

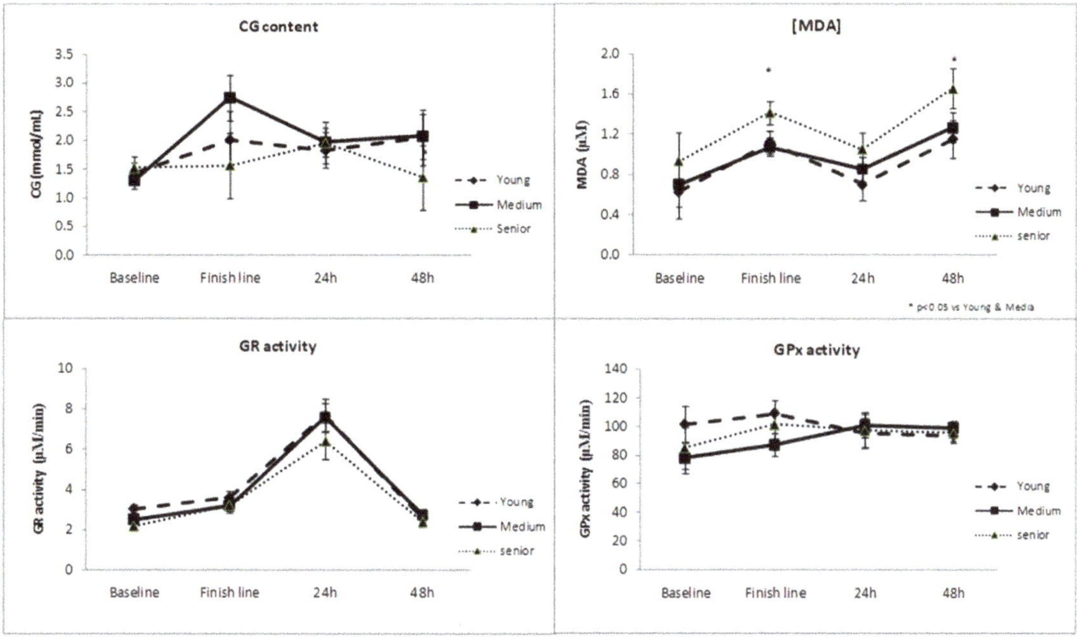

Figure 2. Changes in plasma markers of oxidative status throughout the study according to runners' gender. CG (carbonyl groups) content (**A**), MDA (malondialdehyde) (**B**), GR (glutathione reductase) activity (**C**) and GPx (glutathione peroxidase) activity (**D**).

3.4. Correlation between the Plasma Markers of Oxidative Stress and the Skeletal Muscle Force Production, Muscle Damage and Systemic Inflammatory Response

Regarding the post-race skeletal muscle strength, Table 3 shows that there was significant negative correlation between basal MDA concentration and the fold increase in HG values in the finish line. We also observed a significant positive correlation between basal GR activity and the fold increase in SJ values at the finish line.

Table 3. Significant correlations between baseline plasma markers of oxidative stress and Delta values of muscle strength (SJ and HG) and muscle damage (CK and LDH).

	R Value	*p* Value
GR (UI/mL)/Δ SJ Finish line	0.405	0.027
GR (UI/mL)/Δ CK Finish line	−0.411	0.019
GR (UI/mL)/Δ ck 24/GR (UI/mL)	−0.352	0.048
GR (UI/mL)/Δ LDH Finish line	−0.402	0.023
GR (UI/mL)/Δ LDH 24	−0.418	0.017
GR (UI/mL)/Δ LDH	−0.406	0.021
GPx (μmol/L × min)/Δ CK 48	−0.360	0.043
CG (nmol/mL)/Δ LDH 24	0.358	0.048
CG (nmol/mL)/Δ LDH 48	0.363	0.045
MDA (μM)/Δ HG Finish line	−0.379	0.032

In addition, basal GR activity negatively correlated with the delta values of LDH at the finish line, 24 h and 48 h post-race. Moreover, basal GR activity negatively correlated with the fold increase of CK at the finish line. It was also demonstrated a significant positive correlation between basal CG content (oxidative damage to proteins) and the rise in LDH observed 24 h and 48 h post-race (Table 3). The inflammatory response observed after the race did not correlate significantly with any of the plasma markers of oxidative stress parameters measured.

3.5. Multiple Regression Analysis

Results of the multiple regression analysis are listed in Table 4. When performing the multiple linear regression analysis using Age and GR as the predictive variables, a significant regression equation was obtained for the SJ delta Value dependent variable. This analysis would indicate that younger age as main predictor and higher basal GR concentration, a lower SJ delta value after finishing the ultramarathon was obtained. This regression model predicts the 28.3% of the variance. Another multiple linear regression model was obtained in which the dependent variable was the Δ CK finish line value and the main predictive variables were Age and GR. In this scenario, the regression analysis model predicts a 18.3% of the variance.

Table 4. Linear regression models.

Model	R^2 Adjusted	Standardized Coefficients Beta	Standard Error	F (*p*)
Dependent Variable: Δ SJ finish line Covariables: Age, GR.	0.283	0.217	0.16389	0.6521 (0.005)
Dependent Variable: Δ CK finish line Covariables: Age, GR.	0.183	-0.413	0.15558	3.431 (0.002)

Abbreviations: Δ SJ (Fold increase Squat Jump); Δ CK (Fold increase Creatine Kinase); GR (Glutathione reductase).

As we have previously mentioned, according to the criterion proposed by Cohen [38], our regression models might be considered for their predictive value, explaining within a multicausal model context the influence of oxidative stress on muscle damage and fatigue after a severe effort such as running an ultramarathon.

4. Discussion

The present investigation aimed to ascertain the relationship between several athlete's plasma markers of oxidative stress and the degree of muscle strength and damage after ultraendurance exercise. Results of this study were also extended to investigate the possible runner's sex and age influence. It is important to remark that this short of studies present limitations regarding the sample size due to the difficulty to finish the competition by the runners. The demographic/anthropometric characteristics of the present study (age, body composition and resistance training) as well as muscle damage, acute inflammation and muscle strength variables have been previously analyzed and discussed by our group [3] (See Supplementary Table S1).

Briefly, male participants showed significant higher pre-race values for weight, BMR, BMI and percentage of muscular mass when compared to women runners [31,32]. The muscular membrane disruption variables, LDH and CK release, peaked after the race and returned to normal values after 24 h. In the Supplementary material section, we also have included results about acute inflammatory processes (CRP), muscle damage (CK and LDH) and muscle strength (SJ and HG), previously published by our research group [4]. No changes were observed in the performance of SJ and HG before the race when compared with the finish line. Both acute inflammation and muscle damage were observed in the finish line, as well as 24 h and 48 h post race.

Regarding the plasma markers of oxidative stress, a time-course analysis of GPx and GR activity, and the oxidative damage to lipids and proteins (MDA and CG, respectively) was performed at the finish line, 24 h and 48 h post-race. It has been showed that ultra endurance exercise is associated with a notably enhanced rate of oxygen utilization and the generation and accumulation of ROS [6]. Moreover, the glutathione system increases its activity to restore the cell redox balance when the formation of ROS is enhanced. Our data showed an increase of the oxidative damage to macromolecules (lipids and proteins) indicating an increase of ROS cellular levels. Thus, lipid peroxidation damage appears to be increased in the finish line and 48 h post-race as confirmed by the MDA levels, proving the presence of oxidative damage to lipids two days after the extreme exercise. A partially recovering effect was observed 24 h after the race as can be seen in Table 2. At this respect, previous studies have shown controversial results regarding the blood MDA levels after ultraendurance exercise. Several studies showed an increase on the cellular MDA levels [14,16,39] compared to data supporting no lipid peroxidation effect after extreme endurance exercise [12,13]. In addition, Skenderi et al., [9] demonstrated an MDA levels decrease 48 h post-race when compared to control and post-race values. However, in this study, the sport type (running, swimming), the distance, the accumulated altitude or the anthropometric characteristics of the sample might be influencing these results. It is also noteworthy to mention that these differences can be attributed to methodological aspects. Authors used the TBARS (thiobarbituric acid-reacting substances) technique which, appear to be a less robust measure of lipid peroxidation [6].

The analysis of the CG content remained significantly elevated for all the time points of the study (finish line, 24 h post-race and 48 h post-race). Interestingly, Spanidis et al., [12] did not find significative differences for this parameter after an ultramarathon mountain race. This discrepancy is explained due to a sample size effect. In contrast, Turner et al., [16] demonstrated an increase in plasma CG content after an ultramarathon race, thus confirming oxidative damage to proteins after ultraendurance exercise.

Regarding the glutathione system enzymes, no significant differences were observed for the GPx activity. Conversely, the GR activity showed a significant increase in the finish line that was even greater 24 h post race, interestingly concurring with the partial MDA concentration recovery. Although both enzymes are mainly located in the intracellular compartment, their plasmatic activity have been broadly used as a measurement of the antioxidant status [40–43]. In the case of ultraendurance exercise, previous studies show contradictory results to what concern to the antioxidant enzymes [7,12,14,39], but our results support the hypothesis of a compensatory mechanism based on a temporary increase of the antioxidant defense to compensate an oxidative insult [44,45]. Although the activity of these two enzymes has been used to evaluate the presence of oxidative stress, GR is considered the limiting factor of these antioxidative system [46]. It is plausible that the increase of GR activity is not accompanied by an enhanced GPx activity because of other antioxidant enzymes such as catalase or paraoxonase also able to degrade hydrogen peroxide that could be affected by the intense exercise [10,47].

Moreover, it is remarkable that we report a significant negative correlation between the basal GR activity of the runners and the degree of muscle membrane disruption after the race. Thus, the basal activity of this enzyme correlated with [LDH] in the finish line, after 24 h and 48 h and with the [CK] in the finish line and after 24 h. In addition, the resting levels of oxidative damage to proteins (CG content) also showed a significant positive correlation with the magnitude of post-race muscle injury ([LDH] after 24 h and 48 h). It is important to notice that we assume that serum CK and LDH assess for muscle membrane disruption and do not necessarily correlate with muscle structural damage. Furthermore, we have also reported a significant positive correlation between basal GR activity and the improvement in the SJ performance as well as, a significant negative correlation between the basal levels of lipid peroxidation (MDA concentration) and the enhancement in the HG execution. Although there is a limitation on the correlation's coefficient power to assume an evident causality between oxidative stress and muscle fatigue, these novel findings, suggest that a stronger basal plasma oxidative status might improve muscle strength during ultraendurance sports practice. However, further studies are necessary to increase the number of research volunteers and validate the present results.

Finally, we have reported sex differences in oxidative damage to macromolecules. Surprisingly, female athletes showed higher CG content and less MDA levels than male athletes, although female athletes have less muscular mass and higher body fat mass percentage. There is almost no literature considering sex differences in type of events, probably due to the difficulty in getting sufficient sample sizes. Although several studies include female runners in their research, they do not describe sex differences in the parameters measured [48,49]. Recently Devrim-Lanpir et al., reported significant interaction between time at exhaustion and dietary antioxidant intake in males, but not in females, who underwent an acute exhaustive exercise test (a cycle ergometer) followed by a treadmill test in a laboratory [50]. Moreover, the results of the multivariate analysis show us the predictive value of basal GR concentration and sex in relation to muscle fatigue and cell damage, after ultramarathon.

5. Conclusions

Interestingly, the study yielded new results regarding the age of the runners. Senior runners (45–53) showed significant higher levels of lipid peroxidation (MDA concentration) than medium (38–44) and young runners (31–37) throughout the study. A plausible explanation would be the higher body fat percentage observed in senior runners, together with the loss of muscular mass compared to younger runners (Table 1). These results are in agreement with Hattori and cols. who reported that ultramarathon runners aging less than 45 years old had lower ROS levels at all race points [51]. Again, larger series would be necessary to study the predictive value of this assay to consider if a personalized antioxidant supplementation might promote the physiological recovery after great physical efforts.

Supplementary Materials: The following are available online at https://www.mdpi.com/2076-392 1/10/3/355/s1, Supplementary Table S1. Evolution of muscle strength (SJ and HG), muscle damage (LDH and CK) and acute inflammation (CRP) biomarkers.

Author Contributions: Conceptualization, C.G., C.H., E.C.-B. and M.M.; methodology, P.B., B.H. and C.H.; formal analysis, C.G., M.M. and C.H.; investigation, Barbara Hernando, I.M.-N. and C.H.; writing—original draft preparation, E.C.-B., C.G., C.H.; writing—M.M., P.B., C.H.; funding acquisition, C.H. and I.M.-N. All authors have read and agreed to the published version of the manuscript.

Funding: This research was funded by Vithas Hospitals group (https://vithas.es), Penyagolosa Trails organization (http://penyagolosatrails.com), catedra Endavant Villarreal CF de l'Esport (https://endavant.villarrealcf.es/) and the following grant code: UJI-B2019-38. The funders had no role in study design, data collection and analysis, decision to publish, or preparation of the manuscript.

Institutional Review Board Statement: The investigation was conducted in accordance with the Declaration of Helsinki and approval for the project was obtained from the research Ethics Committee of the University Jaume I of Castellon (Expedient Number CD/007/2019). This study is enrolled in the ClinicalTrails.gov database, with the code number NCT03990259 (www.clinicaltrials.gov).

Informed Consent Statement: Informed consent was obtained from all volunteers involved in the study.

Data Availability Statement: Data is contained within the article or supplementary material.

Conflicts of Interest: The authors declare no conflict of interest.

References

1. Panizo González, N.; Reque Santivañez, J.E.; Hernando Fuster, B.; Collado Boira, E.J.; Martinez-Navarro, I.; Chiva Bartoll, Ó.; Hernando Domingo, C. Quick Recovery of Renal Alterations and Inflammatory Activation after a Marathon. *Kidney Dis.* **2019**, *5*, 259–265. [CrossRef]
2. Martínez-Navarro, I.; Sánchez-Gómez, J.; Sanmiguel, D.; Collado, E.; Hernando, B.; Panizo, N.; Hernando, C. Immediate and 24-h post-marathon cardiac troponin T is associated with relative exercise intensity. *Eur. J. Appl. Physiol.* **2020**. [CrossRef]
3. Martínez-Navarro, I.; Sánchez-Gómez, J.M.; Collado-Boira, E.J.; Hernando, B.; Panizo, N.; Hernando, C. Cardiac damage biomarkers and heart rate variability following a 118-km mountain race: Relationship with performance and recovery. *J. Sport. Sci. Med.* **2019**, *18*, 615.
4. Martínez-Navarro, I.; Sanchez-Gómez, J.M.; Aparicio, I.; Priego-Quesada, J.I.; Pérez-Soriano, P.; Collado, E.; Hernando, B.; Hernando, C. Effect of mountain ultramarathon distance competition on biochemical variables, respiratory and lower-limb fatigue. *PLoS ONE* **2020**, *15*. [CrossRef] [PubMed]
5. Sies, H. Oxidative stress: Oxidants and antioxidants. *Exp. Physiol.* **1997**, *82*, 291–295. [CrossRef]

6. Turner, J.E.; Bennett, S.J.; Bosch, J.A.; Griffiths, H.R.; Aldred, S. Ultra-endurance exercise: Unanswered questions in redox biology and immunology. *Biochem. Soc. Trans.* **2014**, *42*, 989–995.
7. Kłapcińska, B.; Waåkiewicz, Z.; Chrapusta, S.J.; Sadowska-Krępa, E.; Czuba, M.; Langfort, J. Metabolic responses to a 48-h ultra-marathon run in middle-aged male amateur runners. *Eur. J. Appl. Physiol.* **2013**, *113*, 2781–2793. [CrossRef] [PubMed]
8. Marumoto, M.; Suzuki, S.; Akihiro, H.; Kazuyuki, H.; Kiyoshi Shibata, A.; Fuku, M.; Goto, C.; Tokudome, Y.; Hoshino, H.; Imaeda, N.; et al. Changes in thioredoxin concentrations: An observation in an ultra-marathon race. *Environ. Health Prev. Med.* **2010**, *15*, 129–134. [CrossRef]
9. Skenderi, K.P.; Tsironi, M.; Lazaropoulou, C.; Anastasiou, C.A.; Matalas, A.-L.; Kanavaki, I.; Thalmann, M.; Goussetis, E.; Papassotiriou, I.; Chrousos, G.P. Changes in free radical generation and antioxidant capacity during ultramarathon foot race. *Eur. J. Clin. Investig.* **2008**, *38*, 159–165. [CrossRef] [PubMed]
10. Benedetti, S.; Catalani, S.; Peda, F.; Luchetti, F.; Citarella, R.; Battistelli, S. Impact of the 24-h ultramarathon race on homocysteine, oxidized low-density lipoprotein, and paraoxonase 1 levels in professional runners. *PLoS ONE* **2018**, *13*. [CrossRef]
11. Mrakic-Sposta, S.; Gussoni, M.; Moretti, S.; Pratali, L.; Giardini, G.; Tacchini, P.; Dellanoce, C.; Tonacci, A.; Mastorci, F.; Borghini, A.; et al. Effects of mountain ultra-marathon running on ROS production and oxidative damage by micro-invasive analytic techniques. *PLoS ONE* **2015**, *10*. [CrossRef] [PubMed]
12. Spanidis, Y.; Stagos, D.; Orfanou, M.; Goutzourelas, N.; Bar-Or, D.; Spandidos, D.; Kouretas, D. Variations in Oxidative Stress Levels in 3 Days Follow-up in Ultramarathon Mountain Race Athletes. *J. Strength Cond. Res.* **2017**, *31*, 582–594. [CrossRef]
13. Kabasakalis, A.; Kyparos, A.; Tsalis, G.; Loupos, D.; Pavlidou, A.; Kouretas, D. Blood Oxidative Stress Markers After Ultramarathon Swimming. *J. Strength Cond. Res.* **2011**, *25*, 805–811. [CrossRef]
14. Neubauer, O.; König, D.; Kern, N.; Nics, L.; Wagner, K.H. No indications of persistent oxidative stress in response to an ironman triathlon. *Med. Sci. Sports Exerc.* **2008**, *40*, 2119–2128. [CrossRef] [PubMed]
15. Miyata, T.; Wada, Y.; Cai, Z.; Iida, Y.; Horie, K.; Yasuda, Y.; Maeda, K.; Kurokawa, K.; Van Ypersele De Strihou, C. Implication of an increased oxidative stress in the formation of advanced glycation end products in patients with end-stage renal failure. *Kidney Int.* **1997**, *51*, 1170–1181. [CrossRef]
16. Turner, J.E.; Hodges, N.J.; Bosch, J.A.; Aldred, S. Prolonged depletion of antioxidant capacity after ultraendurance exercise. *Med. Sci. Sports Exerc.* **2011**, *43*, 1770–1776. [CrossRef]
17. Cheng, A.J.; Yamada, T.; Rassier, D.E.; Andersson, D.C.; Westerblad, H.; Lanner, J.T. Reactive oxygen/nitrogen species and contractile function in skeletal muscle during fatigue and recovery. *J. Physiol.* **2016**, *594*, 5149–5160. [CrossRef] [PubMed]
18. Mason, S.A.; Morrison, D.; McConell, G.K.; Wadley, G.D. Muscle redox signalling pathways in exercise. Role of antioxidants. *Free Radic. Biol. Med.* **2016**, *98*, 29–45. [CrossRef]
19. Bernat-Adell, M.D.; Collado-Boira, E.J.; Moles-Julio, P.; Panizo-GonzálezGonz, N.; Martínez-Navarro, I.; Hernando-Fuster, B.; Hernando-Domingo, C. Recovery of Inflammation, Cardiac, and Muscle Damage Biomarkers After Running a Marathon. *J. Strength Cond. Res.* **2019**. [CrossRef]
20. Lawrence, R.A.; Parkhill, L.K.; Burk, R.F. Hepatic cytosolic non selenium-dependent glutathione peroxidase activity: Its nature and the effect of selenium deficiency. *J. Nutr.* **1978**, *108*, 981–987. [CrossRef] [PubMed]
21. Smith, I.K.; Vierheller, T.L.; Thorne, C.A. Assay of glutathione reductase in crude tissue homogenates using 5,5′-dithiobis(2-nitrobenzoic acid). *Anal. Biochem.* **1988**, *175*, 408–413. [CrossRef]
22. Richard, M.J.; Guiraud, P.; Meo, J.; Favier, A. High-performance liquid chromatographic separation of malondialdehyde-thiobarbituric acid adduct in biological materials (plasma and human cells) using a commercially available reagent. *J. Chromatogr. B Biomed. Sci. Appl.* **1992**, *577*, 9–18. [CrossRef]
23. Romero, F.J.; Bosch-Morell, F.; Romero, M.J.; Jareño, E.J.; Romero, B.; Marín, N.; Romá, J. Lipid peroxidation products and antioxidants in human disease. *Environ. Health Perspect.* **1998**, *106*, 1229–1234. [CrossRef]
24. Levin, R.L. Determination of carbonyl content in oxidatively modified proteins. *Methods Enzym.* **1990**, *186*, 464–478.
25. Tiana, L.; Caib, Q.; Wei, H. Alterations of antioxidant enzymes and oxidative damage to macromolecules in different organs of rats during aging. *Free Radic. Biol. Med.* **1998**, *24*, 1477–1484. [CrossRef]
26. Dill, D.B.; Costill, D.L. Calculation of percentage changes in volumes of blood, plasma, and red cells in dehydration. *J. Appl. Physiol.* **1974**, *37*, 247–248. [CrossRef] [PubMed]
27. Samozino, P.; Morin, J.B.; Hintzy, F.; Belli, A. A simple method for measuring force, velocity and power output during squat jump. *J. Biomech.* **2008**, *41*, 2940–2945. [CrossRef]
28. Bohannon, R.W. Muscle strength. *Curr. Opin. Clin. Nutr. Metab. Care* **2015**, *18*, 465–470. [CrossRef] [PubMed]
29. Dourado, V.Z.; de Antunes, L.C.O.; Tanni, S.E.; de Paiva, S.A.R.; Padovani, C.R.; Godoy, I. Relationship of Upper-Limb and Thoracic Muscle Strength to 6-min Walk Distance in COPD Patients. *Chest* **2006**, *129*, 551–557. [CrossRef]
30. Reuter, S.E.; Massy-Westropp, N.; Evans, A.M. Reliability and validity of indices of hand-grip strength and endurance. *Aust. Occup. Ther. J.* **2011**, *58*, 82–87. [CrossRef] [PubMed]
31. Ozkaplan, A.; Rhodes, E.C.; Sheel, A.W.; Taunton, J.E. A comparison of inspiratory muscle fatigue following maximal exercise in moderately trained males and females. *Eur. J. Appl. Physiol.* **2005**, *95*, 52–56. [CrossRef]
32. Tong, T.K.; Wu, S.; Nie, J.; Baker, J.S.; Lin, H. The occurrence of core muscle fatigue during high-intensity running exercise and its limitation to performance: The role of respiratory work. *J. Sport. Sci. Med.* **2014**, *13*, 244–251.

33. Byrne, N.M.; Hills, A.P.; Hunter, G.R.; Weinsier, R.L.; Schutz, Y. Metabolic equivalent: One size does not fit all. *J. Appl. Physiol.* **2005**, *99*, 1112–1119. [CrossRef] [PubMed]
34. Mohd Razali, N.; Bee Wah, Y. Power comparisons of Shapiro-Wilk, Kolmogorov-Smirnov, Lilliefors and Anderson-Darling tests. *J. Stat. Model. Anal.* **2011**, *2*, 21–33.
35. Thomas, J.; Nelson, J.; Silverman, S. *Research Methods in Physical Activity*; Human Kinetics: Champaign, IL, USA, 2015.
36. Stoica, P.; Söderström, T. On the parsimony principle. *Int. J. Control* **1982**, *36*, 409–418. [CrossRef]
37. Williams, M.N.; Grajales, C.A.G.; Kurkiewicz, D. Assumptions of multiple regression: Correcting two misconceptions. *Pract. Assess. Res. Eval.* **2013**, *18*, 1–14. [CrossRef]
38. Cohen, J. The Analysis of Variance and Covariance. In *Statistical Power Analysis for the Behavioral Sciences*, 2nd ed.; Taylor Francis Group: Abingdon, UK, 1988.
39. Knez, W.L.; Jenkins, D.G.; Coombes, J.S. Oxidative stress in half and full Ironman triathletes. *Med. Sci. Sports Exerc.* **2007**, *39*, 283–288. [CrossRef]
40. Irigaray, P.; Caccamo, D.; Belpomme, D. Oxidative stress in electrohypersensitivity self-reporting patients: Results of a prospective in vivo investigation with comprehensive molecular analysis. *Int. J. Mol. Med.* **2018**, *42*, 1885–1898. [CrossRef]
41. Beltrán-Sarmiento, E.; Arregoitia-Sarabia, C.K.; Floriano-Sánchez, E.; Sandoval-Pacheco, R.; Galván-Hernández, D.E.; Coballase-Urrutia, E.; Carmona-Aparicio, L.; Ramos-Reyna, E.; Rodríguez-Silverio, J.; Cárdenas-Rodríguez, N. Effects of valproate monotherapy on the oxidant-antioxidant status in mexican epileptic children: A longitudinal study. *Oxid. Med. Cell. Longev.* **2018**, *2018*. [CrossRef]
42. Arribas, L.; Almansa, I.; Miranda, M.; Muriach, M.; Romero, F.J.; Villar, V.M. Serum Malondialdehyde Concentration and Glutathione Peroxidase Activity in a Longitudinal Study of Gestational Diabetes. *PLoS ONE* **2016**, *11*, e0155353. [CrossRef]
43. Brigelius-Flohé, R.; Maiorino, M. Glutathione peroxidases. *Biochim. Biophys. Acta Gen. Subj.* **2013**, *1830*, 3289–3303. [CrossRef] [PubMed]
44. Sánchez-Vallejo, V.; Benlloch-Navarro, S.; Trachsel-Moncho, L.; López-Pedrajas, R.; Almansa, I.; Romero, F.J.; Miranda, M. Alterations in glutamate cysteine ligase content in the retina of two retinitis pigmentosa animal models. *Free Radic. Biol. Med.* **2016**, *96*, 245–254. [CrossRef]
45. Tsuru-Aoyagi, K.; Potts, M.B.; Trivedi, A.; Pfankuch, T.; Raber, J.; Wendland, M.; Claus, C.P.; Koh, S.-E.; Ferriero, D.; Noble-Haeusslein, L.J. Glutathione peroxidase activity modulates recovery in the injured immature brain. *Ann. Neurol.* **2009**, *65*, 540–549. [CrossRef] [PubMed]
46. Finkler, M.; Lichtenberg, D.; Pinchuk, I. The relationship between oxidative stress and exercise. *J. Basic Clin. Physiol. Pharmacol.* **2014**, *25*, 1–11. [CrossRef]
47. Somani, S.M.; Frank, S.; Rybak, L.P. Responses of antioxidant system to acute and trained exercise in rat heart subcellular fractions. *Pharmacol. Biochem. Behav.* **1995**, *51*, 627–634. [CrossRef]
48. Ginsburg, G.S. Effects of a Single Bout of Ultraendurance Exercise on Lipid Levels and Susceptibility of Lipids to Peroxidation in Triathletes. *JAMA J. Am. Med. Assoc.* **1996**, *276*, 221. [CrossRef]
49. Nieman, D.C.; Dumke, C.I.; Henson, D.A.; Mcanulty, S.R.; Mcanulty, L.S.; Lind, R.H.; Morrow, J.D. Immune and Oxidative Changes During and Following the Western States Endurance Run. *Int. J. Sport. Med.* **2003**, *24*, 541–547. [CrossRef]
50. Devrim-Lanpir, A.; Bilgic, P.; Kocahan, T.; Deliceoğlu, G.; Rosemann, T.; Knechtle, B. Total Dietary Antioxidant Intake Including Polyphenol Content: Is It Capable to Fight against Increased Oxidants within the Body of Ultra-Endurance Athletes? *Nutrients* **2020**, *12*, 1877. [CrossRef]
51. Hattori, N.; Hayashi, T.; Nakachi, K.; Ichikawa, H.; Gotoau, C.; Tokudome, Y.; Kuriki, K.; Hoshino, H.; Shibata, K.; Yamada, N.; et al. Changes of ROS during a two-day ultra-marathon race. *Int. J. Sports Med.* **2009**, *30*, 426–429. [CrossRef] [PubMed]

 *antioxidants*

Article

Eccentric Cycling Training Improves Erythrocyte Antioxidant and Oxygen Releasing Capacity Associated with Enhanced Anaerobic Glycolysis and Intracellular Acidosis

Yu-Chieh Huang [1], Mei-Ling Cheng [2,3,4], Hsiang-Yu Tang [2], Chi-Yao Huang [5], Kuan-Ming Chen [5] and Jong-Shyan Wang [5,6,7,*]

[1] Department of Physical Therapy, College of Medical and Health Science, Asia University, Taichung 413, Taiwan; yuchieh@asia.edu.tw
[2] Metabolomics Core Laboratory, Healthy Aging Research Center, Chang Gung University, Taoyuan 333, Taiwan; chengm@mail.cgu.edu.tw (M.-L.C.); tangshyu@gmail.com (H.-Y.T.)
[3] Clinical Metabolomics Core Laboratory, Chang Gung Memorial Hospital, Taoyuan 333, Taiwan
[4] Department of Biomedical Sciences, College of Medicine, Chang Gung University, Taoyuan 333, Taiwan
[5] Healthy Aging Research Center, Graduate Institute of Rehabilitation Science, Medical Collage, Chang Gung University, Taoyuan 333, Taiwan; kjes9210@hotmail.com (C.-Y.H.); ttlike76@gmail.com (K.-M.C.)
[6] Heart Failure Center, Department of Physical Medicine and Rehabilitation, Keelung Chang Gung Memorial Hospital, Keelung 204, Taiwan
[7] Research Center for Chinese Herbal Medicine, College of Human Ecology, Chang Gung University of Science and Technology, Taoyuan 333, Taiwan
* Correspondence: s5492@mail.cgu.edu.tw; Tel.: +886-3-2118800 (ext. 5748); Fax: +886+886-3-2118700

Citation: Huang, Y.-C.; Cheng, M.-L.; Tang, H.-Y.; Huang, C.-Y.; Chen, K.-M.; Wang, J.-S. Eccentric Cycling Training Improves Erythrocyte Antioxidant and Oxygen Releasing Capacity Associated with Enhanced Anaerobic Glycolysis and Intracellular Acidosis. *Antioxidants* **2021**, *10*, 285. https://doi.org/10.3390/antiox10020285

Academic Editor: Gareth Davison and Conor McClean

Received: 25 January 2021
Accepted: 10 February 2021
Published: 13 February 2021

Publisher's Note: MDPI stays neutral with regard to jurisdictional claims in published maps and institutional affiliations.

Abstract: The antioxidant capacity of erythrocytes protects individuals against the harmful effects of oxidative stress. Despite improved hemodynamic efficiency, the effect of eccentric cycling training (ECT) on erythrocyte antioxidative capacity remains unclear. This study investigates how ECT affects erythrocyte antioxidative capacity and metabolism in sedentary males. Thirty-six sedentary healthy males were randomly assigned to either concentric cycling training (CCT, $n = 12$) or ECT ($n = 12$) at 60% of the maximal workload for 30 min/day, 5 days/week for 6 weeks or to a control group ($n = 12$) that did not receive an exercise intervention. A graded exercise test (GXT) was performed before and after the intervention. Erythrocyte metabolic characteristics and O_2 release capacity were determined by UPLC-MS and high-resolution respirometry, respectively. An acute GXT depleted Glutathione (GSH), accumulated Glutathione disulfide (GSSG), and elevated the GSSG/GSH ratio, whereas both CCT and ECT attenuated the extent of the elevated GSSG/GSH ratio caused by a GXT. Moreover, the two exercise regimens upregulated glycolysis and increased glucose consumption and lactate production, leading to intracellular acidosis and facilitation of O_2 release from erythrocytes. Both CCT and ECT enhance antioxidative capacity against severe exercise-evoked circulatory oxidative stress. Moreover, the two exercise regimens activate erythrocyte glycolysis, resulting in lowered intracellular pH and enhanced O_2 released from erythrocytes.

Keywords: eccentric exercise; redox status; erythrocyte; metabolism

1. Introduction

Endurance training is essential to maximally improve cardiopulmonary fitness and delay the disease process. However, this may be intolerable due to the overload of the cardiopulmonary system to elderly individuals or patients with chronic diseases, traditional concentric work at usual training intensity [1]. Eccentric endurance training has the ability to overcome these limitations because of less respiratory requirement and metabolic oxygen, as well as lower heart rate (H), cardiac index and blood lactate concentration than concentric type at equivalent workload [2]. The benefits of using eccentric cycling training (ECT) in chronic heart failure patients [3], elderly individuals [4] and chronic obstructive pulmonary disease [5] have been confirmed. Conventionally, most studies have focused on the contribution to elicit neuromuscular adaptations of eccentric work [6]; nevertheless, a recent study further demonstrated that either acute bout of concentric or eccentric cycling at moderate intensity induced increased enzymatic antioxidant capacity and decreased oxidative stress markers [7]. Moreover, ECT induces greater fat utilization compared to concentric cycling training (CCT) at a fixed workload [8] and greater fat loss in obese adolescents [9]. Therefore, the different cardiopulmonary loading and metabolic oxygen demands in ECT and CCT may result in distinct changes in antioxidative metabolism and O_2 release adaptations [10]. However, there is very limited evidence regarding these mechanisms of chronic physiological responses to eccentric cycling [11].

Erythrocytes are vital to humans because of their abundance and the irreplaceable function they have of delivering O_2. However, they are susceptible to sustained free radical damage during circulation, which impairs their O_2 release capacity and reduces their lifespan [12]. Previous studies have reported that blood antioxidation capacity is impaired with acute exercise [13]; in contrast, regular exercise may increase antioxidative capacity [14]. The lower oxygen and energy consumed in ECT may avoid repeated, excessive exposure to oxidative stress, which progressively impair the erythrocyte [15]. However, whether this lower metabolic stress in comparable ECT might be enough to elicit physiological adaptations as CCT or not is another concern [16]. To date, the adaptations of the antioxidation capacity and regulatory mechanism of erythrocytes under different exercise regimens remain unclear. Here, we identified the key regulatory mechanisms using metabolomics profiling technology.

When exercising, erythrocytes must accelerate O_2 release into peripheral tissue according to the Bohr effect [17] and enhance the demand for glycolytically derived ATP to restore intracellular ion balances. This process is at a constant rate when ATP consumption is steady, but the activity of the process changes rapidly in response to enhanced ATP utilization [18]. Importantly, erythrocytes are also exposed to dramatically enhanced oxidative stress that must be controlled by accelerated production of reducing equivalents derived from the pentose phosphate pathway (PPP), which is the sole source of NADPH and produces GSH as an antioxidant. In the sickle cells, the impaired antioxidant capacity leaves to a loss of glycolysis and the PPP shifting mechanism control and further homeostasis rupture, contributing to a decreased lifespan of cells [19]. Moreover, altering glycolytic pathway dominance has been demonstrated to limit antioxidation capacity under hypoxia [20]. Therefore, exercise may introduce continuous substrate competition between the main glycolysis pathway and the PPP, although this needs to be further elucidated.

2,3-BPG is a strong allosteric modulator that leads to O_2 unloading [21]. However, the generation of 2,3-BPG bypasses the main phosphoglycerate kinase reaction so that the overall production of ATP per mole of glucose is decreased to zero. GSH de novo synthesis is ATP dependent and is therefore impaired when the stocks of intracellular ATP are depleted. In addition, lactate is the only end product of glycolysis in erythrocytes, and it also helps create a low pH value environment to decrease Hb-O_2 affinity [22] and influence GSH synthesis [23]. Therefore, one of the biggest puzzles regarding erythrocyte metabolism during exercise is how the programming of erythrocyte glucose metabolism, 2,3-BPG production, and antioxidative capacity is regulated.

To address the abovementioned questions, this study elucidated the pathways underlying the regulation of the main glycolysis and the PPP and explored the effects of oxidation and antioxidation capacity in erythrocytes after six weeks of interventions. In addition, we also investigated the capacity for O_2 release under different lactate concentrations under hypoxic and normoxic conditions. The aim of this study was to provide direct evidence that both ECT and CCT induce metabolic adaptations within erythrocytes that counteract the high oxidative stress evoked by vigorous exercise.

2. Materials and Methods

2.1. Subjects

The intervention was performed in accordance with the Declaration of Helsinki and was approved by the Institutional Review Board of Chang Gung Memorial Hospital in Taiwan. A total of 36 sedentary males who were nonsmokers, nonusers of medication/vitamins, and free of any cardiopulmonary/hematological risks were recruited from Chang Gung University, Taiwan. No subject had performed regular exercise (i.e., exercise frequency once per week, duration < 20 min) for at least 1 year before the experiment. All subjects provided informed consent after the experimental procedures were explained. These subjects were randomly divided into three groups: the concentric cycling training (CCT, n = 12), the eccentric cycling training (ECT, n = 12), and the control (CTL, n = 12) groups. All subjects arrived at the testing center at 9:00 AM to eliminate any possible circadian effect. Participants were instructed to fast for at least 8 h and to refrain from strenuous physical exercise for at least 48 h before sampling.

2.2. Protocol and Interventions

Both the CCT and ECT groups performed exercise regimens on a stationary bicycle ergometer (CCT: Corival 400, Lode; ECT: custom-built cycle ergometer) 5 times a week for 6 weeks. For comparison, the CTL group did not perform any exercise, but their physical activities and daily diet were carefully documented.

All subjects reported their daily activities and nutrition intakes via questionnaires throughout the experiment. The participants were instructed to refrain from extra regular exercise until the end of this study. Moreover, the compliance rates for all three interventions were 100%.

The graded exercise test (GXT) was performed 48 h before and after the intervention. Both the CCT and ECT groups had a 3 day familiarization program upon initiation of training. The exercise intensity was set at 20%, 30%, and 40% of the maximal workload (W_{max}) on each day. The first week's intensity was set at 45% W_{max} and progressively increased 5% W_{max} per week until 70% W_{max} was obtained in the sixth week. Each training session contained a 6-min warm-up phase (3 min at 0% and 3 min at 30% W_{max}), 30-min training period and 6-min cool-down phase (3 min at 30% and 3 min at 0% W_{max}) (Figure 1). The training groups were asked to record their daily activities and nutritional intake using the short form of the international physical activity questionnaire and a written diet record, respectively. Subjects were asked to refrain from regular extra exercise until the end of the study. The participant compliance rate was 100% throughout this study.

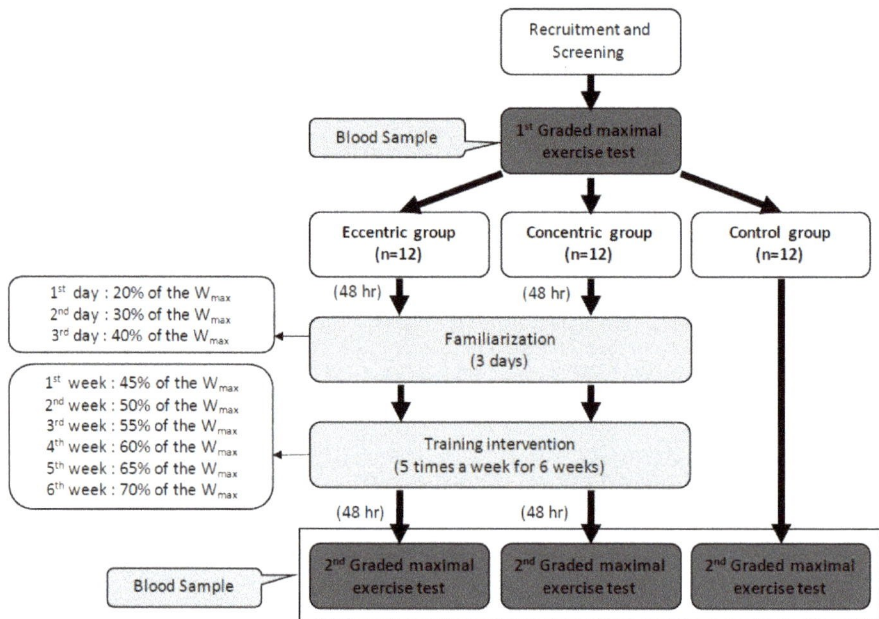

Figure 1. Design of the experiment and the training intensity of eccentric and concentric groups in each week. W_{max}: the maximal workload of the first graded maximal exercise test.

2.3. Graded Exercise Tests

To assess aerobic capacity, a GXT was performed on a cycle ergometer (Corival 400, Lode B.V., Groningen, The Netherlands). After a 5-min baseline resting period, a 2-min warm-up period (60 rpm, unloaded pedaling) was initiated, followed by incremental work (30 Watt increase for each 3 min) until exhaustion (i.e., progressive exercise to VO_{2max}). Minute ventilation (V_E), oxygen consumption (VO_2), and carbonic dioxide production (VCO_2) were measured for each breath by using a computer-based system (MasterScreen CPX, CareFusion, Franklin Lakes, NJ, USA). The criteria used to define VO_{2max} were as follows: (i) the level of VO_2 increased by < 2 mL/kg/min over at least 2 min; (ii) H exceeded its predicted maximum; (iii) the respiratory exchange ratio (RER) exceeded 1.2; and (iv) the venous lactate concentration was >8 mM, which was consistent with the guidelines of the American College of Sports Medicine for exercise testing [24]. Additionally, the ventilation threshold (VT) was determined when VE/VO_2 increased without a corresponding increase in the V_E-to-VCO_2 ratio and when end-tidal PO_2 increased without a decrease in end-tidal PCO_2 or a deviation from linearity for V_E.

2.4. Erythrocyte Isolation and Blood Collection

At rest and immediately after the GXT, a 10 mL blood sample was collected from the antecubital vein via clean venipuncture (20-gauge needle) and added to a tube with ethylenediaminetetraacetic acid (EDTA, 4 mM). Blood cells were counted by using a Sysmex SF-3000 cell counter (GMI Inc., Ramsey, MN, USA), and the blood pH and lactate concentration were tested by an i-STAT handheld blood analyzer (Abbott Point of Care Inc., Princeton, NJ, USA). Erythrocytes were isolated from whole blood by centrifugation ($1000\times g$ for 15 min at RT), the supernatant was discarded, and the buffy coat was discarded, followed by three washing steps in PBS with 0.1% (*w/v*) glucose (Sigma). The erythrocyte count was adjusted to 1×10^4 cells/µL using PBS.

2.5. Measurement of Reactive Oxygen Species (ROS) Production

2′, 7′-Dichlorofluorescin diacetate (DCFDA) is a fluorogenic dye that measures hydroxyl, peroxyl and other ROS activities within the cell. This study used a DCFDA Cellular ROS Detection Assay Kit (ab113851, Abcam) to measure intracellular ROS according to the manufacturer's protocol. DCFDA-labeled erythrocytes were treated with different concentrations of tert-butyl hydroperoxide (TBHP) (5 mM, 10 mM, 50 mM and 100 mM) at 37 °C for 30 min. TBHP mimics ROS activity to oxidize DCFDA to fluorescent DCF. Finally, the mean fluorescence intensities obtained from 50,000 erythrocytes were measured using FACSCalibur (Becton Dickinson, NJ, USA). All samples were analyzed in triplicate, and the intraassay CV was 4.1 ± 0.7%.

2.6. Erythrocyte Intracellular pH

As our previous study presented [25], erythrocytes were loaded with the fluorescent pH indicator carboxy SNARF-1 (1 μM, Invitrogen) at 37 °C for 30 min in the dark and then washed with HBSS (Sigma) at 2500× g for 5 min. SNARF-1-loaded cells were incubated with pH-controlled normal K$^+$-balanced buffer (137.9 mM NaCl, 5.33 mM KCl, 0.441 mM KH2PO4, 4.17 mM NaHCO3, 0.338 mM Na2HPO4, 5.56 mM glucose, and 20 mM HEPES, pH = 7.5) and high K$^+$-balanced buffer at different pH values (43.7 mM NaCl, 100 mM KCl, 0.441 mM KH2PO4, 4.17 mM NaHCO3, 0.338 mM Na2HPO4, 5.56 mM glucose, and 20 mM HEPES, pH = 6.8, 7.0, 7.2, 7.4, 7.6, 7.8, and 8.0) containing 10 μM nigericin (Invitrogen). The pH was always adjusted at RT prior to use. The pH-dependent spectral shifts exhibited by SNARF-1 allowed calibration of the pH response in terms of the ratio of fluorescence intensities measured at two different wavelengths, FL2 and FL3, in a FACSCalibur (λ1 = 580 nm and λ2 = 600~640 nm and fixed excitation at 514 nm), as described in the manufacturer's protocol (Invitrogen). All samples were analyzed in triplicate, and the intraassay CV was 3.7 ± 0.9%. The equation is as follows:

$$ \mathrm{pH} = \mathrm{pK_a} - \log_{10}\left[\frac{R - R_B}{R_A - R} \times \frac{F_{B(\lambda 2)}}{F_{A(\lambda 2)}} \right] \tag{1} $$

pKa values: 7.5 for carboxy SNARF-1;
R: The ratio Fλ1/Fλ2 of fluorescence intensities (F) measured at λ1 and λ2;
The subscript A represents the limiting values at the acidic endpoints of the titration;
The subscript B represents the limiting values at the basic endpoints of the titration.

2.7. Oxygen Release Efficacy in Erythrocytes

High-resolution respirometry, an Oroboros Oxygraph-2 K (Oroboros Instruments, Innsbruck, Austria), was used to measure the O$_2$ pressure (mmHg) and O$_2$ flux per volume (pmol·s^{-1}·mL^{-1}) of erythrocytes at 0-, 1-, and 4 mM lactate during hypoxia (PO$_2$ = 20 ± 3 mmHg) and normoxia (PO$_2$ = 147 ± 3 mmHg) in HBSS medium, respectively. Hypoxic conditions were prepared by gassing with nitrogen (N$_2$) gas. After heating at 37 °C and equilibration and calibration for the target PO$_2$ conditions, 2 × 10^6 isolated erythrocytes were added, and after the signaling stabilized, 2 and 6 μL of 1 M lactic acid (Sigma) were sequentially added to form 1 and 4 mM lactic acid environments to simulate the rest and lactate threshold conditions, respectively (Figure S1). All samples were analyzed in triplicate, and the intraassay CV was 3.9 ± 0.8%.

2.8. Sample Preparation for Targeted Metabolite Identification and Quantification

Fifteen randomly selected samples (ECT = 5, CCT = 5, and CTL = 5) were quantified for the target metabolite. Erythrocytes (6×10^8) were lysed in 200 μL of ddH$_2$O containing 100 ppb of debrisoquine sulfate (Sigma) as an internal control. The lysate was extracted in 800 μL of warmed methanol. The sample was incubated at RT for 15 min to precipitate proteins and centrifuged at 16,000× *g* for 30 min at 4 °C. The supernatant was transferred to a new tube, dried under nitrogen gas, and stored at −80 °C. Prior to analysis, the sample was dissolved in 200 μL of ddH$_2$O containing 0.1% formic acid. The procedure was carried out according to the method of Tang et al. [26].

2.9. Target Metabolite Analysis of Glycolysis Intermediates

All samples were analyzed using ultrahigh-performance liquid chromatography (UH-PLC) coupled with Xevo TQ-S MS (Waters Corp., MA USA) as previously described with modifications [27]. MS was performed in negative-ion multiple-reaction-monitoring (MRM) mode. For tuning purposes, a single analysis standard dissolved in a mixture of water/methanol 50:50 (*v/v*) was infused at a flow rate of 10 μL/min. The desolvation gas flow was set at 1000 L/h at a temperature of 500 °C, and the source temperature was set at 150 °C. The capillary voltage and cone voltage were set to 1.3 kV and 25 V, respectively. For chromatographic separation, a BEH C18 (100 mm × 2.1 mm, 1.7 μm; Waters Corp, MA, USA) was used with eluent A (10 mM tributylamine aqueous solution with 15 mM acetic acid) and eluent B (50% acetonitrile containing 10 mM tributylamine and 15 mM acetic acid), the flow rate was 0.3 mL/min, and the column temperature was set at 25 °C. The gradient profile was as follows: linear-gradient 99–98% solvent B, 8 min; 12% solvent B, 2 min; 55% solvent B, 2 min; and 99% solvent B, 2 min. The column was then re-equilibrated for 4 min. QC samples were analyzed during the analytical runs. All samples were analyzed in triplicate, and the intraassay CV was 3.4 ± 0.6%.

2.10. Senescence-Related Biological Markers and Methemoglobin Concentrations in Erythrocytes

The erythrocyte suspension (1×10^4 cells/μL) was incubated with saturating concentrations of monoclonal anti-human CD47 antibody (BioLegend) or monoclonal anti-human CD147 antibody (eBioscience) conjugated with fluorescein isothiocyanate (FITC) in the dark for 30 min at 37 °C. The mean fluorescence intensity (MFI) obtained from 50,000 erythrocytes was measured by a FACSCalibur (Becton Dickinson, NJ USA). Human methemoglobin (met-Hb) ELISA kit (CSB-E09493 h, CUSABIO, Huston, TX USA) obtained from CUSABIO was used according to the manufacturers' instructions. All samples were analyzed in triplicate, and the intraassay CV was 4.4 ± 0.7%.

2.11. Statistical Analysis

Data were analyzed using the statistical software SPSS 22.0 (SPSS, Chicago, IL USA), and continuous data are expressed as the means ± SEM. Nonparametric results were examined by the Mann–Whitney U test and Wilcoxon signed ranked test. Parametric results were tested by two-way repeated-measures ANOVA (group × time points) and the Newman–Keuls post hoc test to identify significant changes pre- vs. postintervention and pre- vs. post-graded exercise tests. Correlations were measured by Pearson's correlation coefficient. The level of statistical significance was $p < 0.05$.

3. Results

3.1. Cardiopulmonary Fitness and Hematological and Blood Gas Parameters

There were no differences in anthropometric characteristics, hematological parameters, blood pH, lactate concentration or exercise performance among the groups at baseline (Table 1). Following 6 weeks of training, both the CCT and ECT groups demonstrated increases in work rate, VE, and VO_2 at the ventilation threshold (VT). Moreover, CCT was superior to ECT for enhancing the work rate and VO_2 at VT. At the peak performance, only CCT enhanced the VEmax and VO_2 max, while ECT only resulted in an improvement in the work rate (Table 1). However, 6 weeks of the CTL did not influence hematological parameters or cardiopulmonary responses to a GXT (Table 1).

Table 1. Anthropometric data and ventilatory responses to graded exercise test in concentric and eccentric training groups.

		CCT		ECT		CTL	
		Pre	Post	Pre	Post	Pre	Post
Anthropometrics Characteristics							
Age, year		21.3 ± 0.5	—	21.7 ± 0.4	—	21.6 ± 0.6	—
Height, cm		174 ± 1	—	173 ± 2	—	175 ± 1	—
Weight, kg		67.5 ± 2.3	68.4 ± 1.9	68.1 ± 1.3	67.4 ± 1.5	67.2 ± 2.2	68.0 ± 2.2
Hematological Parameters							
Red blood cells, $10^6/\mu L$		5.10 ± 0.08	5.05 ± 0.05	5.13 ± 0.06	5.06 ± 0.07	5.13 ± 0.06	5.13 ± 0.07
Hb, g/dL		14.9 ± 0.2	14.5 ± 0.2	15.0 ± 0.4	14.9 ± 0.3	15.0 ± 0.3	14.7 ± 0.3
Hematocrit, %		45.4 ± 0.7	44.5 ± 0.6	46.2 ± 0.6	45.3 ± 0.6	46.2 ± 0.6	45.2 ± 0.5
i-STAT Parameters							
Blood pH, unit	Rest	7.37 ± 0.02	7.37 ± 0.01	7.36 ± 0.01	7.37 ± 0.01	7.36 ± 0.02	7.35 ± 0.01
	Ex	$7.23 \pm 0.02\ \#$	$7.21 \pm 0.02\ \#$	$7.19 \pm 0.01\ \#$	$7.19 \pm 0.01\ \#$	$7.19 \pm 0.03\ \#$	$7.19 \pm 0.02\ \#$
Blood lactate, mM	Rest	0.88 ± 0.11	0.87 ± 0.11	0.87 ± 0.06	0.98 ± 0.08	0.89 ± 0.09	0.93 ± 0.11
	Ex	$13.00 \pm 0.59\ \#$	$12.66 \pm 0.64\ \#$	$13.99 \pm 0.51\ \#$	$13.9 \pm 0.49\ \#$	$13.16 \pm 0.69\ \#$	$12.38 \pm 0.73\ \#$
Ventilation Threshold							
Work-rate, watt		125 ± 6	151 ± 6 * †	120 ± 4	136 ± 5 *	121 ± 4.3	122 ± 6.8
V_E, L/min		44.8 ± 2.3	52.3 ± 2.7 *	43.5 ± 1.9	51.8 ± 4.3 *	45.2 ± 1.8	46.2 ± 3.9
VO_2, mL/min/kg		21.3 ± 0.8	26.4 ± 1.0 * †	21.3 ± 0.6	23.3 ± 0.5 *	21.6 ± 1.1	21.2 ± 1.2
% of VO_{2max}, %		59.8 ± 2.0	66.1 ± 2.0 *	60.6 ± 1.7	67.5 ± 1.2 *	60.92 ± 2.1	61.16 ± 1.9
Maximal Exercise Performance							
Work-rate, watt		191 ± 3	223 ± 5 * †	189 ± 4	201 ± 5 *	188 ± 5	190 ± 5
V_E, L/min		107.4 ± 3.2	118.8 ± 2.5 *	111.9 ± 3.7	115.3 ± 2.2	109.95 ± 4.7	108.3 ± 4.3
VO_2, mL/min/kg		35.7 ± 1.1	40.0 ± 0.8 *	35.2 ± 0.7	34.6 ± 0.7	34.1 ± 1.0	34.6 ± 1.5
OUES, unit		814 ± 23	886 ± 20 *	817 ± 16	829 ± 24	816 ± 19	825 ± 20
V_E-VCO_2 slope, unit		36.8 ± 1.5	36.8 ± 1.6	37.6 ± 1.9	38.5 ± 2.4	35.7 ± 1.6	35.7 ± 1

Values were mean $\pm$ SEM. **Hb**, hemoglobin; **V_E**, minute ventilation; **VO_2**, oxygen consumption; **OUES**, oxygen uptake efficiency slope; **CCT**, concentric cycling training; **ECT**, eccentric cycling training; **CTL**, control group. **Pre**, pre-intervention; **Post**, post-intervention; **Rest**, at rest; **Ex**, immediately after the GXT; # $p < 0.05$, Rest vs. Ex; * $p < 0.05$, Pre vs. Post; † $p < 0.05$, CCT vs. ECT.

3.2. Pain Scale Scores, Heart Rate and Systolic Blood Pressure during the Training Period

The CCT group had significantly higher levels of pain, H, and SBP than the ECT group throughout the 6 week training period (Figure 2). Concerning the assessment of pain or soreness, the specific pain scale score was close to zero in both groups before each training session (Figure 2A).

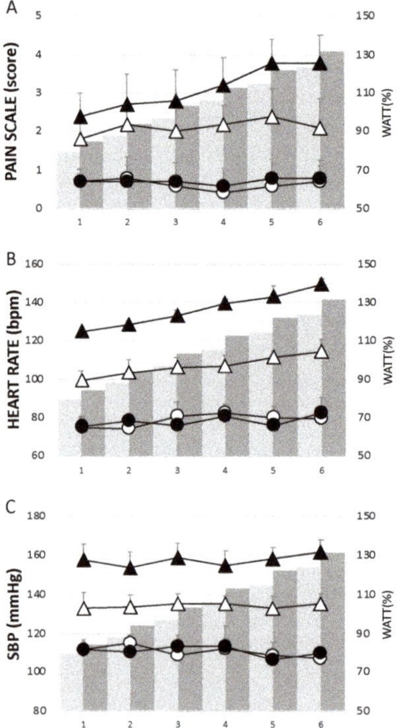

Figure 2. The physiological responses in each week during the training period. (**A**) pain scale score, (**B**) heart rate, and (**C**) systolic blood pressure. Whites— eccentric cycling training (ECT); Blacks— concentric cycling training (CCT); Dots—before training; Triangles—after training. Values were mean ± SEM.

3.3. Erythrocyte Senescence-Related Markers and Antioxidation Capacity

The ratios of Ex to Rt in CD147 and CD47 cells were less than 1 before training, indicating enhanced senescence in erythrocytes due to an acute GXT. After the interventions, these ratios significantly increased to nearly 1 in response to a GXT (Figure 3A,B). Intracellular ROS levels were significantly increased after an acute GXT among the three groups; however, both training groups had lower ROS production related to the GXT after training (Figure 3C). Furthermore, as Figure 3D,F shows, a higher TBHP concentration induced greater ROS generation. Nevertheless, the two exercise regimens significantly diminished the extent of ROS generation under 50 and 100 mM TBHP conditions (Figure 3D,E). No alteration was observed in the CTL group (Figure 3F).

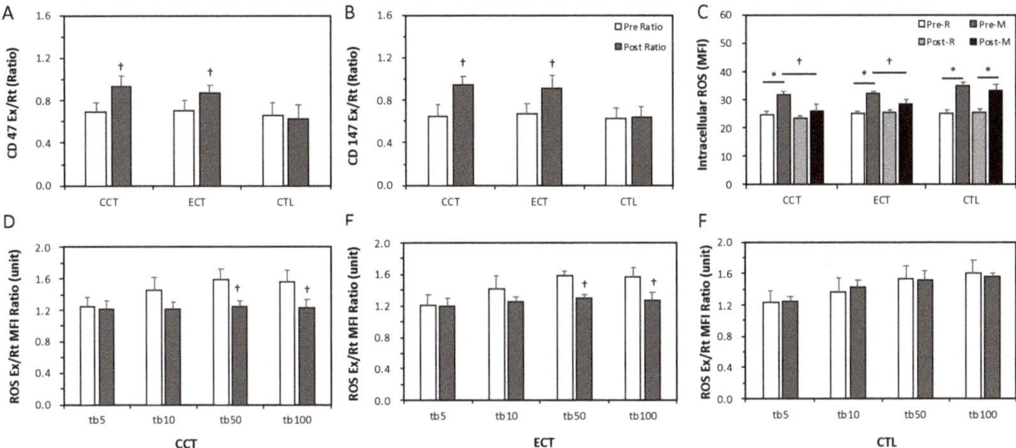

Figure 3. Effects of various ECT and CCT on the erythrocyte senescence-related biomarkers, intracellular reactive oxygen species (ROS) level and the ROS dose-response. (**A**) the ratio of post-graded exercise test (GXT) to pre-GXT in CD47, (**B**) the ratio of post-GXT to pre-GXT in CD147, (**C**) the intracellular ROS level among three groups; the ratio of Ex to Rt ROS response in erythrocytes treated with different concentrations of tert-butyl hydroperoxide (tb): (**D**) the CCT group, (**E**) the ECT group, and (**F**) the control (CTL) group. **Pre**, pre-intervention; Post, post-intervention; M or Ex, immediately after a GXT; R or Rt, at rest. * $p < 0.05$, R vs. M; † $p < 0.05$, Pre vs. Post. Values were mean $\pm$ SEM.

3.4. Target Metabolite Analysis of Glycolysis and Pentose Phosphate Pathway Intermediates

At baseline, the acute GXT showed greater glucose consumption and was accompanied by a series of unchanged levels of downstream metabolites until G3P (Figure 4A). The GSSG/GSH ratio significantly increased following an acute GXT, indicating an accumulation of oxidative stress. In addition, 2,3-BPG was depleted after the GXT.

Both glycolysis and the PPP were upregulated due to exercise training. Although no change was observed in the glucose level, the downstream metabolites were markedly elevated even under resting conditions. Interestingly, although 2,3-BPG was lower after training, the end product of glycolysis, lactate, was dramatically increased. The PPP flux was also facilitated, as evidenced by a decrease in 6PG and a constant GSSG/GSH ratio. Furthermore, the higher levels of Ru5P and GSH suggested a significant enhancement of GSH biosynthesis under stress. X5P and E4P are intermediates between the PPP and glycolysis, which was dramatically enhanced in both training groups.

In addition, the level of met-Hb was significantly increased by the acute GXT compared to the rest among the three groups, whereas the extent of elevation was diminished after training. The intracellular pH in erythrocytes did not change relative to the GXT before training, whereas both CCT and ECT lowered the pH value after 6 weeks (Figure 4B).

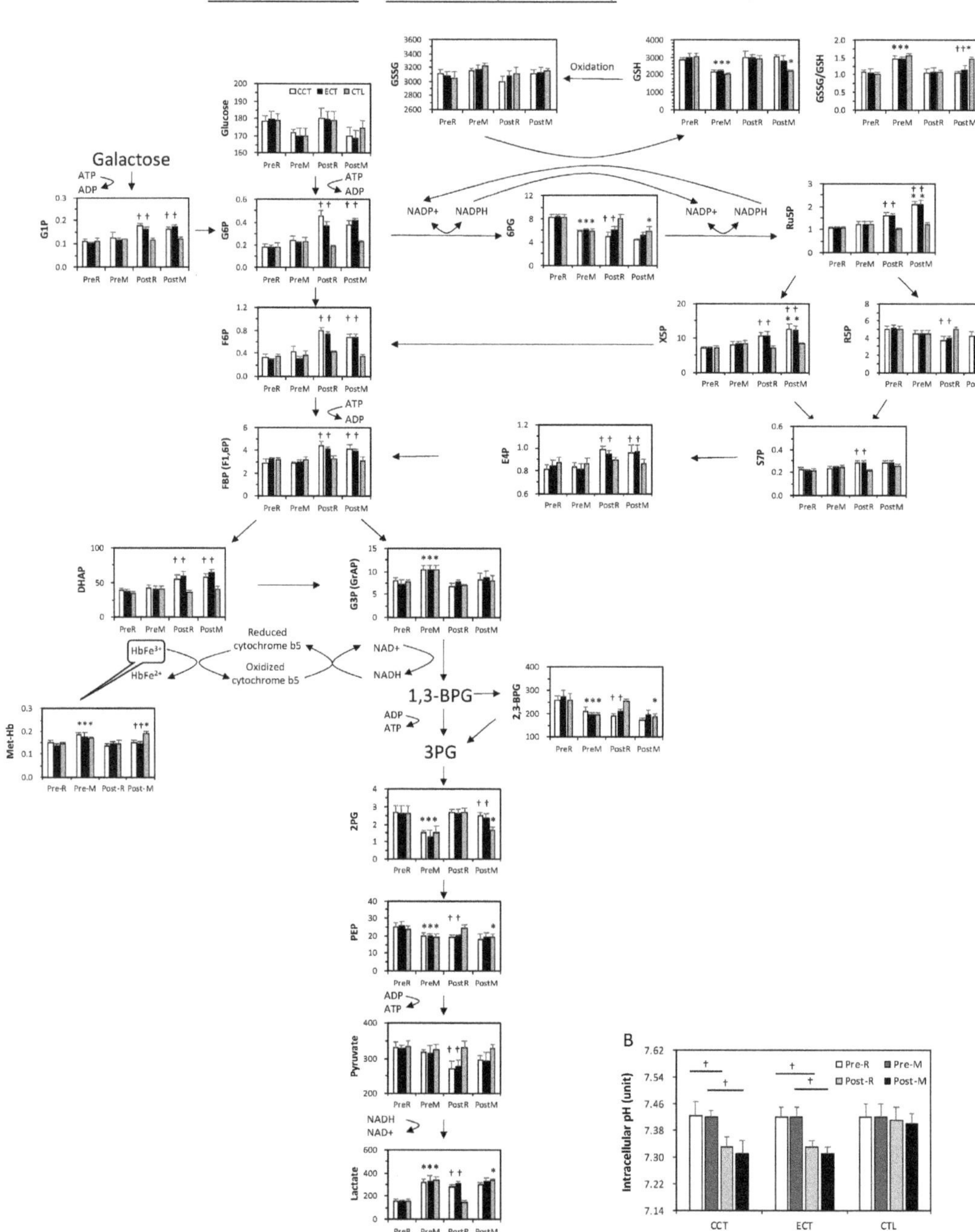

Figure 4. The target metabolite analysis of glycolysis and the pentose phosphate pathway intermediates in erythrocytes in ECT and CCT. (**A**) Levels of metabolites in pentose phosphate pathway and glycolytic pathway (n = 5) (**B**) the intracellular pH of erythrocytes. **Pre**, pre-intervention; **Post**, post-intervention; R, at rest; M, immediately after a GXT; * $p < 0.05$, R vs. M; † $p < 0.05$, Pre vs. Post. Values were mean ± SEM.

3.5. Erythrocyte O_2 Release Capacity in Normal Conditions

As Figure 5A,F shows, oxygen was absorbed into the cell when isolated erythrocytes were added to the normoxia chamber, which produced a negative oxygen pressure difference (O_2 pressure-diff). Furthermore, this oxygen was released at lactate acid concentrations of 1 and 4 mM. After an acute GXT, the magnitudes of oxygen absorption and release were diminished (Figure 5A,C) and coupled with a reduced flux velocity (Figure 5D,F). Following the interventions, both training groups exhibited a diminished O_2 pressure-diff and velocity in 0 and 1 mM lactate acid conditions at rest and even after the GXT. However, both CCT and ECT induced maintenance of these parameters at resting levels in the 4 mM lactate acid condition (Figure 5C,F).

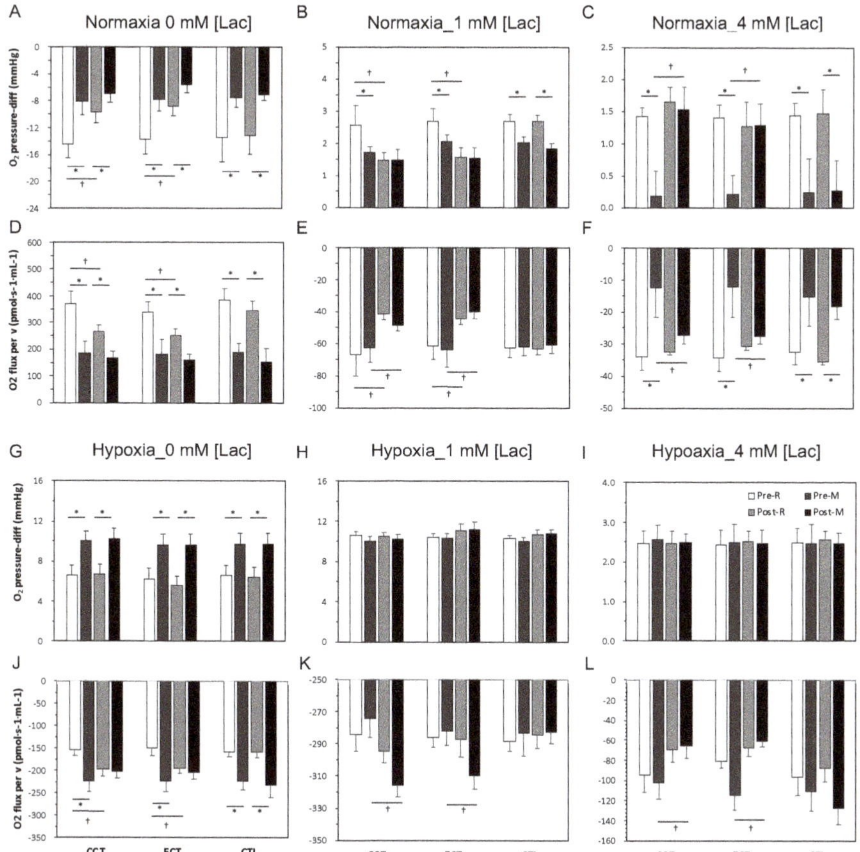

Figure 5. Measurement of oxygen release capacity of erythrocytes among three groups in normoxia and hypoxia conditions. Levels of O_2 pressure-diff in normoxia condition: (**A**) at 0 mM [lac], (**B**) at 1 mM [lac], and (**C**) at 4 mM [lac]; oxygen flux per volume in normoxia condition: (**D**) at 0 mM [lac], (**E**) at 1 mM [lac], and (**F**) at 4 mM [lac]; levels of O_2 pressure-diff in hypoxia condition: (**G**) at 0 mM [lac], (**H**) at 1 mM [lac], and (**I**) at 4 mM [lac]; oxygen flux per volume in hypoxia condition: (**J**) at 0 mM [lac], (**K**) at 1 mM [lac], and (**L**) at 4 mM [lac]; Pre, pre-intervention; Post, post-intervention; R, at rest; M, immediately after a GXT; * $p < 0.05$, R vs. M; † $p < 0.05$, Pre vs. Post. Values were mean ± SEM.

3.6. Erythrocyte O_2 Release Capacity in Hypoxia Conditions

O_2 was released immediately (O_2 pressure-diff was positive) from erythrocytes when erythrocytes were added to the hypoxia chamber, and the magnitude of release was further augmented due to the GXT (Figure 5G,L). An enhanced release velocity was also noticed (Figure 5J). After training, although no alternation in the oxygen release amount was observed at rest or even after the GXT, a faster velocity was observed in both the CCT and ECT groups under 0, 1 and 4 mM lactate acid (Figure 5J,K).

3.7. Relationships between GSSG/GSH and Lactate/Pyruvate and between Intracellular pH and Lactate Concentration

Figure 6A shows that the increase in the GSSG/GSH ratio caused by the GXT was significantly linearly related to the augmented lactate/pyruvate ratio ($r = 0.72$, $p < 0.05$). However, this correlation was blunted after training in both the CCT and ECT groups (Figure 6B, $r = 0.29$, $p = 0.12$). Furthermore, Figure 6C demonstrates that the lowered intracellular pH was moderately related to the greater lactate concentration ($r = -0.50$, $p < 0.05$).

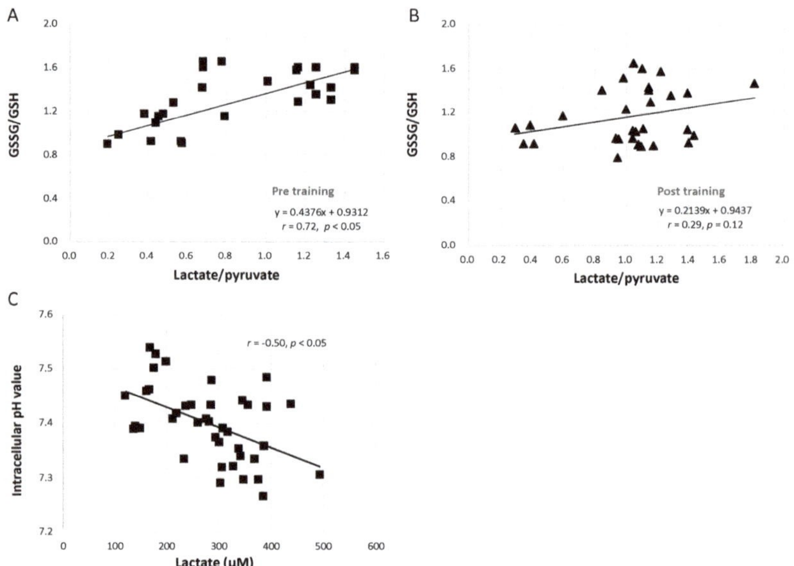

Figure 6. Correlation analysis between GSSG/GSH, lactate/pyruvate, intracellular pH value and lactate concentration. (**A**) relationship between GSSG/GSH and lactate/pyruvate before interventions, (**B**) relationship between GSSG/GSH and lactate/pyruvate after interventions, and (**C**) the correlation between intracellular pH value and lactate concentration. GSH, glutathione; GSSG, glutathione disulfide.

4. Discussion

Erythrocyte metabolism includes glycolytic pathways producing both energy and oxidation–reduction intermediates that support O2 transport and antioxidative capacity. This study is the first to demonstrate that both CCT and power-matched ECT not only ameliorate antioxidation capacity in erythrocytes but also significantly increase the flux of anaerobic glycolysis to facilitate oxygen release efficacy. We further elucidated that the reduced oxygen affinity is due to greater lactate synthesis and not to the production of 2,3-BPG. Although ECT did not result in significant improvement in VO2 max, the im-proved VT performance indicates the positive effect ECT has on the aerobic capacity of young and sedentary men.

Several studies have reported that blood GSSG and thus the GSH/GSSG ratio decrease in response to acute exercise, and regular exercise may increase antioxidative capacity [28]. After both CCT and ECT, neither the GSSG/GSH ratio nor GSH decreased due to the

GXT, while enhanced anaerobic glycolysis provided more precursors to activate the PPP. In addition, the downstream X5P and E4P, from the PPP back to glycolysis, also increased after training, thus suggesting significant enhancement of GSH biosynthesis under stress. The linear relationship between GSSG/GSH and lactate/pyruvate was disrupted after training, which may be due to a changed dominance of energy or/and antioxidant production [29]. In addition, the oxidative environment leads to the production of $Fe3^+$ (met-Hb). To restore Hb function, met-Hb must be reduced mainly by NADH-dependent cytochrome b5 reductase [30].

An interesting aspect of the metabolic pathways is that intracellular pH (pHi) regulates both the glycolytic pathway and the PPP. As with glycolysis, the optimum activity of the enzymes driving the PPP occurs at an alkaline pHi [31]. Generally, the presence of NADPH blocks PPP negative feedback control and shifts metabolism from the PPP to glycolysis, thus increasing the formation of NAD^+ [32]. Although NADH does not directly participate in the reduction of $Fe3^+$ to $Fe2^+$ in hemoglobin, it has the ultimate responsibility of providing the reducing power needed for such a reaction [33]. Under normal homeostasis in general, and especially in the case of the high glycolytic flux that is required during high-intensity exercise, lactate dehydrogenase oxidizes NADH back to NAD^+ in the conversion of pyruvate to lactate, thereby maintaining necessary levels of the cofactor for the continuation of glycolysis. Cyclists in the high-class group had a higher posttest lactate/pyruvate ratio, which is proportional to $NADH/NAD^+$ and a marker of glycolytic capacity [34]. Additionally, it has been confirmed that high glucose levels induce in-creases in lactate and 6PG production in vitro and ensure a longer supply of energy sources, preventing the loss of GSH [35].

The blood lactate that was progressively elevated with exercise intensity further reduces local blood pH and thus enhances the Bohr effect to attenuate O_2 affinity and facilitate O_2 release [17]. In contrast, pulmonary O_2 uptake is enhanced, but muscle unloading is hindered with high-affinity hemoglobin [36]. The capillary transit times were very limited; thus, the exchange speed is critical for evaluating the physiological fitness of erythrocytes [37]. Therefore, we developed a novel method for quantifying gas exchange in a constant number of erythrocytes and used it to assess the quality and quantity of O_2 releasing capacity.

To clarify the oxygen release efficacy of erythrocytes, we measured the PO_2, oxygen release/absorption velocity and acceleration under 0, 1, and 4 mM lactic acid concentrations to mimic resting and near AT conditions, respectively. The PO_2 increased (oxygen release) when lactic acid was added. Even with a smaller amount of O_2 being supplied to the tissue, the efficacy was enhanced with faster acceleration and velocity, which indicates better efficiency for release. The improved release of oxygen efficacy at 4 mM [lac] might be associated with improved cycling performance before reaching the anaerobic threshold. In addition, a diminished magnitude and velocity of O_2 absorption after the acute GXT were noticed. We first speculated that this was a consequence of sufficiently oxygenated Hb after high oxygen demand activity or that this impaired quantity and quality may also be related to increases in oxidative and met-Hb levels after exhaustive exercise [38]. We further demonstrated an enhanced, strong correlation between lactate concentration and oxygen release magnitude and velocity under hypoxic conditions. The O_2 affinity of athletes is lower than that of untrained subjects, which is consistent with our results [39]. Slow VO_2 kinetics incur a high O_2 deficit, usually resulting in poor exercise tolerance [40].

Lactic acid plays a vital indirect role in tissue O_2 delivery apart from the direct allosteric interaction of lactate ions with Hb [41]. Lactic acid increases the Bohr shift via acidification as well as via liberation of CO_2 [42]. Therefore, the lower affinity of hemoglobin for the O_2 of erythrocytes in athletes at rest is maintained by the factor(s) dominating pH and lactate-driven regulation. Under heavy exercise (above the lactic acidosis threshold), acidification of muscle capillary blood by lactic acid accounts for virtually all of the oxygen unloaded from Hb [43].

Erythrocytes must be considered a potential storage site of lactate, storage of which leads to a greater gradient from the interstitial fluid to plasma. This mechanism improves the rate of release from muscle and ameliorates exercise performance [44]. However, a previous study demonstrated that the lactate distribution in erythrocytes and plasma after high-intensity running was not different between trained and untrained subjects. Hence, lactate uptake by erythrocytes cannot or can only in part be seen as a contributor to aerobic athletic performance [45]. Traditionally, a higher Bohr effect is supposed to be related to a higher 2,3-BPG in erythrocytes [46]. However, the generation of 2,3-BPG results in the overall production of ATP per mole of glucose is decreased to zero. Therefore, accumulation of 2,3-BPG leads to decreased production of 2,3-BPG by competitive feedback inhibition of di-phosphoglycerate mutase [47]. The relative ratio of 2,3-BPG synthase to 2,3-BPG phosphatase decreased dramatically with decreasing pH value [48]. The lactate effect even increased after 2,3-BPG depletion [49]. In this study, the presence of large lactate concentrations leading to lower pH values may effectively limit the production of 2,3-BPG [50]. When the downstream enzymes of 2,3-BPG, such as pyruvate kinase and lactate dehydrogenase, maintained higher activities, the enzyme activities of other pathways were significantly repressed [48]. Therefore, this reversed flux of the 2,3-BPG shunt is crucial in maintaining the activities of the latter part of glycolysis and the production of ATP in the latter half of the storage period.

The results in this study clearly presented that, although both ECT and CCT ameliorate the erythrocyte antioxidant and oxygen releasing capacity, thus further delays the anaerobic threshold, yet only ECT has significantly less cardiopulmonary stress without undesirable fatigue or pain impact during the whole training period. Therefore, we suggest that ECT is preferred to those who have exercise intolerance or low physical activity, whereas CCT may be more feasible for those who have general physical activity to increase the ability to cope with the physical demands of daily activity. These findings provide a new suggestion on why ECT is worthy to further developed as a suitable training strategy in cardiopulmonary rehabilitation or the elderly.

A small sample size (n = 12 in each group) is a major limitation of this study. However, the results for aerobic capacity and the novel interpretation of the O_2 release and antioxidative mechanisms in the metabolic pathways obtained from this investigation have high statistical power (0.862 to 1.000). We speculate that metabolic alteration of erythrocytes generates higher ATP concentrations followed by lactate production. We did not directly detect the ATP concentration because of the fast rate of ATP hydrolysis. Therefore, in this study, we indirectly inferred ATP demand by the lactate/pyruvate ratio. We did not see potential model alterations in Hb affinity as blood traverses the exercising muscles in accordance with local changes in temperature, pHi, or CO_2. There are certainly differences between and challenges with in vivo vs. in vitro measurements of O_2 dissociation curve dynamics in both the lungs and muscles in response to variables such as temperature, pHi, and 2,3-BPG. A few studies have shown that the additive effects of temperature and pH are responsible for shifting the O_2 dissociation curve affinity, especially with prolonged exercise [51]. Additionally, the subjects tended to be young and healthy; thus, further clinical evidence is still required to extrapolate the present results to patients with hemorheological or hemodynamic disorders.

Although the glutathione system is a principal nonenzymic antioxidant system in erythrocytes yet, GSSG is rapidly formed, but it quickly disappears once the oxidative stimulus is interrupted; conversely, S-glutathionylated proteins (PSSGs) may be produced more slowly but are more durable [52]. Therefore, the PSSGs is a worthy parameter for further investigation [53]. In addition, although previously study suggested that exercise-induced changes in the nonenzymatic glutathione system seem to be more effective in erythrocytes [54]. Nevertheless, many studies have indicated the activity of glutathione peroxidase (GPx) plays a key component in the antioxidant experiment [55], and regular cardiovascular training increased GPx activity in skeletal muscle [56]. Taking together, both PSSGs and the immunoblotting for GPx are very used to supply important information on the state of this antioxidant network in the future. To assess the reliabilities of biomarkers, metabolites and oxygen releasing capacity to exercise, the subjects ($n = 5$) in a prior study were tested twice at two-day intervals. Results of responses to exercise were highly reproducible from day to day. The intraclass correlation coefficients were from 0.811 to 0.954. Additionally, it requires separate analytical measurements for GSH and GSSG for accurate analysis and specific methodological procedures needed to detect samples [57]. Although the use of classical and well-validated in previous studies [26,58], techniques to perform our measurements, requiring immediate and complex processing of blood samples [59]. This limits the possibility of receiving samples from different centers to be analyzed. In the present study, we tested our participants at the same time of the day and asked them to record their nutritional intake and to maintain the same diet (data not shown). Thus, we assume that our results well represent the physical adaptations after exercise training. Importantly, other more adequate and precise methodologies should be considered in future studies [53].

5. Conclusions

This study presented evidence that both ECT and CCT simultaneously promote flux into the pentose phosphate pathway and anaerobic glycolysis pathway in response to overcoming accumulated oxidative stress and regulating internal O_2 dissociation, respectively. The adaptations of the metabolite process not only increased the synthesis of GSH but also enhanced the production of lactate in glucose metabolism in trained erythrocytes. The lower intracellular pH value related to lactate, instead of 2,3-BPG, ameliorated the O_2 release efficacy of erythrocytes under different O_2 gradients. In addition, the reduced amount of met-Hb also contributed to the O_2 release. The above experimental findings reflect many positive effects of both interventions and provide a novel interpretation of delayed anaerobic threshold by ameliorated O_2 release and antioxidative mechanism in the metabolic pathway (Figure 7). Therefore, ECT is a feasible and promising exercise regimen that promotes a less cardiovascular stress way to exercise without undesirable fatigue impact and provides important implications for those who have exercise intolerance.

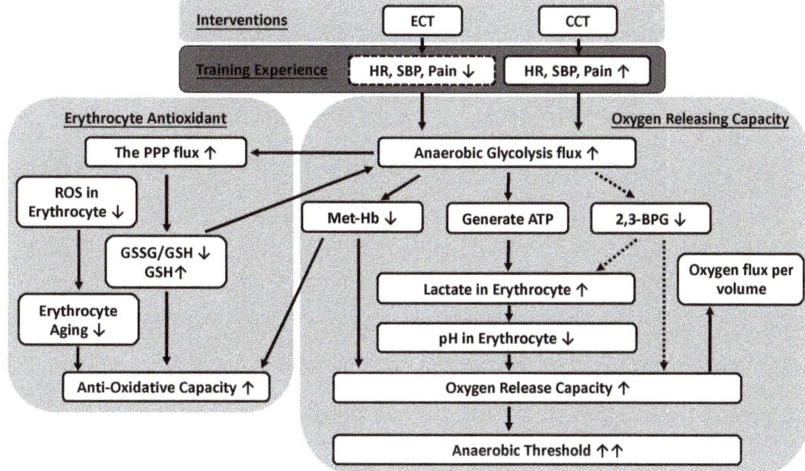

Figure 7. Possible mechanisms of anaerobic glycolytic pathways producing both oxygen releasing and antioxidative capacity caused by eccentric (ECT) and concentric cycling training (CCT). Both CCT and ECT at a given workload enhance an anaerobic glycolysis flux to ameliorate antioxidative capacity in erythrocytes, as well as significantly facilitate oxygen release efficacy. The reduced oxygen affinity is due to greater lactate synthesis and lower intracellular pH, instead of the production of 2,3-BPG. Solid line: positive regulation; dotted line: negative regulation.

Supplementary Materials: The following are available online at https://www.mdpi.com/2076-392 1/10/2/285/s1, Figure S1: the scheme of testing oxygen release capacity in erythrocytes by using high-resolution respirometry.

Author Contributions: Conceptualization, J.-S.W. and Y.-C.H.; methodology, J.-S.W., Y.-C.H., H.-Y.T. and M.-L.C.; formal analysis, J.-S.W. and Y.-C.H.; investigation, Y.-C.H. and H.-Y.T.; writing—original draft preparation, Y.-C.H.; writing—review and editing, J.-S.W., Y.-C.H., H.-Y.T., M.-L.C., C.-Y.H. and K.-M.C.; supervision, J.-S.W.; funding acquisition, J.-S.W. and Y.-C.H. All authors have read and agreed to the published version of the manuscript.

Funding: This research was funded by the National Science Council of Taiwan, grant number NSC 108-2314-B-182 -039 -MY3 and 109-2314-B-468 -008 -MY2, and by the Chang Gung Medical Research Program, grant number CMRPD1J0222 and CMRPG2F0193.

Institutional Review Board Statement: The study was conducted according to the guidelines of the Declaration of Helsinki and approved by the Institutional Review Board of Chang Gung Memorial Hospital in Taiwan (protocol code 201900415A3 and 2019/5/17 approval).

Informed Consent Statement: Informed consent was obtained from all subjects involved in the study.

Data Availability Statement: All data is contained within the article.

Acknowledgments: The authors would like to thank the volunteers for their enthusiastic participation and assistance from medical writers, proof-readers and editors.

Conflicts of Interest: The authors declare no conflict of interest. The funders had no role in the design of the study; in the collection, analyses, or interpretation of data; in the writing of the manuscript, or in the decision to publish the results.

References

1. Hody, S.; Croisier, J.L.; Bury, T.; Rogister, B.; Leprince, P. Eccentric Muscle Contractions: Risks and Benefits. *Front. Physiol.* **2019**, *10*. [CrossRef]
2. Lastayo, P.C.; Reich, T.E.; Urquhart, M.; Hoppeler, H.; Lindstedt, S.L. Chronic eccentric exercise: Improvements in muscle strength can occur with little demand for oxygen. *Am. J. Physiol.* **1999**, *276*, 611–615. [CrossRef] [PubMed]
3. Haynes, A.; Linden, M.D.; Chasland, L.C.; Nosaka, K.; Maiorana, A.; Dawson, E.A.; Dembo, L.H.; Naylor, L.H.; Green, D.J. Acute impact of conventional and eccentric cycling on platelet and vascular function in patients with chronic heart failure. *J. Appl. Physiol.* **2017**, *122*, 1418–1424. [CrossRef]
4. Gault, M.L.; Willems, M.E. Aging, functional capacity and eccentric exercise training. *Aging Dis.* **2013**, *4*, 351–363. [CrossRef] [PubMed]
5. Nickel, R.; Troncoso, F.; Flores, O.; Gonzalez-Bartholin, R.; Mackay, K.; Diaz, O.; Jalon, M.; Peñailillo, L. Physiological response to eccentric and concentric cycling in patients with chronic obstructive pulmonary disease. *Appl. Physiol. Nutr. Metab.* **2020**, *45*, 1232–1237. [CrossRef] [PubMed]
6. Douglas, J.; Pearson, S.; Ross, A.; McGuigan, M. Chronic Adaptations to Eccentric Training: A Systematic Review. *Sports Med.* **2017**, *47*, 917–941. [CrossRef]
7. Peñailillo, L.; Mackay, K.; Gonzalez, R.; Valladares, D.; Contreras-Ferrat, A.; Zbinden-Foncea, H.; Nosaka, K. Effects of Eccentric and Concentric Cycling on Markers of Oxidative Stress and Inflammation in Elderly. *Med. Sci. Sports Exerc.* **2018**, *50*, 518. [CrossRef]
8. Peñailillo, L.; Blazevich, A.; Nosaka, K. Energy expenditure and substrate oxidation during and after eccentric cycling. *Eur. J. Appl. Physiol.* **2014**, *114*, 805–814. [CrossRef]
9. Julian, V.; Thivel, D.; Miguet, M.; Pereira, B.; Costes, F.; Coudeyre, E.; Duclos, M.; Richard, R. Eccentric cycling is more efficient in reducing fat mass than concentric cycling in adolescents with obesity. *Scand. J. Med. Sci. Sports* **2019**, *29*, 4–15. [CrossRef] [PubMed]
10. Aoi, W.; Naito, Y.; Yoshikawa, T. Role of oxidative stress in impaired insulin signaling associated with exercise-induced muscle damage. *Free Radic. Biol. Med.* **2013**, *65*, 1265–1272. [CrossRef] [PubMed]
11. Barreto, R.V.; de Lima, L.C.R.; Denadai, B.S. Moving forward with backward pedaling: A review on eccentric cycling. *Eur. J. Appl. Physiol.* **2021**, *121*, 381–407. [CrossRef] [PubMed]
12. Mohanty, J.G.; Nagababu, E.; Rifkind, J.M. Red blood cell oxidative stress impairs oxygen delivery and induces red blood cell aging. *Front. Physiol.* **2014**, *5*, 84. [CrossRef]
13. Laaksonen, D.E.; Atalay, M.; Niskanen, L.; Uusitupa, M.; Hanninen, O.; Sen, C.K. Blood glutathione homeostasis as a determinant of resting and exercise-induced oxidative stress in young men. *Redox Rep. Commun. Free Radic. Res.* **1999**, *4*, 53–59. [CrossRef]

14. Marzatico, F.; Pansarasa, O.; Bertorelli, L.; Somenzini, L.; Della Valle, G. Blood free radical antioxidant enzymes and lipid peroxides following long-distance and lactacidemic performances in highly trained aerobic and sprint athletes. *J. Sports Med. Phys. Fit.* **1997**, *37*, 235–239.

15. González-Bartholin, R.; Mackay, K.; Valladares, D.; Zbinden-Foncea, H.; Nosaka, K.; Peñailillo, L. Changes in oxidative stress, inflammation and muscle damage markers following eccentric versus concentric cycling in older adults. *Eur. J. Appl. Physiol.* **2019**, *119*, 2301–2312. [CrossRef] [PubMed]

16. Elmer, S.J.; McDaniel, J.; Martin, J.C. Alterations in neuromuscular function and perceptual responses following acute eccentric cycling exercise. *Eur. J. Appl. Physiol.* **2010**, *110*, 1225–1233. [CrossRef] [PubMed]

17. Böning, D.; Hollnagel, C.; Boecker, A.; Göke, S. Bohr shift by lactic acid and the supply of O2 to skeletal muscle. *Respir. Physiol.* **1991**, *85*, 231–243. [CrossRef]

18. Mairbaurl, H. Red blood cells in sports: Effects of exercise and training on oxygen supply by red blood cells. *Front. Physiol.* **2013**, *4*, 332. [CrossRef]

19. Chaves, N.A.; Alegria, T.G.P.; Dantas, L.S.; Netto, L.E.S.; Miyamoto, S.; Bonini Domingos, C.R.; da Silva, D.G.H. Impaired antioxidant capacity causes a disruption of metabolic homeostasis in sickle erythrocytes. *Free Radic. Biol. Med.* **2019**, *141*, 34–46. [CrossRef]

20. Rogers, S.C.; Said, A.; Corcuera, D.; McLaughlin, D.; Kell, P.; Doctor, A. Hypoxia limits antioxidant capacity in red blood cells by altering glycolytic pathway dominance. *FASEB J.* **2009**, *23*, 3159–3170. [CrossRef] [PubMed]

21. Brown, S.P.; Keith, W.B. The effects of acute exercise on levels of erythrocyte 2,3-bisphosphoglycerate: A brief review. *J. Sports Sci.* **1993**, *11*, 479–484. [CrossRef]

22. Guesnon, P.; Poyart, C.; Bursaux, E.; Bohn, B. The binding of lactate and chloride ions to human adult hemoglobin. *Respir. Physiol.* **1979**, *38*, 115–129. [CrossRef]

23. Roth, S.; Gmünder, H.; Dröge, W. Regulation of intracellular glutathione levels and lymphocyte functions by lactate. *Cell. Immunol.* **1991**, *136*, 95–104. [CrossRef]

24. American College of Sports Medicine. *ACSM's Guidelines for Exercise Testing and Prescription*, 10th ed.; Lippincott Williams & Wilkins: Philadelphia, PA, USA, 2017.

25. Chou, S.L.; Huang, Y.C.; Fu, T.C.; Hsu, C.C.; Wang, J.S. Cycling Exercise Training Alleviates Hypoxia-Impaired Erythrocyte Rheology. *Med. Sci. Sports Exerc.* **2016**, *48*, 57–65. [CrossRef]

26. Tang, H.Y.; Ho, H.Y.; Wu, P.R.; Chen, S.H.; Kuypers, F.A.; Cheng, M.L.; Chiu, D.T.L. Inability to Maintain GSH Pool in G6PD-Deficient Red Cells Causes Futile AMPK Activation and Irreversible Metabolic Disturbance. *Antioxid. Redox Sign.* **2015**, *22*, 744–759. [CrossRef]

27. McDonald, J.G.; Smith, D.D.; Stiles, A.R.; Russell, D.W. A comprehensive method for extraction and quantitative analysis of sterols and secosteroids from human plasma. *J. Lipid Res.* **2012**, *53*, 1399–1409. [CrossRef] [PubMed]

28. Kawamura, T.; Muraoka, I. Exercise-Induced Oxidative Stress and the Effects of Antioxidant Intake from a Physiological Viewpoint. *Antioxidants* **2018**, *7*, 119. [CrossRef]

29. Sastre, J.; Asensi, M.; Gasco, E.; Pallardo, F.V.; Ferrero, J.A.; Furukawa, T.; Vina, J. Exhaustive physical exercise causes oxidation of glutathione status in blood: Prevention by antioxidant administration. *Am. J. Physiol.* **1992**, *263*, 992–995. [CrossRef] [PubMed]

30. Maeda, S.; Kobori, H.; Tanigawa, M.; Sato, K.; Yubisui, T.; Hori, H.; Nagata, Y. Methemoglobin reduction by NADH-cytochrome b(5) reductase in Propsilocerus akamusi larvae. *Comp. Biochem. Physiol. Part B Biochem. Mol. Biol.* **2015**, *185*, 54–61. [CrossRef]

31. Alfarouk, K.O.; Ahmed, S.B.M.; Elliott, R.L.; Benoit, A.; Alqahtani, S.S.; Ibrahim, M.E.; Bashir, A.H.H.; Alhoufie, S.T.S.; Elhassan, G.O.; Wales, C.C.; et al. The Pentose Phosphate Pathway Dynamics in Cancer and Its Dependency on Intracellular pH. *Metabolites* **2020**, *10*, 285. [CrossRef]

32. Alfarouk, K.O.; Muddathir, A.K.; Shayoub, M.E.A. Tumor Acidity as Evolutionary Spite. *Cancers* **2011**, *3*, 408–414. [CrossRef]

33. Ho, H.Y.; Cheng, M.L.; Chiu, D.T. Glucose-6-phosphate dehydrogenase-from oxidative stress to cellular functions and degenerative diseases. *Redox Rep. Commun. Free Radic. Res.* **2007**, *12*, 109–118. [CrossRef]

34. San-Millán, I.; Stefanoni, D.; Martinez, J.L.; Hansen, K.C.; D'Alessandro, A.; Nemkov, T. Metabolomics of Endurance Capacity in World Tour Professional Cyclists. *Front. Physiol.* **2020**, *11*, 578. [CrossRef]

35. Viskupicova, J.; Blaskovic, D.; Galiniak, S.; Soszyński, M.; Bartosz, G.; Horakova, L.; Sadowska-Bartosz, I. Effect of high glucose concentrations on human erythrocytes in vitro. *Redox Biol.* **2015**, *5*, 381–387. [CrossRef] [PubMed]

36. Dominelli, P.B.; Wiggins, C.C.; Baker, S.E.; Shepherd, J.R.A.; Roberts, S.K.; Roy, T.K.; Curry, T.B.; Hoyer, J.D.; Oliveira, J.L.; Joyner, M.J. Influence of high affinity haemoglobin on the response to normoxic and hypoxic exercise. *J. Physiol.* **2020**, *598*, 1475–1490. [CrossRef]

37. Richardson, S.L.; Hulikova, A.; Proven, M.; Hipkiss, R.; Akanni, M.; Roy, N.B.A.; Swietach, P. Single-cell O$_2$ exchange imaging shows that cytoplasmic diffusion is a dominant barrier to efficient gas transport in red blood cells. *Proc. Natl. Acad. Sci. USA* **2020**, *117*, 10067. [CrossRef]

38. Xiong, Y.; Xiong, Y.; Wang, Y.; Zhao, Y.; Li, Y.; Ren, Y.; Wang, R.; Zhao, M.; Hao, Y.; Liu, H.; et al. Exhaustive-exercise-induced oxidative stress alteration of erythrocyte oxygen release capacity. *Canad. J. Physiol. Pharmacol.* **2018**, *96*, 953–962. [CrossRef]

39. Makhro, A.; Haider, T.; Wang, J.; Bogdanov, N.; Steffen, P.; Wagner, C.; Meyer, T.; Gassmann, M.; Hecksteden, A.; Kaestner, L.; et al. Comparing the impact of an acute exercise bout on plasma amino acid composition, intraerythrocytic Ca2+ handling, and red cell function in athletes and untrained subjects. *Cell Calcium* **2016**, *60*, 235–244. [CrossRef]

40. Calbet, J.; González-Alonso, J.; Helge, J.; Søndergaard, H.; Munch-Andersen, T.; Saltin, B.; Boushel, R. Central and peripheral hemodynamics in exercising humans: Leg vs arm exercise. *Scand. J. Med. Sci. Sports* **2015**, *25*, 144–157. [CrossRef] [PubMed]
41. Robergs, R.A.; Ghiasvand, F.; Parker, D. Biochemistry of exercise-induced metabolic acidosis. *Am. J. Physiol. Regul. Integr. Comp. Physiol.* **2004**, *287*, 502–516. [CrossRef]
42. Jensen, F.B. Red blood cell pH, the Bohr effect, and other oxygenation-linked phenomena in blood O_2 and CO_2 transport. *Acta Physiol. Scand.* **2004**, *182*, 215–227. [CrossRef] [PubMed]
43. Stringer, W.; Wasserman, K.; Casaburi, R.; Porszasz, J.; Maehara, K.; French, W. Lactic acidosis as a facilitator of oxyhemoglobin dissociation during exercise. *J. Appl. Physiol.* **1994**, *76*, 1462–1467. [CrossRef] [PubMed]
44. Connes, P.; Caillaud, C.; Mercier, J.; Bouix, D.; Casties, J.F. Injections of recombinant human erythropoietin increases lactate influx into erythrocytes. *J. Appl. Physiol.* **2004**, *97*, 326–332. [CrossRef] [PubMed]
45. Tomschi, F.; Bizjak, D.A.; Predel, H.-G.; Bloch, W.; Grau, M. Lactate distribution in red blood cells and plasma after a high intensity running exercise in aerobically trained and untrained subjects. *J. Hum. Sport Exerc.* **2018**, *13*, 384–392. [CrossRef]
46. Mairbaurl, H.; Humpeler, E.; Schwaberger, G.; Pessenhofer, H. Training-dependent changes of red cell density and erythrocytic oxygen transport. *J. Appl. Physiol.* **1983**, *55*, 1403–1407. [CrossRef] [PubMed]
47. Oslund, R.C.; Su, X.; Haugbro, M.; Kee, J.-M.; Esposito, M.; David, Y.; Wang, B.; Ge, E.; Perlman, D.H.; Kang, Y.; et al. Bisphosphoglycerate mutase controls serine pathway flux via 3-phosphoglycerate. *Nat. Chem. Biol.* **2017**, *13*, 1081–1087. [CrossRef]
48. Nishino, T.; Yachie-Kinoshita, A.; Hirayama, A.; Soga, T.; Suematsu, M.; Tomita, M. Dynamic simulation and metabolome analysis of long-term erythrocyte storage in adenine-guanosine solution. *PLoS ONE* **2013**, *8*, e71060. [CrossRef] [PubMed]
49. Böning, D.; Schünemann, H.J.; Maassen, N.; Busse, M.W. Reduction of oxylabile CO2 in human blood by lactate. *J. Appl. Physiol.* **1993**, *74*, 710–714. [CrossRef]
50. Mairbäurl, H.; Schobersberger, W.; Hasibeder, W.; Schwaberger, G.; Gaesser, G.; Tanaka, K.R. Regulation of red cell 2,3-DPG and Hb-O_2-affinity during acute exercise. *Eur. J. Appl. Physiol. Occup. Physiol.* **1986**, *55*, 174–180. [CrossRef] [PubMed]
51. Dempsey, J.A.; Wagner, P.D. Exercise-induced arterial hypoxemia. *J. Appl. Physiol.* **1999**, *87*, 1997–2006. [CrossRef]
52. Rossi, R.; Milzani, A.; Dalle-Donne, I.; Giannerini, F.; Giustarini, D.; Lusini, L.; Colombo, R.; Di Simplicio, P. Different metabolizing ability of thiol reactants in human and rat blood: Biochemical and pharmacological implications. *J. Biol. Chem.* **2001**, *276*, 7004–7010. [CrossRef]
53. Giustarini, D.; Colombo, G.; Garavaglia, M.L.; Astori, E.; Portinaro, N.M.; Reggiani, F.; Badalamenti, S.; Aloisi, A.M.; Santucci, A.; Rossi, R.; et al. Assessment of glutathione/glutathione disulphide ratio and S-glutathionylated proteins in human blood, solid tissues, and cultured cells. *Free Radic. Biol. Med.* **2017**, *112*, 360–375. [CrossRef] [PubMed]
54. Unt, E.; Kairane, C.; Vaher, I.; Zilmer, M. Red blood cell and whole blood glutathione redox status in endurance-trained men following a ski marathon. *J. Sports Sci. Med.* **2008**, *7*, 344–349. [PubMed]
55. Li, X.D.; Sun, G.F.; Zhu, W.B.; Wang, Y.H. Effects of high intensity exhaustive exercise on SOD, MDA, and NO levels in rats with knee osteoarthritis. *Genet. Mol. Res. GMR* **2015**, *14*, 12367–12376. [CrossRef] [PubMed]
56. Erbs, S.; Höllriegel, R.; Linke, A.; Beck Ephraim, B.; Adams, V.; Gielen, S.; Möbius-Winkler, S.; Sandri, M.; Kränkel, N.; Hambrecht, R.; et al. Exercise Training in Patients with Advanced Chronic Heart Failure (NYHA IIIb) Promotes Restoration of Peripheral Vasomotor Function, Induction of Endogenous Regeneration, and Improvement of Left Ventricular Function. *Circ. Heart Fail.* **2010**, *3*, 486–494. [CrossRef] [PubMed]
57. Stempak, D.; Dallas, S.; Klein, J.; Bendayan, R.; Koren, G.; Baruchel, S. Glutathione stability in whole blood: Effects of various deproteinizing acids. *Ther. Drug Monit.* **2001**, *23*, 542–549. [CrossRef]
58. Cheng, M.-L.; Lin, J.-F.; Huang, C.-Y.; Li, G.-J.; Shih, L.-M.; Chiu, D.T.-Y.; Ho, H.-Y. Sedoheptulose-1,7-bisphospate Accumulation and Metabolic Anomalies in Hepatoma Cells Exposed to Oxidative Stress. *Oxid. Med. Cell. Longev.* **2019**, 5913635. [CrossRef]
59. Gutteridge, B.H.J.M.C. *Free Radicals in Biology and Medicine*, 5th ed.; Oxford University Press Inc.: New York, NY, USA, 2015.

antioxidants

Article

Differences between Professional and Amateur Cyclists in Endogenous Antioxidant System Profile

Francisco Javier Martínez-Noguera [1,*], Pedro E. Alcaraz [1], Raquel Ortolano-Ríos [1], Stéphane P. Dufour [2,3] and Cristian Marín-Pagán [1]

[1] Research Center for High Performance Sport, Campus de los Jerónimos, Catholic University of Murcia, 30107 Murcia, Spain; palcaraz@ucam.edu (P.E.A.); rortolano@ucam.edu (R.O.-R.); cmarin@ucam.edu (C.M.-P.)

[2] Faculty of Medicine, Translational Medicine Federation (FMTS) UR 3072, University of Strasbourg, 67000 Strasbourg, France; sdufour@unistra.fr

[3] Faculty of Sport Sciences, University of Strasbourg, 67084 Strasbourg, France

[*] Correspondence: fjmartinez3@ucam.edu; Tel.: +34-96-827-8566

Abstract: Currently, no studies have examined the differences in endogenous antioxidant enzymes in professional and amateur cyclists and how these can influence sports performance. The aim of this study was to identify differences in endogenous antioxidants enzymes and hemogram between competitive levels of cycling and to see if differences found in these parameters could explain differences in performance. A comparative trial was carried out with 11 professional (PRO) and 15 amateur (AMA) cyclists. All cyclists performed an endogenous antioxidants analysis in the fasted state (visit 1) and an incremental test until exhaustion (visit 2). Higher values in catalase (CAT), oxidized glutathione (GSSG) and GSSG/GSH ratio and lower values in superoxide dismutase (SOD) were found in PRO compared to AMA ($p < 0.05$). Furthermore, an inverse correlation was found between power produced at ventilation thresholds 1 and 2 and GSSG/GSH (r = -0.657 and r = -0.635; $p < 0.05$, respectively) in PRO. Therefore, there is no well-defined endogenous antioxidant enzyme profile between the two competitive levels of cyclists. However, there was a relationship between GSSG/GSH ratio levels and moderate and submaximal exercise performance in the PRO cohort.

Keywords: catalase; superoxide dismutase; oxidized glutathione; reduced glutathione; hemoglobin; power output

Citation: Martínez-Noguera, F.J.; Alcaraz, P.E.; Ortolano-Ríos, R.; Dufour, S.P.; Marín-Pagán, C. Differences between Professional and Amateur Cyclists in Endogenous Antioxidant System Profile. *Antioxidants* **2021**, *10*, 282. https://doi.org/10.3390/antiox10020282

Academic Editors: Gareth Davison and Conor McClean

Received: 16 January 2021
Accepted: 10 February 2021
Published: 12 February 2021

Publisher's Note: MDPI stays neutral with regard to jurisdictional claims in published maps and institutional affiliations.

1. Introduction

Competitive cycling is highly stressful on both aerobic and anaerobic metabolisms. Road cycling races require the riders to produce high relative power output (W/kg) for short duration (i.e., less than 1 min at the start, during steep climb and at the end of the race) while also sustain efforts that last for several minutes to several hours [1]. Overall, professional cyclists (PRO) perform high training volumes (~32,500 Km) during the competitive season, which include 90–100 race days [2]. On the other hand, amateur competitive cyclists (AMA) can be defined as cyclists that train 3–7 times per week, with daily training volumes of 60–120 min and that compete about 20 times in a year [3]. During training sessions and competitions (aerobic and anaerobic exercise), there is a rise in reactive oxygen species (ROS) and subsequent oxidative stress, which can lead to a favorable adaptation in the body's antioxidant defense system [4]. This improvement in the endogenous antioxidant system is generally associated with lower levels of oxidative stress biomarkers [5].

Within the endogenous antioxidant system, superoxide dismutase (SOD) is the first line of enzymatic defense that transforms the superoxide radical($O_2^{\bullet-}$) into hydrogen peroxide (H_2O_2) [6]. Then, H_2O_2, which is also harmful to cells, can be metabolized in a couple of ways: (1) conversion into water by glutathione peroxidase (GPx) with the reduced glutathione consumption (GSH) being converted into oxidized glutathione (GSSG) and (2) when the production of H_2O_2 exceeds the capacity of glutathione peroxidase, catalase (CAT) takes over to remove the excess H_2O_2 [7]. However, un-neutralized H_2O_2 can interact with transition metals, such as Fe^{2+} and Cu^+, and result in the production of the hydroxyl radical ($^{\bullet}OH$; i.e., Fenton reaction), which is an extremely powerful oxidizing agent that reacts with all biological macromolecules. However, currently it is not clear what the end product of the Fenton reaction $^{\bullet}OH$ or FeO^{2+} is [8]. Finally, $^{\bullet}OH$ can elicit damage to DNA, oxidation of the thiol group of proteins and peroxidation of lipids [9].

Exercise increases oxygen uptake and almost 0.15% of the oxygen consumed can be converted into ROS, which can be harmful to muscle and mitochondrial function [10]. Recent findings show that the main source of ROS during exercise is nicotinamide adenine dinucleotide phosphate oxidases [11]. It is known that long-duration strenuous exercise and extensive sprint training can exceed our ability to detoxify the action of reactive oxygen species within the blood cells, as well as at the muscle level [12]. Conversely, prevention of oxidative stress to enhance performance in professional athletes can be done by the adaptation mechanisms (hormesis) and detoxifying function of antioxidant enzymes (SOD, CAT, GPx, glutathione reductase (GR), glutathione-s-transferase), as well as via non-enzymatic antioxidants (such as vitamins E, A, C, and GSH and GSSG [13]. Therefore, it has been suggested that higher levels of the endogenous antioxidant system may improve performance of skeletal muscle contraction [5]. Cordova et al. [14] analyzed antioxidant markers and showed an average GSH of 3.24 $\mu mol \cdot g^{-1}$ Hb, GSSG of 1.54 $\mu mol \cdot g^{-1}$ Hb, GSSG/GSH ratio of 0.56 %, catalase (CAT) of 172.0 $mmol \cdot min^{-1} \cdot g^{-1}$ Hb and superoxide dismutase (SOD) of 1983.0 $U \cdot g^{-1}$ Hb activity. In a previous study [15], average values of GSH of ~4.7 $\mu mol \cdot g^{-1}$ Hb, GSSG of ~0.7 $\mu mol \cdot g^{-1}$ Hb and GSSG/GSH ratio of 0.15% were observed in PRO cyclists in February in of the same competitive season, suggesting that training season can modify the levels of these components of the endogenous antioxidant system [5,12,15].

At rest, endogenous antioxidant enzymes (EAE) levels are generally lower in athletes than in sedentary subjects, although higher or unchanged levels have been observed [6,16]. Several factors may explain this discrepancy, the most important being differences in the methods used to estimate the state of oxidative stress, the characteristics of the sample population (high-level athlete, sedentary, etc.) and the time of measurement (period of the sport's season) [13]. However, what is clearly evident is that acute exercise can lead to an imbalance between ROS and endogenous antioxidants, causing what is known as oxidative stress [7]. A recent study has shown that acute exercise at low, moderate or high intensity has the capacity to reduce GSH concentration and increase SOD and CAT activity compared to baseline, in addition to increasing F_2-isoprostanes (markers of oxidative stress) at all levels of exercise. [17]. In addition, it is generally known that chronic exercise causes an increase in enzymatic and non-enzymatic antioxidant defense, leading to adaptations to the training response and improving the protection against ROS [18]. Chronic exercise (6-week cycling training) has the ability to increase the concentration and activity of GSH, SOD and CAT, while maintaining the levels of F_2-isoprostanes [17]. This same study also found that moderate and high intensity exercise promoted greater adaptations in antioxidant markers than low intensity exercise at baseline.

Despite the large amount of information on the EAE status and their relationship to the effects of acute and chronic exercise, to our knowledge, there are no studies that have compared the status of antioxidant enzymes between professional and amateur cyclists and their relationship to performance. Therefore, the main objective of this research was to determine the differences in endogenous antioxidants enzymes and hemogram levels between PRO and AMA, and whether these might be related to differences in performance (power output at VT1, VT2 and VO_{2MAX}) in an incremental test. Finally, the secondary objective was to assess whether differences between endogenous antioxidant enzymes and hematological were associated with differences in performance between PRO and AMA.

2. Methodology

2.1. Selection of Participants

A total of 26 male cyclists (11 PRO, 15 AMA) were recruited and completed the study. The PRO were competing at the *Union Cycliste Internationale* (UCI) PRO TOUR level and have participated in UCI major stage races (*Vuelta a España, Giro d'Italia, Tour de France*). The 15 AMA were from the southeast region of Spain. The PRO riders were selected based on the following criteria: (1) 20 to 40 years of age, (2) enrolled in a professional licensed team and (3) competed in at least one of the main 3-week stage races in the last years. Subjects for the AMA group had to meet the following inclusion criteria: (1) 20 to 40 years of age, (2) had at least 3 years of cycling experience and (3) performed specific training 6–12 h/week.

All subjects signed the informed consent document before their participation. The study was performed following the guidelines of the Helsinki Declaration for Human Research [19] and was approved by the Ethics Committee of the Catholic University of Murcia (CE091802).

2.2. Study Protocol

The experimental design of the study required each rider to visit the laboratory twice between the end of October and December (i.e., post-season period). In the first visit, a medical exam and blood analysis were completed to check their state of health. In the second visit (post-48 h), the cyclists performed a maximal incremental test. The 2 h prior to this latter test, they ingested a standardized breakfast, which was based relative to body mass (557.7 kcal) and composed of 95.2 g of carbohydrates (68%), 19.0 g of protein (14%) and 11.3 g of lipids (18%), established by a sports nutritionist. All subjects were instructed to refrain from high-intensity training 48 h before each visit.

2.3. Incremental Test

An incremental step test with final ramp until exhaustion was performed on a cycle ergometer (Cyclus 2TM, RBM elektronik- automation GmbH, Germany) using a metabolic cart (Metalyzer 3B. Leipzig, Germany) to determine VT1, VT2 and VO_{2max}, as well as the associated levels of power output. The testing protocol started with 35 W and increased by 35 W every 2 min until RER > 1.05 was reached, from which the final ramp ($+35\ W\cdot min^{-1}$) until exhaustion was initiated [20]. To ensure that VO_{2MAX} was achieved, at least 2 of the following criteria had to be met: plateau in the final VO_2 values (increase $\leq 2.0\ mL\cdot kg^{-1}\cdot min^{-1}$ in the two last loads), maximal theoretical HR (220-age)·0.95) [21], RER ≥ 1.15 and lactate $\geq 8.0\ mmol\cdot L^{-1}$ [22,23]. Ventilatory thresholds were obtained using the ventilatory equivalents method described by Wasserman [24].

2.4. Blood Analysis

A total of 21.5 mL of blood were withdrawn from the antecubital vein for analyses: one 3.0 mL tube with ethylenediaminetetraacetic acid (EDTA) for hemogram and another 3.5 mL tube with polyethylene terephthalate (PET) for biochemical parameters. For the measurement of antioxidant parameters, five 3.0 mL EDTA tubes were obtained, where one tube was immediately centrifuged at 3500 rpm at 4 °C for 10 min. All tubes were temporarily stored at 2–4 °C and then sent to an external laboratory for analysis. Red blood cell count was carried out in an automated Cell-Dyn 3700 analyser (Abbott Diagnostics, Chicago, IL, USA) using internal (Cell-Dyn 22) and external (Program of Excellence for Medical Laboratories-PEML) controls. Values of erythrocytes, hemoglobin, hematocrit and hematometra indexes (mean cell volume (MCV), mean cell hemoglobin (MCH) and mean cell hemoglobin concentration (MCHC)) were estimated.

2.5. Oxidative Stress and Antioxidant Status Markers

2.5.1. Catalase

The activity of catalase was measured in the whole blood using a UV-VIS spectrophotometer. The catalase enzyme extracts the peroxides from the region of the gel it occupies, following the isolation of the native protein. The removal of peroxide does not cause potassium ferricyanide (yellow substance) to be reduced to potassium ferrocyanide, which reacts with ferric chloride to form a blue Prussian precipitate. The catalase positive control activity is defined in international unit equals (1 unit) to the amount of catalase necessary to decompose 1.0 μM of H_2O_2 per minute at pH 7.0 at 25 °C while H_2O_2 concentration falls from $\approx$ 10.3 mM to 9.2 mM. The absorbance of H_2O_2 decreases at 240 nm proportional to its decomposition so that the concentration of H_2O_2 is critical in this determination. The decrease in absorbance per time unit is the measure of catalase activity [25]. Results were expressed in U/g of Hb.

2.5.2. SOD

Superoxide dismutase (SOD) activity was measured using an SD125 Ransod kit (Randox Ltd. Crumlin, Reino Unido) in whole blood. Xanthine and xanthine oxidase were used to produce superoxide anion ($O_2^{\bullet-}$), which responded with the 2-(4-iodophenyl)-3-(4-nitrophenol)-5-phenyltetrazolium chloride (INT) reactive and formed a red complex that is detectable at 420 nm. The SOD activity was measured as the inhibition degree of this reaction [26]. Results were expressed in U/g of Hb.

2.5.3. Glutathione

The analysis of reduced glutathione (GSH) was performed using the glutathione-S-transferase assay described by Akerboom and Sies [27]. Calculation of GSH was performed from lymphocytes treated with perchloric acid at a final concentration of 6%, collecting the supernatant after vortexing and subsequent centrifugation for 10 min at 10,000 rpm. Following collection of the supernatants in vials, high-performance liquid chromatography (HPLC) coupled to a Waters NH2 ODS S5 column (0.052, 25 cm) was conducted. Oxidized glutathione (GSSG) was analyzed using a similar method described by Asensi. [28].

2.6. Statistical Analyses

The statistical analysis was performed using the Statistical Package for Social Sciences (SPSS 21.0, International Business Machines Chicago, IL, USA). Descriptive statistics are presented as mean $\pm$ standard deviation (SD). Levene and Shapiro–Wilks tests were performed to check for homogeneity and normality of the data, respectively. A Student's t-test for unpaired data was used to evaluate differences between groups. Additionally, the standardized mean differences were calculated using Cohen's effect size (ES) (95% confidence interval) for all comparisons. Threshold values for ES statistics were as follows: >0.2 small, >0.5 moderate, >0.8 large [29]. The different correlations between the parameters were evaluated using Pearson's correlation (r). Significance level was set at $p \leq 0.05$.

3. Results

3.1. Subject Characteristics

The general characteristics and hemogram results are presented in Table 1. Age, body mass and height were not different between PRO and AMA groups. Interestingly, PRO had higher MCH (4.8%, $p < 0.001$) and MCHC (3.6%, $p < 0.001$) compared to AMA. There were no group differences in RBC, Hb, HCT and MCV (Table 1). No correlation was found between age and antioxidant markers in both groups. However, there were correlations found between age and Hb, HCT and MCHC ($r \leq -0.597$, $p < 0.05$).

3.2. Antioxidant Parameters

Table 2 shows the outcomes of CAT, SOD, GSSG, GSH, %GSSG/GSH and GSSG+GSH, which were measured at baseline before the incremental tests. Higher levels in CAT (30.0%, $p < 0.001$), GSSG (63.2%, $p < 0.001$) and %GSSG/GSH (70.1%, $p < 0.001$), and lower levels in SOD (-16.2%, $p = 0.009$) were found in PRO compared to AMA. However, no differences in GSH (-4.3%, $p = 0.216$) and GSSG+GSH (-3.5%, $p = 0.317$) values were observed between PRO and AMA.

Table 1. Baseline general characteristics and hemogram variables of professional and amateur cyclists.

	PRO	AMA	*p*-Value	Cohen's d	Effect Size
Age (years)	28.3 (4.65)	29.3 (6.54)	0.671	0.17	Trivial
Body mass (kg)	68.5 (4.43)	69.9 (5.50)	0.488	0.28	Small
Height (cm)	178.0 (6.93)	175.0 (6.71)	0.274	0.44	Small
HEMOGRAM					
RBC ($10^6 \cdot \mu L^{-1}$)	5.06 (0.281)	5.15 (0.260)	0.441	0.08	Trivial
Hb ($g \cdot dl^{-1}$)	15.6 (0.827)	15.1 (0.676)	0.107	0.49	Small
HCT (%)	44.5 (2.28)	44.6 (1.57)	0.866	0.13	Trivial
MCV (fl)	87.9 (2.19)	86.8 (2.92)	0.305	1.10	Large
MCH (pg)	30.8 (0.35)	29.4 (1.03)	<0.001	1.44	Large
MCHC (%)	35.0 (0.74)	33.8 (0.60)	<0.001	1.19	Large

Values are expressed as mean (SD). Abbreviations: RBC = red blood cell; Hb = hemoglobin; HCT = hematocrit; MCV = mean corpuscular volume; MCH = mean corpuscular hemoglobin; MCHC = mean corpuscular hemoglobin concentration and SD = standard deviation.

Table 2. Endogenous antioxidant enzymes from professional and amateur cyclists.

	PRO	AMA	*p*-Value	Cohen's d	Effect Size
CAT (U/g Hb)	32.5 (5.34)	25.0 (4.51)	<0.001	1.55	Large
SOD (U/g Hb)	1213 (233.0)	1447 (184.4)	0.009	1.13	Large
GSSG (nmol/mg protein)	0.524 (0.103)	0.321 (0.077)	<0.001	2.28	Large
GSH (nmol/mg protein)	24.4 (2.00)	25.5 (2.17)	0.216	0.50	Moderate
GSSG/GSH	2.16 (0.436)	1.27 (0.279)	<0.001	2.52	Large
GSSG+GSH (nmol/mg protein)	24.9 (2.02)	25.8 (2.19)	0.317	0.41	Small

Values are expressed as mean (SD). Abbreviations: CAT = catalase; SOD = superoxide dismutase; GSH = reduced glutathione; GSSG = oxidized glutathione; % GSSG/GSH = oxidized/reduced glutathione ratio and SD = standard deviation.

3.3. Physiological and Metabolic Parameters at VT1

VO_2, W, WR, $\%VO_{2MAX}$, HR and RER values at VT1 are shown in Table 3. Significant group differences in VO_2 (76.0%, $p < 0.001$), W (90.4%, $p < 0.001$), WR (92.5%, $p < 0.001$), $\%VO_{2MAX}$ (53.3%, $p < 0.001$) and HR (12.9%, $p = 0.004$), but not for RER (0.78%, $p = 0.707$) were observed.

Table 3. Metabolic and performance variables of professional and amateur cyclists.

	PRO	AMA	*p*-Value	Cohen's d	Effect Size
VT1					
VO_2 ($mL \cdot min^{-1}$)	3593 (271.0)	2041 (401.0)	<0.001	4.40	Large
W	299 (32.9)	157 (36.1)	<0.001	4.07	Large
WR ($W \cdot kg^{-1}$)	4.37 (0.42)	2.27 (0.56)	<0.001	4.14	Large
$\%VO_{2max}$	76.2 (3.91)	49.7 (5.58)	<0.001	5.36	Large
HR ($beats \cdot min^{-1}$)	149 (14.7)	132 (13.2)	0.004	1.25	Large
RER	0.906 (0.05)	0.899 (0.04)	0.707	0.15	Trivial

Table 3. *Cont.*

	PRO	AMA	*p*-Value	Cohen's d	Effect Size
		VT2			
VO_2 $(mL \cdot min^{-1})$	4259 (234.0)	3389 (505.0)	<0.001	2.10	Large
W	379 (34.0)	286 (45.1)	<0.001	2.28	Large
WR $(W \cdot kg^{-1})$	5.54 (0.41)	4.13 (0.74)	<0.001	2.28	Large
$\%VO_{2max}$	90.3 (2.36)	84.7 (5.67)	0.005	1.24	Large
HR $(beats \cdot min^{-1})$	168 (11.1)	171 (9.4)	0.467	0.29	Small
RER	1.01 (0.05)	1.03 (0.03)	0.323	0.40	Small
		VO_{2max}			
VO_2 $(mL \cdot min^{-1})$	4714 (241.0)	4066 (580.7)	0.002	1.38	Large
VO_2/R $(mL \cdot kg^{-1} \cdot min^{-1})$	69.0 (3.94)	58.7 (9.58)	0.003	1.34	Large
W	474 (31.5)	383 (49.2)	<0.001	2.13	Large
WR $(W \cdot kg^{-1})$	6.93 (0.44)	5.51 (0.81)	<0.001	2.09	Large
HR $(beats \cdot min^{-1})$	186 (7.42)	186 (7.62)	0.966	0.02	Trivial
RER	1.22 (0.04)	1.14 (0.06)	0.001	1.49	Large

Values are expressed as mean (SD). Abbreviations: VO_2 = oxygen uptake; VO_{2max} = maximum oxygen consumption; VO_2/R = maximum oxygen consumption relative to weight; W = power output; WR = power output relative to weight; $\%VO_{2max}$ = percentage of VO_{2max}; HR = heart rate $(beats \cdot min^{-1})$; RER = respiratory exchange ratio; VT1 = ventilatory threshold 1; VT2 = ventilatory threshold 2 and SD = standard deviation.

GSSG/GSH was significantly correlated with W_{VT1} and VO_{2VT1} (r = −0.657 and r = −0.651; *p* < 0.05, respectively) in PRO (Table 4) (Figure 1).

Table 4. Correlation between endogenous antioxidant enzymes and performance-metabolic variables from professional and amateur cyclists.

		CAT	SOD	GSSG	GSH	%GSSG/GSH	GSSG + GSH
				PRO (*n* = 11)			
W_{VT1}	r	−0.120	0.305	−0.449	0.425	**−0.657**	0.397
	p-value	0.72	0.36	0.17	0.19	**0.03**	0.23
VO_{2VT1}	r	0.001	0.378	−0.442	0.457	**−0.651**	0.429
	p-value	0.998	0.252	0.173	0.157	**0.030**	0.188
W_{VT2}	r	−0.253	0.183	−0.575	0.116	**−0.635**	0.085
	p-value	0.45	0.59	0.06	0.73	**0.04**	0.80
VO_{2VT2}	r	−0.319	0.423	−0.518	0.277	−0.622	0.247
	p-value	0.34	0.20	0.10	0.41	0.04	0.46
W_{MAX}	r	−0.045	0.186	−0.342	0.239	−0.443	0.219
	p-value	0.90	0.58	0.30	0.48	0.17	0.52
VO_{2MAX}	r	−0.375	0.422	−0.312	0.304	−0.414	0.284
	p-value	0.26	0.20	0.35	0.36	0.21	0.40

Table 4. *Cont.*

		CAT	SOD	GSSG	GSH	%GSSG/GSH	GSSG + GSH
		AMA (*n* = 15)					
W_{VT1}	r	0.181	0.172	0.206	−0.102	0.256	−0.098
	p-value	0.52	0.54	0.46	0.72	0.36	0.73
VO_{2VT1}	r	0.360	0.159	0.182	−0.108	0.230	−0.105
	p-value	0.19	0.57	0.52	0.70	0.41	0.71
W_{VT2}	r	0.414	0.113	−0.002	0.047	−0.046	0.040
	p-value	0.13	0.69	0.99	0.87	0.87	0.89
VO_{2VT2}	r	0.358	0.234	−0.104	−0.068	−0.097	−0.077
	p-value	0.19	0.69	0.71	0.81	0.73	0.78
W_{MAX}	r	0.180	0.173	−0.136	−0.379	0.009	−0.386
	p-value	0.52	0.54	0.63	0.16	0.97	0.16
VO_{2MAX}	r	0.289	0.278	−0.118	−0.334	0.001	−0.339
	p-value	0.30	0.32	0.66	0.22	0.10	0.22

Values are expressed as mean (SD). Abbreviations: CAT = catalase (U/g Hb); SOD = superoxide dismutase (U/g Hb); GSH = reduced glutathione (nmol/mg protein); GSSG = oxidized glutathione (nmol/mg protein); % GSSG/GSH = oxidized/reduced glutathione ratio; VO_2 = oxygen uptake; VO_{2MAX} = maximum oxygen consumption; VT1 = ventilatory threshold 1; VT2 = ventilatory threshold 2 and W = power output.

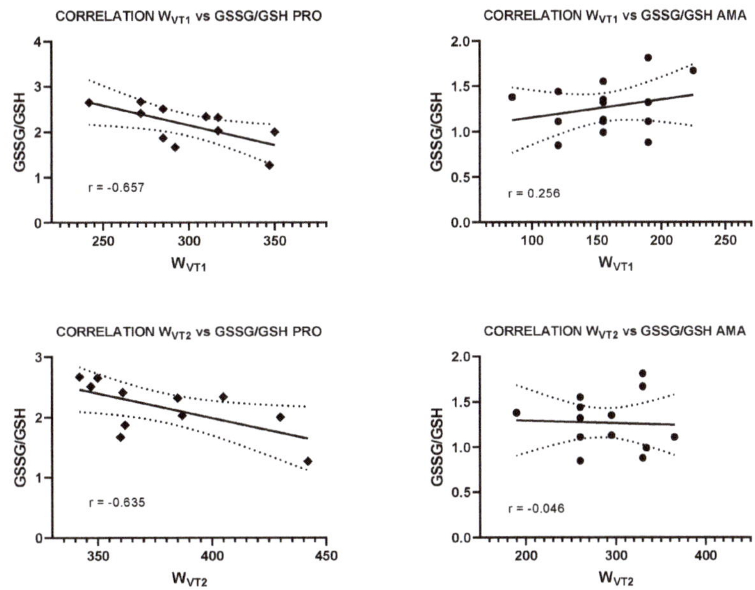

Figure 1. Correlations between the power generated at the ventilation threshold 1 and 2 between the GSSS/GSH ratio in PRO and AMA.

3.4. Physiological and Metabolic Parameters at VT2

Table 3 demonstrates the VT2 results of VO_2, W, WR, %VO_{2MAX}, HR and RER. Significant group differences in VO_{2VT2} (25.6%, p = <0.001), W_{VT2} (32.5%, p = <0.001), WR_{VT2} (34.1%, p < 0.001) and %$VO_{2MAXVT2}$ (6.6%, p = 0.005) were observed.

GSSG/GSH was significantly correlated with W_{VT2} and VO_{2VT2}(r = −0.635 and r = −0.622; p < 0.05, respectively) in PRO (Figure 1). GSSG tended to correlate with W_{VT2} (r = −0.575; p = 0.06) in PRO (Table 4).

3.5. Physiological and Metabolic Parameters at VO$_{2max}$

Maximal values of VO$_2$, VO$_2$/R, W, WR, HR and RER are presented in Table 3. Significant group differences in VO$_{2MAX}$ (15.9%, $p = 0.002$), VO$_2$/R$_{MAX}$ (17.5%, $p = 0.002$), W$_{MAX}$ (23.8%, $p < 0.001$), WR$_{MAX}$ (25.8%, $p < 0.001$), and RER$_{MAX}$ (7.0%, $p = 0.001$), but not for HR$_{MAX}$ ($p = 0.966$) were found.

In VO$_{2MAX}$, no correlation with any antioxidant marker was observed (Table 4).

4. Discussion

This study provides the first direct comparison of endogenous antioxidant, hematological, performance and metabolic biomarkers (VT1, VT2 and VO$_{2MAX}$) between PRO and AMA cyclists. Our results demonstrate that: (i) PRO have higher values in MCH, MCHC, CAT, GSSG and GSSG/GSH but lower values in SOD than AMA; (ii) PRO have higher levels of absolute and relative power output and oxygen consumption in all intensity zones (VT1, VT2 and VO$_{2MAX}$) than in AMA, with the largest differences found at VT1; (iii) inverse correlations were identified in W$_{VT1}$, VO$_{2VT1}$, W$_{VT2}$ and VO$_{2VT2}$ with GSSG/GSH in PRO.

Differences in Antioxidant Enzymes and Hemogram

When intense physical exercise is performed (especially in untrained or those not familiar with the exercise), there is an increase in the production of reactive oxygen species, which are neutralized by our complex endogenous antioxidant defense system (GSH, GSSG, CAT, SOD, GPx and GR) and by exogenous antioxidants (vitamin C, vitamin E, carotenes) [30].

Regarding EAE, we observed higher levels of CAT activity, GSSG and GSSG/GSH, but lower levels of SOD activity in PRO versus AMA. Mena et al. [31] found higher resting levels of SOD, CAT and GPx in a sample of PRO cyclists compared to sedentary people. Tauler et al. [32] also showed differences in antioxidant enzyme activity in erythrocyte between PRO and AMA at rest. In the same study, a decrease in CAT (-12%), GPx (-14%) and GR activity (-16%) but an increase in SOD activity of about 25% after a submaximal test (80% VO$_{2MAX}$; 1 h 30 min) was reported [32].

Long distance runners have been shown to have a three-fold higher CAT activity compared to short distance runners [18]. Similarly, it was observed that marathon runners had twice as high catalase activity compared to sprinters [12]. In this study, we also demonstrated higher levels of CAT in PRO than in AMA, and this may be largely explained by the fact that PRO perform greater volume, intensity and competitions (higher aerobic load and prolonged periods of exercise) than AMA, which induces higher levels of exposure to ROS and, consequently, adaptations of EAE [6]. When CAT levels increase, it is possible that GPx activity is not sufficient to neutralize high levels H$_2$O$_2$ (endurance exercise) [7].

Regarding SOD, Mena et al. [31] observed lower levels of SOD activity (-32.1%) in PRO than in elite cyclist, but in the case of CAT (80.0%) and GPx (149.0%), the levels were higher, reporting an ascending behavior of SOD during a stage race (2800 km in 17 stages) in PRO. Tauler et al. [32] has also found lower levels of SOD activity in PRO (-19.8%) than in AMA at baseline, which are in line with the results of our study. Antioxidant enzyme activity can be modified either by an initial increase (adaptation) or a decrease if the oxidative stress of long duration (utilization) [33]. Therefore, the low basal levels of SOD activity in professional cyclists could be overwhelmed and the high concentration of superoxide anions could activate CAT, allowing compensated metabolization of H$_2$O$_2$. This may be the reason why PRO has lower levels of SOD activity than AMA, as PRO have higher levels of exercise exigency that sometimes get close to exhaustion, which can lead a decrease in the working capacity of SOD.

On the other hand, there is evidence to suggest that GSH or GSH/GSSG decreases during exercise because of its utilization against ROS [33]. Ultra-endurance exercise depletes erythrocyte GSH levels by $\sim$66% for 24 h and levels remain $\sim$33% lower than normal 1 month later [34]. PRO frequently compete in longer distance events than AMA, which can lead to lower levels of GSH in PRO than in AMA, although no differences were observed in GSH between PRO vs. AMA in our study. In addition, the muscle can import GSH from plasma during exercise, and as a result, there is a change in the GSH/GSSG ratio after exercise with a decrease in the GSH/GSSG ratio at the time of exhaustion [35]. Furthermore, it is important to mention that tissues are not only capable of importing GSH but also exporting GSSG under oxidative stress [35]. Moreover, GSH is a molecule that is key in cellular redox status regulation, and consequences of prolonged GSH depletion may include a compromise in immunity, where lower GSH is associated with decreased lymphocyte proliferation and increased viral reactivation [34].

GSSG levels are a biomarker of cellular oxidative stress, since GSH is an important antioxidant in many tissues and oxidizes in the catalyzed reduction of H_2O_2 to H_2O to become GSSG [36]. The increase in GSH (mainly) and GSSG in plasma after exercise could be explained by an efflux from the liver to other tissues, including skeletal muscle [37]. GSSG levels in skeletal muscle have previously been shown to increase by ~50% in rats after running on a treadmill at moderate intensity [38] and by ~20% after cycling in humans (workload corresponding to 90% of VO_2peak; 10×4 min) [39]. Leonardo et al. [40] observed an increase in both GSSG and GSSG/GSH after a period of intense PRO training, which returned to their baseline levels after a period of tapering. We found similar baseline values of GSSG in our study. In addition, we found higher levels of GSSG and GSSG/GSH in PRO than in AMA.

The efforts made during cycling competitions produce oxidative stress in lymphocytes, leading to a reduction in GSH levels and an increase in GSSG levels. The decrease in GSH and increase in GSSG during exercise may be explained by an increase in H_2O_2 formation, as reported by Wang et al. who found that high-intensity exercise (80% VO_{2MAX}) decreased GSH levels while lipid peroxidation increased immediately and after 24h of exercise [41]. Furthermore, in this study, lymphocytes were incubated with H_2O_2 for 2 and 4 h, promoting an increase in DNA fragmentation immediately and 24 h after high intensity exercise. Thus, H_2O_2 would cause a failure of the endogenous antioxidant system leading to DNA damage in lymphocytes. Ferrer et al. [42] found that high intensity exercise (swimming) increased GPx activity (converts GSH to GSSG) in lymphocytes, in the same way as other authors found after a cycling stage [43,44]. This supports the decrease in GSH and increase in GSSG after high intensity exercise. Therefore, the higher levels of GSSG and GSSG/GSH in PRO vs. AMA in our study may be due to a higher production of ROS, which leads to a higher production of GSSG and, consequently, of GSSG/GSH together with a decrease in GSH.

In addition, our study is the first to show correlations between GSSG/GSH with W_{VT1} (r = -0.657) and W_{VT2} (r = -0.635) in PRO. This is also supported by a trend towards a significant correlation between GSSG and WVT2 (r = -0.575; $p = 0.06$) in PRO. These relationships suggest that cyclists who generate more power at VT1 and VT2 have lower GSSG/GSH levels, and therefore, less oxidative stress, as GSSG/GSH ratio is known to be a marker of antioxidant status [20].

In response to strenuous physical working conditions, the body's antioxidant capacity may be temporarily diminished, as its components are used to scavenge the harmful radicals that are produced [45]. It is well known that exercise-induced ROS are detrimental to physiological function, including decreased performance and immune function and increased fatigue [45]. Moreover, it has been shown that the response of antioxidant capacity to exercise responds in a similar way to the activity of EAE [45]. Therefore, the antioxidant defense system may be temporarily reduced in response to increased ROS production but may increase during the recovery period as a result of the initial prooxidant insult [46]. However, contradictory findings have been reported where increases in GPx, SOD, and CAT, as well as decreases in GPx, GR, SOD have been observed [45]. Evidently, this controversy may depend on the moment of sampling (i.e., period of the season), as well as on the duration and intensity of the exercise, which varies considerably between studies.

It could be that there is an undefined optimal level of ROS production and oxidative damage required for adaptations in antioxidant defenses and other physiological parameters, leading to health and performance improvements [45]. However, overproduction of ROS and oxidative damage due to chronic long-term exercise and/or overtraining may exceed the above-mentioned optimal level, resulting in irreparable oxidative damage, which can lead to the development or progression of poor health and/or disease [47]. Therefore, the measurement of the antioxidant capacity (CAT, SOD, GSH, GSSG and GSGG/GSH) of the body is used as a marker of oxidative stress and can provide us insight on how it affects performance. Given the results of our research and the evidence shown in the scientific literature, there is no endogenous antioxidant profile defined in PRO compared to AMA.

There are also other antioxidant proteins, such as peroxiredoxin (PRX) and thioredoxin (TRX) containing thiol groups, with a high capacity to neutralize reactive oxygen and nitrogen species and decrease oxidative stress [48]. One study showed how moderate and high-intensity exercise and a low volume high intensity interval training trial increased TRX (85%, 64% and 206%, respectively); however, PRX only increased during high intensity exercise (moderate: -6229%; high: 203% and low volume high intensity interval: -23%, respectively) in peripheral blood mononuclear cells [48]. In addition, an increase in nuclear transcription factor kappa B was found during all exercises, suggesting an activation of the inflammatory system, probably due to increased oxidative stress. Future studies should examine whether there are differences in these antioxidant proteins between PRO and AMA and their relationship with performance.

Regarding hematological parameters, no significant differences were found except for MCH and MCHC between PRO vs. AMA. Schumacher et al. found hematological values in elite cyclists from the German national team (blood samples collected between November and January) and the values were similar to ours in Hb (~15.5 g/dL), Hct (~45.0%) and RBC ($\sim5.0 \times 10^6/\text{mm}^3$) in PRO [49]. In addition, other studies have found hematological values of approximately 15.0 g/dL of Hb and 45% of Hct in professional cyclists [50–52]. Well-trained cyclists have found values of 14.3 g/dL in Hb and 43.1% in Hct, values lower than PRO [53]. However, Bejder et al. [54] observed amateur competitive cyclist values of 14.8 g/dL Hb, 42.8% Hct, $4.92 \times 10^6 \cdot \mu\text{L}^{-1}$ RBC, 87.1 fl MCV, 30.1 pg MCH and 34.6 g/dL MCHC, lower than those reported in PRO.

MCH indicates the amount of hemoglobin contained in an erythrocyte and MCHC is the average hemoglobin concentration [55]. Therefore, the red blood cells of PROs will have a higher oxygen transport capacity due to the higher levels of MCH and MCHC. Currently, no study on cyclists has examined the differences in MCH and MCHC, so we cannot draw many conclusions in this regard. These hematological parameters have mainly been used as markers of anemia both in athletes and in the general population [56], but so far, they are not associated with an athlete's performance level in this study.

5. Limitations

Our study had limitations with regards to the sample number, since it was more difficult to recruit PRO athletes than lower-level athletes (AMA).

Differences in endogenous antioxidant marker between this study and previous works may be influenced by the instrumentation and methodology used, the timing of the season at which the measurements were made, and the training status of the cyclists.

6. Conclusions

Regarding the endogenous antioxidants profile, PRO had higher values of CAT, GSSG and GSSG/GSH compared to AMA. An inverse correlation was found for the first time between W_{VT1} and W_{VT2} with GSSG/GSH at rest only in PRO. This indicates better antioxidant status that allow for higher performance with regard to power output. Future studies should examine how training adaptations affect the studied variables and how antioxidant enzymes evolve during a race stage (e.g. Tour de France), in order to see their association with performance, recovery and fatigue, thereby helping to develop monitoring tools for medical doctors, nutritionists and coaches.

Author Contributions: Conceptualization, F.J.M.-N., C.M.-P., R.O.-R. and P.E.A.; methodology, F.J.M.-N., C.M.-P., R.O.-R. and P.E.A.; formal analysis, F.J.M.-N. and C.M.-P.; investigation, F.J.M.-N., C.M.-P. and R.O.-R.; resources, F.J.M.-N. and C.M.-P.; data curation, F.J.M.-N., C.M.-P. and R.O.-R.; writing—original draft preparation, F.J.M.-N.; writing—review and editing, F.J.M.-N., C.M.-P. and S.P.D.; visualization, F.J.M.-N. and C.M.-P.; supervision, C.M.-P., S.P.D. and P.E.A.; project administration, F.J.M.-N. and C.M.-P. All authors have read and agreed to the published version of the manuscript.

Institutional Review Board Statement: The study was conducted according to the guidelines of the Declaration of Helsinki, and approved by the Ethics Committee of the Catholic University of Murcia (CE091802).

Informed Consent Statement: Informed consent was obtained from all subjects involved in the study.

Data Availability Statement: All data is contained within the article.

Acknowledgments: This study was supported by the Research Center in High-Performance Sport of the Catholic University of Murcia (Murcia, Spain). We would like to acknowledge Linda H. Chung for her help in this project.

Conflicts of Interest: The authors declare no conflict of interest.

References

1. Faria, E.W.; Parker, D.L.; Faria, I.E. The science of cycling: Physiology and training—Part 1. *Sports Med.* **2005**, *35*, 285–312. [CrossRef]
2. Lucia, A.; Hoyos, J.; Chicharro, J.L. Physiology of professional road cycling. *Sports Med.* **2001**, *31*, 325–337. [CrossRef]
3. Jeukendrup, A.E.; Craig, N.P.; Hawley, J.A. The bioenergetics of World Class Cycling. *J. Sci. Med. Sport* **2000**, *3*, 414–433. [CrossRef]
4. Margaritelis, N.V.; Paschalis, V.; Theodorou, A.A.; Kyparos, A.; Nikolaidis, M.G. Redox basis of exercise physiology. *Redox Biol.* **2020**, *35*, 101499. [CrossRef] [PubMed]
5. Bentley, D.J.; Ackerman, J.; Clifford, T.; Slattery, K.S. Acute and chronic effects of antioxidant supplementation on exercise performance. In *Antioxidants in Sport Nutrition*; CRC Press: Boca Raton, FL, USA, 2014; Volume 141.
6. Bloomer, R.J.; Fisher-Wellman, K.H. Blood oxidative stress biomarkers: Influence of sex, exercise training status, and dietary intake. *Gend Med.* **2008**, *5*, 218–228. [CrossRef]
7. Urso, M.L.; Clarkson, P.M. Oxidative stress, exercise, and antioxidant supplementation. *Toxicology* **2003**, *189*, 41–54. [CrossRef]
8. Koppenol, W.H.; Hider, R.H. Iron and redox cycling. Do's and don'ts. *Free Radic. Biol. Med.* **2019**, *133*, 3–10. [CrossRef] [PubMed]
9. Mylonas, C.; Kouretas, D. Lipid peroxidation and tissue damage. *In Vivo* **1999**, *13*, 295–309.
10. St-Pierre, J.; Buckingham, J.A.; Roebuck, S.J.; Brand, M.D. Topology of superoxide production from different sites in the mitochondrial electron transport chain. *J. Biol. Chem.* **2002**, *277*, 44784–44790. [CrossRef]
11. Sakellariou, G.K.; Jackson, M.J.; Vasilaki, A. Redefining the major contributors to superoxide production in contracting skeletal muscle. The role of NAD(P)H oxidases. *Free Radic. Res.* **2014**, *48*, 12–29. [CrossRef]
12. Marzatico, F.; Pansarasa, O.; Bertorelli, L.; Somenzini, L.; Della Valle, G. Blood free radical antioxidant enzymes and lipid peroxides following long-distance and lactacidemic performances in highly trained aerobic and sprint athletes. *J. Sports Med. Phys. Fitness* **1997**, *37*, 235–239.
13. Pingitore, A.; Lima, G.P.; Mastorci, F.; Quinones, A.; Iervasi, G.; Vassalle, C. Exercise and oxidative stress: Potential effects of antioxidant dietary strategies in sports. *Nutrition* **2015**, *31*, 916–922. [CrossRef] [PubMed]
14. Cordova, A.; Sureda, A.; Albina, M.L.; Linares, V.; Belles, M.; Sanchez, D.J. Oxidative stress markers after a race in professional cyclists. *Int. J. Sport Nutr. Exerc. Metab.* **2015**, *25*, 171–178. [CrossRef]
15. Serrano, E.; Venegas, C.; Escames, G.; Sanchez-Munoz, C.; Zabala, M.; Puertas, A.; de Haro, T.; Gutierrez, A.; Castillo, M.; Acuna-Castroviejo, D. Antioxidant defence and inflammatory response in professional road cyclists during a 4-day competition. *J. Sports Sci.* **2010**, *28*, 1047–1056. [CrossRef] [PubMed]
16. Falone, S.; Mirabilio, A.; Pennelli, A.; Cacchio, M.; Di Baldassarre, A.; Gallina, S.; Passerini, A.; Amicarelli, F. Differential impact of acute bout of exercise on redox- and oxidative damage-related profiles between untrained subjects and amateur runners. *Physiol. Res.* **2010**, *59*, 953–961. [CrossRef] [PubMed]
17. Margaritelis, N.V.; Theodorou, A.A.; Paschalis, V.; Veskoukis, A.S.; Dipla, K.; Zafeiridis, A.; Panayiotou, G.; Vrabas, I.S.; Kyparos, A.; Nikolaidis, M.G. Adaptations to endurance training depend on exercise-induced oxidative stress: Exploiting redox interindividual variability. *Acta Physiol. (Oxford)* **2018**, *222*. [CrossRef]
18. Kostaropoulos, I.A.; Nikolaidis, M.G.; Jamurtas, A.Z.; Ikonomou, G.V.; Makrygiannis, V.; Papadopoulos, G.; Kouretas, D. Comparison of the blood redox status between long-distance and short-distance runners. *Physiol. Res.* **2006**, *55*, 611–616. [PubMed]

19. World Medical Association. WMA Declaration of Helsinki-Ethical principles for medical research involving human subjects. *JAMA J. Am. Med. Assoc.* **2013**, *310*, 2191–2194. [CrossRef] [PubMed]
20. Martínez-Noguera, F.J.; Marín-Pagán, C.; Carlos-Vivas, J.; Rubio-Arias, J.A.; Alcaraz, P.E. Acute Effects of Hesperidin in Oxidant/Antioxidant State Markers and Performance in Amateur Cyclists. *Nutrients* **2019**, *11*, 1898. [CrossRef]
21. Millet, G.P.; Vleck, V.E.; Bentley, D.J. Physiological Differences Between Cycling and Running. *Sports Med.* **2009**, *39*, 179–206. [CrossRef]
22. Edvardsen, E.; Hem, E.; Anderssen, S.A. End criteria for reaching maximal oxygen uptake must be strict and adjusted to sex and age: A cross-sectional study. *PLoS ONE* **2014**, *9*, e85276. [CrossRef] [PubMed]
23. Howley, E.T.; Bassett, D.R., Jr.; Welch, H.G. Criteria for maximal oxygen uptake: Review and commentary. *Med. Sci. Sports Exerc.* **1995**, *27*, 1292–1301. [CrossRef]
24. Wasserman, K.; Beaver, W.L.; Whipp, B.J. Gas exchange theory and the lactic acidosis (anaerobic) threshold. *Circulation* **1990**, *81*, Ii14–Ii30.
25. Weydert, C.J.; Cullen, J.J. Measurement of superoxide dismutase, catalase and glutathione peroxidase in cultured cells and tissue. *Nat. Protoc.* **2010**, *5*, 51–66. [CrossRef]
26. Randox Laboratories Ltd. *Radicales Libres*; Randox Laboratories Ltd.: Crumlin, UK, 1996; pp. 1–16.
27. Akerboom, T.P.; Sies, H. Assay of glutathione, glutathione disulfide, and glutathione mixed disulfides in biological samples. In *Methods in Enzymology*; Elsevier: Amsterdam, The Netherlands, 1981; Volume 77, pp. 373–382.
28. Asensi, M.; Sastre, J.; Pallardo, F.V.; Estrela, J.M.; Viña, J. Determination of oxidized glutathione in blood: High-performance liquid chromatography. In *Methods in Enzymology*; Elsevier: Amsterdam, The Netherlands, 1994; Volume 234, pp. 367–371.
29. Cohen, J. Statistical power. *Anal. Behav. Sci.* **1988**, *2*, 273–406.
30. Evans, W.J. Vitamin E, vitamin C, and exercise. *Am. J. Clin. Nutr.* **2000**, *72*, 647S–652S. [CrossRef] [PubMed]
31. Mena, P.; Maynar, M.; Gutierrez, J.M.; Maynar, J.; Timon, J.; Campillo, J.E. Erythrocyte Free Radical Scavenger Enzymes in Bicycle Professional Racers. Adaptation to Training. *Int. J. Sports Med.* **1991**, *12*, 563–566. [CrossRef]
32. Tauler, P.; Aguiló, A.; Guix, P.; Jiménez, F.; Villa, G.; Tur, J.A.; Córdova, A.; Pons, A. Pre-exercise antioxidant enzyme activities determine the antioxidant enzyme erythrocyte response to exercise. *J. Sports Sci.* **2005**, *23*, 5–13. [CrossRef]
33. Finaud, J.; Lac, G.; Filaire, E. Oxidative stress: Relationship with exercise and training. *Sports Med.* **2006**, *36*, 327–358. [CrossRef]
34. Turner, J.E.; Hodges, N.J.; Bosch, J.A.; Aldred, S. Prolonged depletion of antioxidant capacity after ultraendurance exercise. *Med. Sci. Sports Exerc.* **2011**, *43*, 1770–1776. [CrossRef] [PubMed]
35. Banerjee, A.K.; Mandal, A.; Chanda, D.; Chakraborti, S. Oxidant, antioxidant and physical exercise. *Mol. Cell Biochem.* **2003**, *253*, 307–312. [CrossRef] [PubMed]
36. Morrison, D.; Hughes, J.; Della Gatta, P.A.; Mason, S.; Lamon, S.; Russell, A.P.; Wadley, G.D. Vitamin C and E supplementation prevents some of the cellular adaptations to endurance-training in humans. *Free Radic. Biol. Med.* **2015**, *89*, 852–862. [CrossRef]
37. Lew, H.; Pyke, S.; Quintanilha, A. Changes in the glutathione status of plasma, liver and muscle following exhaustive exercise in rats. *FEBS Lett.* **1985**, *185*, 262–266. [CrossRef]
38. Wadley, G.D.; Nicolas, M.A.; Hiam, D.S.; McConell, G.K. Xanthine oxidase inhibition attenuates skeletal muscle signaling following acute exercise but does not impair mitochondrial adaptations to endurance training. *Am. J. Physiol. Endocrinol. Metab.* **2013**, *304*, E853–E862. [CrossRef] [PubMed]
39. Zhang, S.J.; Sandström, M.E.; Lanner, J.T.; Thorell, A.; Westerblad, H.; Katz, A. Activation of aconitase in mouse fast-twitch skeletal muscle during contraction-mediated oxidative stress. *Am. J. Physiol. Cell Physiol.* **2007**, *293*, C1154–C1159. [CrossRef] [PubMed]
40. Leonardo-Mendonça, R.C.; Concepción-Huertas, M.; Guerra-Hernández, E.; Zabala, M.; Escames, G.; Acuña-Castroviejo, D. Redox status and antioxidant response in professional cyclists during training. *Eur. J. Sport Sci.* **2014**, *14*, 830–838. [CrossRef]
41. Wang, J.-S.; Huang, Y.-H. Effects of exercise intensity on lymphocyte apoptosis induced by oxidative stress in men. *Eur. J. Appl. Physiol.* **2005**, *95*, 290–297. [CrossRef]
42. Ferrer, M.D.; Tauler, P.; Sureda, A.; Tur, J.A.; Pons, A. Antioxidant regulatory mechanisms in neutrophils and lymphocytes after intense exercise. *J. Sports Sci.* **2009**, *27*, 49–58. [CrossRef]
43. Tauler, P.; Aguiló, A.; Gimeno, I.; Guix, P.; Tur, J.A.; Pons, A. Different effects of exercise tests on the antioxidant enzyme activities in lymphocytes and neutrophils. *J. Nutr. Biochem.* **2004**, *15*, 479–484. [CrossRef] [PubMed]
44. Sureda, A.; Tauler, P.; Aguiló, A.; Cases, N.; Fuentespina, E.; Córdova, A.; Tur, J.A.; Pons, A. Relation between oxidative stress markers and antioxidant endogenous defences during exhaustive exercise. *Free Radic. Res.* **2005**, *39*, 1317–1324. [CrossRef]
45. Fisher-Wellman, K.; Bloomer, R.J. Acute exercise and oxidative stress: A 30 year history. *Dyn. Med.* **2009**, *8*, 1. [CrossRef] [PubMed]
46. Steinberg, J.G.; Delliaux, S.; Jammes, Y. Reliability of different blood indices to explore the oxidative stress in response to maximal cycling and static exercises. *Clin. Physiol. Funct. Imag.* **2006**, *26*, 106–112. [CrossRef] [PubMed]
47. Lee, I.M.; Hsieh, C.C.; Paffenbarger, R.S., Jr. Exercise intensity and longevity in men. The Harvard Alumni Health Study. *JAMA* **1995**, *273*, 1179–1184. [CrossRef] [PubMed]
48. Wadley, A.J.; Chen, Y.W.; Bennett, S.J.; Lip, G.Y.; Turner, J.E.; Fisher, J.P.; Aldred, S. Monitoring changes in thioredoxin and over-oxidised peroxiredoxin in response to exercise in humans. *Free Radic. Res.* **2015**, *49*, 290–298. [CrossRef]

49. Schumacher, Y.O.; Jankovits, R.; Bültermann, D.; Schmid, A.; Berg, A. Hematological indices in elite cyclists. *Scand. J. Med. Sci. Sports* **2002**, *12*, 301–308. [CrossRef]
50. Aguiló, A.; Tauler, P.; Pilar Guix, M.; Villa, G.; Córdova, A.; Tur, J.A.; Pons, A. Effect of exercise intensity and training on antioxidants and cholesterol profile in cyclists. *J. Nutr. Biochem.* **2003**, *14*, 319–325. [CrossRef]
51. Guglielmini, C.; Casoni, I.; Patracchini, M.; Manfredini, F.; Grazzi, G.; Ferrari, M.; Conconi, F. Reduction of Hb levels during the racing season in nonsideropenic professional cyclists. *Int. J. Sports Med.* **1989**, *10*, 352–356. [CrossRef] [PubMed]
52. Saris, W.H.; Senden, J.M.; Brouns, F. What is a normal red-blood cell mass for professional cyclists? *Lancet* **1998**, *352*, 1758. [CrossRef]
53. Heuberger, J.; Rotmans, J.I.; Gal, P.; Stuurman, F.E.; van't Westende, J.; Post, T.E.; Daniels, J.M.A.; Moerland, M.; van Veldhoven, P.L.J.; de Kam, M.L.; et al. Effects of erythropoietin on cycling performance of well trained cyclists: A double-blind, randomised, placebo-controlled trial. *Lancet Haematol.* **2017**, *4*, e374–e386. [CrossRef]
54. Bejder, J.; Andersen, A.B.; Goetze, J.P.; Aachmann-Andersen, N.J.; Nordsborg, N.B. Plasma volume reduction and hematological fluctuations in high-level athletes after an increased training load. *Scand. J. Med. Sci. Sports* **2017**, *27*, 1605–1615. [CrossRef]
55. López-Santiago, N. La biometría hemática. *Acta Pediátr. México* **2016**, *37*, 246–249. [CrossRef]
56. Wintrobe, M.M. Classification of the Anemias on the Basis of Differences in the Size and Hemoglobin Content of the Red Corpuscles. *Proc. Soc. Exp. Biol. Med.* **1930**, *27*, 1071–1073. [CrossRef]

 *antioxidants*

Article

Antioxidant Effect of a Probiotic Product on a Model of Oxidative Stress Induced by High-Intensity and Duration Physical Exercise

Maravillas Sánchez Macarro [1], Vicente Ávila-Gandía [1], Silvia Pérez-Piñero [1], Fernando Cánovas [1], Ana María García-Muñoz [1], María Salud Abellán-Ruiz [1], Desirée Victoria-Montesinos [1], Antonio J. Luque-Rubia [1], Eric Climent [2], Salvador Genovés [2], Daniel Ramon [2], Empar Chenoll [2] and Francisco Javier López-Román [1,3,*]

[1] Department of Exercise Physiology, San Antonio Catholic University of Murcia (UCAM), 30107 Murcia, Spain; msanchez4@ucam.edu (M.S.M.); vavila@ucam.edu (V.Á.-G.); sperez2@ucam.edu (S.P.-P.); fcanovas@ucam.edu (F.C.); amgarcia13@ucam.edu (A.M.G.-M.); msabellan@ucam.edu (M.S.A.-R.); dvictoria@ucam.edu (D.V.-M.); ajluque@ucam.edu (A.J.L.-R.)
[2] Research and Development Department, ADM-Biopolis, ADM, Parc Cientific Universitat de Valencia, Paterna, 46980 Valencia, Spain; Eric.Climent@adm.com (E.C.); salvador.genoves@adm.com (S.G.); Daniel.RamonVidal@adm.com (D.R.); Maria.Chenoll@adm.com (E.C.)
[3] Primary Care Research Group, Biomedical Research Institute of Murcia (IMIB-Arrixaca), 30120 Murcia, Spain
* Correspondence: jlroman@ucam.edu

Citation: Sánchez Macarro, M.; Ávila-Gandía, V.; Pérez-Piñero, S.; Cánovas, F.; García-Muñoz, A.M.; Abellán-Ruiz, M.S.; Victoria-Montesinos, D.; Luque-Rubia, A.J.; Climent, E.; Genovés, S.; et al. Antioxidant Effect of a Probiotic Product on a Model of Oxidative Stress Induced by High-Intensity and Duration Physical Exercise. *Antioxidants* **2021**, *10*, 323. https://doi.org/10.3390/antiox10020323

Academic Editors: Gareth Davison and Conor McClean

Received: 30 January 2021
Accepted: 18 February 2021
Published: 22 February 2021

Publisher's Note: MDPI stays neutral with regard to jurisdictional claims in published maps and institutional affiliations.

Abstract: This randomized double-blind and controlled single-center clinical trial was designed to evaluate the effect of a 6-week intake of a probiotic product (1 capsule/day) vs. a placebo on an oxidative stress model of physical exercise (high intensity and duration) in male cyclists (probiotic group, n = 22; placebo, n = 21). This probiotic included three lyophilized strains (*Bifidobacterium longum* CECT 7347, *Lactobacillus casei* CECT 9104, and *Lactobacillus rhamnosus* CECT 8361). Study variables were urinary isoprostane, serum malondialdehyde (MDA), serum oxidized low-density lipoprotein (Ox-LDL), urinary 8-hydroxy-2′-deoxiguanosine (8-OHdG), serum protein carbonyl, serum glutathione peroxidase (GPx), and serum superoxide dismutase (SOD). At 6 weeks, as compared with baseline, significant differences in 8-OHdG (Δ mean difference −10.9 (95% CI −14.5 to −7.3); p < 0.001), MDA (Δ mean difference −207.6 (95% CI −349.1 to −66.1; p < 0.05), and Ox-LDL (Δ mean difference −122.5 (95% CI −240 to −4.5); p < 0.05) were found in the probiotic group only. Serum GPx did not increase in the probiotic group, whereas the mean difference was significant in the placebo group (477.8 (95% CI 112.5 to 843.2); p < 0.05). These findings suggest an antioxidant effect of this probiotic on underlying interacting oxidative stress mechanisms and their modulation in healthy subjects. The study was registered in ClinicalTrials.gov (NCT03798821).

Keywords: oxidative stress; probiotics; physical exercise; male cyclists; oxidative stress biomarkers; antioxidative enzymes

1. Introduction

Oxidative stress is characterized by the inability of the organism to detoxify reactive oxygen species (ROS) caused by a disequilibrium in the balance between their production and accumulation in cells and tissues. ROS generated by biological systems as metabolic by-products include hydrogen peroxide, superoxide and hydroxyl radicals, and singlet oxygen [1]. The oxidation products or nitrosylated products linked to ROS have a variety of detrimental effects on crucial cellular functions. Cell enzymatic antioxidant defensive systems include superoxide dismutase (SOD), catalase (CAT), glutathione reductase, and glutathione peroxidase (GPx) as the most important scavengers [2,3]. On the other hand, overproduction of ROS may result in cell and tissue injury and contribute to oxidative stress and chronic inflammation as the underlying pathophysiological mechanisms of a wide spectrum of pathological conditions related to neurodegeneration, atherosclerosis, metabolic diseases, carcinogenesis, or ageing [4–8].

The relationship between oxidative stress and microbiota dysbiosis has been a focus of increasing interest. The intestinal microbiota performs multiple functions related to signaling pathways and maintenance of homeostasis, interacting with nutrients and drug metabolism, performing intestinal barrier functions, protecting against pathogen colonization, and also working together with the immune system [9,10]. Excessive bioavailability of ROS may result from a disturbance of gut microbiota, contributing to an increase of oxidative stress. It has been shown that microbial-elicited ROS modulates innate immune signaling and mediates motility and increased cellular proliferation [11]. It has been hypothesized that at least partially-mediated ROS-dependent mechanisms are involved in potential beneficial effects of candidate probiotic bacteria as well as in many of the known effects of the normal microbiota on intestinal physiology [12]. Recent studies have shown fecal microbiota transplantation to be effective in the modulation of oxidative stress and reduced inflammation. A variety of mechanisms has been identified for the antioxidant action induced by probiotic bacteria in the gut. These include release of antioxidant molecules (e.g., glutathione) and secretion of antioxidant enzymes, direct ROS scavenging action, and their role as strong chelators of free copper or iron ions to prevent metal ion-catalyzed oxidation [13,14]. Probiotic exposure has also been associated with reduction of the activity of ROS-releasing enzyme systems such as NADPH oxidases and induction of cellular antioxidant signaling pathways such Nrf2-Keap1-ARE [15]. Altogether, it seems plausible that strategies able to impact the microbiome could potentially have an effect on oxidative stress.

On the other hand, intense physical exercise has been shown to be associated with different physiological changes, some of which include glucose and fatty acid oxidation, oxidative phosphorylation, and increased production of ROS and reactive oxygen nitrogen species (RONS) [16,17]. Additionally, gastrointestinal hypoxia and hypoperfusion during endurance exercise may increase intestinal permeability and oxidative stress in the gastrointestinal tract. Exercise-induced oxidative stress is affected by important factors, such as duration and intensity of exercise, training status, and nutritional intake. The effects of antioxidant intake (e.g., vitamin C, vitamin E, polyphenols, resveratrol, β-carotene, *N*-acetylcysteine) on exercise-induced oxidative stress have also been assessed in numerous experimental and human studies [18–20]. However, evidence of improvement of exercise performance or reduced muscle damage is inconsistent due to differences in the conditions of the exercise protocol and the administration of the antioxidant product (i.e., type, dose, timing, duration, etc.).

Based on the potential effects of probiotics as inducers of an antioxidant action and the increased production of ROS elicited by intense physical exercise, this study was conducted to test the hypothesis that supplementation with a probiotic product may be associated with beneficial effects in an oxidative stress model induced by high-intensity and duration physical exercise in male cyclists. Changes in gut bacterial microbiome were also examined.

2. Materials and Methods

2.1. Design

Between July 2018 and January 2019, a randomized, parallel-group, double-blind, placebo-controlled, and single-center trial was conducted at the Health Sciences Department of the Saint Anthony Catholic University (UCAM) in Murcia, Spain. The primary objective of this study was to evaluate the effect of the administration for 6 weeks of a daily regimen of a probiotic product, obtained from the mixture of three lyophilized probiotic strains, on an oxidative stress model based on the performance of physical exercise of high intensity and duration. The secondary objective was the evaluation of changes in bacterial microbiome from fecal samples. The study protocol was approved by the Ethics Committee of UCAM. Written informed consent was obtained from all participants. The study was registered in ClinicalTrials.gov (accessed on 18 February 2021) (NCT03798821).

2.2. Eligibility Criteria and Randomization

Caucasian healthy male volunteers aged 18–45 years who performed aerobic physical exercise between 2 and 4 times a week were eligible provided that they gave the written informed consent and none of the following exclusion criteria were present: history of chronic disease, particularly gastrointestinal disorders; abdominal surgery in last 3 months; asthma; chronic obstructive pulmonary disease (COPD); hypertension; sinus bradycardia; heart failure or cardiogenic shock; current smoking (>10 cigarettes/day); body mass index (BMI) > 30 kg/m^2; alcohol or drug abuse; and poor tolerance or hypersensitivity to any component of the study product. The database of the Health Sciences Department of UCAM was used for the recruitment of participants.

Randomization (1:1) to supplementation with the probiotic product (probiotic group) or placebo (placebo group) was performed by an independent researcher using a random sequence of computer-generated numbers.

2.3. Intervention

Participants were given the probiotic product (300 mg capsules with 100 mg probiotic and maltodextrin and sucrose as carriers, 200 mg) or placebo (300 mg capsules with maltodextrin and sucrose) during 6 weeks. The probiotic product obtained from ADM-Biopolis (Valencia, Spain) was based on a mixture of three lyophilized probiotic strains: *Bifidobacterium longum* CECT 7347, *Lactobacillus casei* CECT 9104, and *Lactobacillus rhamnosus* CECT 8361 (in a ratio 1:4.5:4.5, 1×10^9 total colony-forming units (cfu) per capsule). Participants were recommended to take one daily capsule, at breakfast, for 6 weeks. For all the strains a safety study including in vivo acute oral toxicity was previously evaluated, following the method described by Chenoll et al. [21] for *L. rhamnosus* CECT 8361 and *L. casei* CECT 9104 (data not shown).

2.4. Physical Exercise Oxidative Stress Model

The model was a high-intensity and long-lasting physical activity (90 min) on a bicycle roller. Participants underwent a preliminary test and two subsequent tests (test #1 after a 7-day washout period and test #2 at the end of the study at 6 weeks). The preliminary test was performed to calculate the intensity of tests #1 and #2 for each individual, using a bicycle roller with electromagnetic resistance (Technogym Spin Trainer) with an initial speed load of 12 km/h, with a 2 km/h load increase every minute, maintaining a constant slope of 2%. The cyclists employed free development. In order to calculate the intensity of tests #1 and #2, participants were monitored by ECG and gas analyzer (Jaeger Oxicom Pro$^®$, CareFusion Respiratory Care, Germany) to determine maximal heart rate (MHR) and monitor heart rate above anaerobic threshold and during maximum oxygen uptake (VO$_2$ max). Tests #1 and #2 lasted 90 min, and the maximum maintained load was equivalent to a heart rate corresponding to 75% of VO$_2$ max calculated in the preliminary test. A constant slope of 2% was also used. The water consumption was ad libitum. After test #1, participants were given the assigned supplement (probiotic or placebo). Forty-eight hours before each test participants did not make any intense physical or psychological effort.

2.5. Study Procedures

The study included three visits, one at baseline during the time of the preliminary test, one at the time of test #1, and a final visit after test #2 at 6 weeks. At baseline, participants signed the informed consent, when eligibility criteria were checked, and the study product was given. Clinical evaluations included detailed medical history and measurement of anthropometric variables. Compliance with the intake of the probiotic product was assessed by counting the remaining capsules in the medication container. Adverse events were ascertained by directly asking participants how they were feeling after taking the product and from abnormal changes of laboratory results. During the study period, there were no dietary restrictions, but medications that may affect the microbiome (e.g., antioxidants, statins) were not allowed.

Peripheral blood samples (12 mL) after 12 h fasting were extracted at 30 min before and after each test, and 24 h urine samples were collected one day before and after the test. From the total urine volume, a 9 mL sample was frozen at $-80\,^{\circ}$C for further analysis.

Stool samples were collected during 24 h before test #1 and at 6 weeks during the 24 h before test #2, preserved with REAL stock buffer (Durviz S.L., Paterna, Valencia, Spain), and stored at $-80\,^{\circ}$C until analysis.

2.6. Study Variables

Body weight, BMI, and free fat mass were measured using bioelectrical impedance analysis (BIA) on a whole body BIA analyzer (Tanita BC-420MA, Tanita Corporation, Tokyo, Japan). Biochemical analyses included urinary isoprostanes (8-iso-PGF2α, ELISA kit, Oxford Biomedical Research, Rochester Hills, MI, USA), serum malondialdehyde (MDA) (MDA ELISA kit, Elabscience, Houston, TX, USA), and serum oxidized low-density lipoprotein (Ox-LDL) (Human OxLDL ELISA kit, Elabscience) as lipid-related oxidative stress biomarker; urinary 8-hydroxy-2′-deoxiguanosine (8-OHdG) (ELISA kit, Elabscience) as DNA-related oxidative stress biomarker, and serum protein carbonyl (Protein Carbonyl ELISA kit, Enzo Life Sciences, Lausanne, Switzerland) as protein-related oxidative stress biomarker; and serum glutathione peroxidase (GPx) (ELISA kit, Elabscience) and serum superoxide dismutase (SOD) (ELISA kit, Elabscience) as endogenous antioxidative enzymes. Safety analyses included complete blood count, liver function tests (bilirubin, alanine and aspartate aminotransferases, gamma-glutamyl transpeptidase), and renal function tests (blood urea nitrogen and serum creatinine levels).

For microbiome analysis, DNA was isolated with the aid of a QIAmp Power Fecal Pro DNA kit (Qiagen, Hilden, Germany), with bead beating and enzymatic lysis steps prior to extraction to avoid bias in DNA purification toward misrepresentation of Gram-positive bacteria. Massive genome sequencing of the hypervariable region V3–V4 of the bacterial 16s rRNA gene was conducted to evaluate the bacterial composition of the gut microbiome. Samples were amplified using key-tagged eubacterial primers [22] and sequenced with a MiSeq Illumina Platform, following the Illumina recommendations for library preparation and sequencing for metagenomic studies. The resulting sequences were split per patient, considering the barcode introduced during the PCR reaction. R1 and R2 reads were overlapped using PEAR program version 0.9.1, with an overlap of 50 nucleotides and a quality of overlap with a minimum of Q20, providing a single FASTQ file for each of the samples. Quality control of the sequences was performed by initial quality filtering (minimum threshold of Q20) using fastx tool kit version 0.013, followed by primer (16s rRNA primers) trimming and length selection (reads over 300 nts) with cutadapt version 1.4.126. These FASTQ files were then converted to FASTA files, and chimeras that could arise during the amplification and sequencing steps were removed by the UCHIME program, version 7.0.1001. Those clean FASTA files were BLAST against the National Center for Biotechnology Information (NCBI) 16s rRNA database using blastn version 2.2.29+. The resulting XML files were processed using a python script developed by ADM-Biopolis; (Valencia, Spain) to annotate each sequence at different phylogenetic levels.

2.7. Statistical Analysis

Analyses were performed in the per-protocol (PP) data set, which included all participants who completed the 6-week study period and underwent tests #1 and #2. The sample size was calculated for an expected mean difference between groups in serum levels of MDA of 1.34 nmol/mL with a standard deviation of 1.6 nmol/L according to data of Krotkiewsky et al. [23], so that for a significance level of 5% and statistical power of 80% assuming a drop-out rate of 10% since the primary analysis was performed in the PP data set, 20 evaluable participants for each treatment group were required. Categorical variables were expressed as frequencies and percentages, and continuous variables as mean and standard error (SE). Mean differences and 95% confidence intervals (CI) were calculated for changes between data at 6 weeks as compared with baseline. The chi-square (χ^2) test or the Fisher's exact probability test was used for the comparison of categorical variables between the probiotic and placebo groups. Quantitative variables were assessed using the analysis of variance (ANOVA) for repeated measures with three factors: time (baseline and final), test (test #1 and test #2) as within-subject factors and intervention (probiotic and placebo) as between-subject factor, with Bonferroni's correction for pairwise comparisons.

In the case of microbiome analysis, alpha diversity was conducted using the vegan package, and statistical significance analyzed with the ANOVA test. The DESeq2 package from R (R Core Team, 2012) was used to generate a generalized linear model with fixed effects with negative binomial family, and the Wald test was used to compare operational taxonomic unit (OTU) counts between groups.

Statistical significance was set at $p < 0.05$. The SPSS software version 21.0 (IMB Corp., Armonk, NY, USA) was used for statistical analysis.

3. Results

3.1. Study Population

Of a total of 45 eligible subjects, 1 declined to participate. The remaining 44 were randomized to the study groups (22 in each group), but 1 subject assigned to the placebo group did not receive the assigned intervention and was lost to follow-up. The final study sample included 22 subjects in the probiotic group (25.3 ± 7.2 years) and 21 (27.1 ± 8.4 years) in the placebo group (Figure 1). Baseline BMI was 23.6 (2.6) kg/m^2 and VO$_2$ max 51.1 (8.8) mL/kg/min. Significant differences after randomization were not observed.

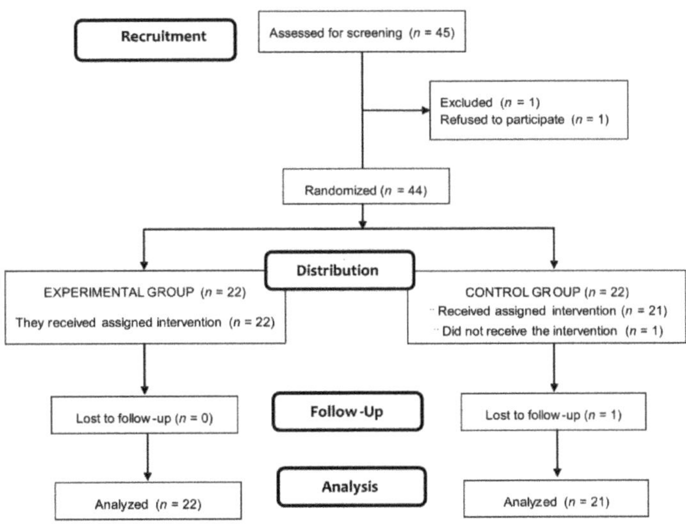

Figure 1. Flow chart of the study population.

3.2. Lipid, Protein, and DNA-Related Oxidative Stress Biomarkers and Antioxidative Enzymes

The oxidative stress model based on the performance of high intensity exercise and duration (test 1) produced statistically significant increases in biomarkers of oxidative stress and enzymes.

As shown in Table 1, urinary isoprostanes increased significantly in both groups after tests #1 and #2 as compared with baseline, but the difference between tests #1 and #2 (Δ mean difference) and between-group differences were not significant. Serum MDA showed a significant Δ mean difference of -207 ng/mL (95% CI -349.1 to 66.1) ($p < 0.05$) in the probiotic group only, with between-group differences also statistically significant ($p < 0.05$). Serum Ox-LDL showed a significant Δ mean difference of -122.5 pg/mL (95% CI -240 to -4.5) ($p < 0.05$) in the probiotic group only, but between-group differences almost reached statistical significance ($p = 0.063$). Urinary 8-OHdG increased significantly in both groups after tests #1 and #2, although the Δ mean difference (-10.9 pg/day, 95% CI -14.5 to -7.3; $p < 0.001$) was only significant in the probiotic group; moreover, between-group differences were also significant ($p < 0.001$). Serum protein carbonyl increased significantly after test #1 and test #2 in both groups, but neither Δ mean difference nor between-group differences were statistically significant. Serum GPx increased significantly in both groups after test #1 and in the placebo group only after test #2; however, neither Δ mean difference nor between-group differences were statistically significant. Serum SOD increased significantly in both groups after test #2, but again neither Δ mean difference nor between-group differences were statistically significant.

Table 1. Results of lipid, protein, and DNA-related oxidative stress biomarkers and antioxidative enzymes.

Variables	Test #1				Test #2 (6-Week Probiotic/Placebo Intake)				Test #1 vs. Test #2	Between-Group Difference p Value F Snedecor
	Baseline Mean (SE)	After Exercise Mean (SE)	Mean Difference (95% CI), p Value		Baseline Mean (SE)	After Exercise Mean (SE)	Mean Difference (95% CI), p Value	Δ Mean Difference (95% CI) p Value	
Urinary isoprostane, pg/day									
Placebo group	1.3 (0.5)	2.5 (0.7)	1.2 (0.5 to 1.9) $p = 0.05$		1.2 (0.5)	2.1 (0.7)	0.9 (0.3 to 1.5) $p < 0.05$	-0.3 (-0.8 to 0.2) $p = 0.292$	$p = 0.213$ F = 1.601
Probiotic group	2.1 (0.5)	3.3 (0.7)	1.3 (0.6 to 2.0) $p < 0.05$		2.2 (0.5)	3.6 (0.7)	1.4 (0.9 to 2.0) $p < 0.001$	0.1 (-0.3 to 0.7) $p = 0.476$	
Serum MDA, ng/mL									
Placebo group	347.4 (84.8)	491.1 (145.3)	143.7 (-25.8 to 313.2) $p = 0.094$		312.9 (64.3)	454.4 (113.3)	141.5 (-52.8 to 335.8) $p = 0.149$	-2.2 (-147 to 142.6) $p = 0.975$	$p < 0.05$ F = 4.195
Probiotic group	433.2 (82.9)	687.4 (142.0)	254 (88 to 419.8) $p < 0.05$		358 (62.9)	404.6 (110.7)	46.6 (-143 to 236.4) $p = 0.623$	-207.6 (-0.341 to -66.1) $p < 0.05$	
Serum Ox-LDL, pg/mL									
Placebo group	740.3 (82.9)	899.6 (64.1)	159.3 (81.9 to 236.7) $p < 0.001$		779.9 (64.2)	977.4 (78.4)	196.6 (83.0 to 310.2) $p < 0.05$	37.3 (-83.5 to 158.0) $p = 0.536$	$p < 0.063$ F = 3.653
Probiotic group	646.2 (60.1)	809.0 (62.6)	162.9 (87.2 to 238.5) $p < 0.001$		772.9 (67.6)	813.3 (77.1)	40.4 (-70.6 to 151.4) $p = 0.467$	-122.5 (-240 to -4.5) $p < 0.05$	
Urinary 8-OHdG, pg/day									
Placebo group	10.7 (0.2)	23.1 (3.8)	12.4 (8.3 to 16.6) $p < 0.001$		11.8 (2.4)	23.4 (3.3)	11.5 (8.1 to 15.0) $p < 0.001$	-0.9 (-4.6 to 2.8) $p = 0.620$	$p < 0.001$ F = 15.144
Probiotic group	13.3 (2.0)	29.0 (3.7)	15.7 (11.6 to 19.7) $p < 0.001$		13.6 (2.4)	18.4 (3.2)	4.8 (1.4 to 8.1) $p < 0.001$	-10.9 (-14.5 to -7.3) $p < 0.01$	
Serum protein carbonyl, pmol/mg protein									

Table 1. *Cont.*

| Variables | Test #1 | | | Test #2 (6-Week Probiotic/Placebo Intake) | | | Test #1 vs. Test #2 | Between-Group Difference *p* Value F Snedecor |
	Baseline Mean (SE)	After Exercise Mean (SE)	Mean Difference (95% CI), *p* Value	Baseline Mean (SE)	After Exercise Mean (SE)	Mean Difference (95% CI), *p* Value	Δ Mean Difference (95% CI) *p* Value	
Placebo group	124.0 (16.3)	160.0 (18.0)	36.0 (18.4 to 53.6) $p < 0.001$	112.4 (19.9)	162.0 (20.6)	49.6 (32.6 to 66.2) $p < 0.001$	13.6 (−4.4 to 31.6) $p = 0.135$	$p = 0.434$ F = 0.625
Probiotic group	166.8 (15.9)	204.2 (17.6)	37.4 (20.1 to 54.6) $p < 0.001$	162.9 (17.5)	204 (20.1)	41.1 (24.8 to 57.3) $p < 0.001$	3.7 (−13.9 to 21.3) $p = 0.671$	
Serum GPx, pg/mL								
Placebo group	526.9 (84.9)	788.0 (92.1)	261.1 (162.2 to 360.0) $p < 0.001$	633.8 (80.3)	1111.7 (214.5)	477.8 (112.5 to 843.2) $p < 0.05$	216.7 (−156.4 to 598.9) $p = 0.248$	$p = 0.253$ F = 1.598
Probiotic group	473.4 (83.0)	594.8 (90.0)	121.4 (24.7 to 218.0) $p < 0.05$	598.5 (78.4)	610.0 (209.6)	11.6 (−345.4 to 368.5) $p = 0.948$	−109.9 (−474.4 to 254.7) $p = 0.546$	
Serum SOD, ng/mL								
Placebo group	24.1 (2.6)	34.5 (3.3)	10.5 (5.9 to 15.1) $p < 0.001$	22.1 (2.2)	29.9 (2.7)	7.8 (3.7 to 11.9) $p < 0.001$	−2.9 (−8.6 to 3.2) $p = 0.358$	$p = 0.267$ F = 1.274
Probiotic group	29.2 (2.6)	33.1 (3.3)	3.9 (−0.7 to 8.5) $p = 0.094$	24.9 (2.2)	30.7 (2.7)	5.8 (1.7 to 10) $p < 0.05$	2 (−4 to 7.8) $p = 0.511$	

SE: standard error; CI: confidence interval; MDA: malondialdehyde; Ox-LDL: oxidized low-density lipoprotein; GPx: glutathione peroxidase; SOD: superoxide dismutase; F: F-Snedecor.

3.3. Microbiome Analysis

A total of 86 samples were included in the microbiome analysis (44 samples from participants in the probiotic group before test #1 (n = 22) and at 6 weeks before test #2 (n = 22), and 42 samples from participants in the placebo group before test #1 (n = 21) and at 6 weeks before test #2 (n = 21)). The local contributions to beta diversity (LCBD) at family and genus levels from taxonomic identification of the samples sequenced is shown in Figure 2.

Bacterial composition of samples was grouped, and both groups (placebo and probiotic) were compared at baseline and at 6 weeks at the end of the study. Richness, Simpson diversity index, and Shannon diversity index did not change after probiotic consumption (ANOVA test, $p > 0.05$ for all comparisons) (Figure 3).

Differences in bacterial population were measured with a Wald test using DESeq2 analysis. After 6 weeks of ingestion of the probiotic product or placebo (end of study), families Rhodospirillaceae (placebo vs. probiotic, log2 fold = 2.71, adjusted p value = 0.019) and Streptococcaceae (placebo vs. probiotic, log2 fold = 2.20, adjusted p value = 0.019) showed lower values in the probiotic group (Figure 4, left panel), considering a minimum threshold value of 10 counts (total average). There were statistically significant changes in seven genera, *Rhodospirillum* and *Streptococcus* being higher in the placebo group (Figure 4, right panel). However, within-group differences in the probiotic group showed an increase in specific genera, *Methanobrevibacter* (*M. smithii*), *Holdemanella* (*H. biformis*), and *Blautia* being the most remarkable, although *Lactobacillus* and *Lachnospira* decreased at the end of the study. Within-group differences in the placebo group revealed increases in *Bifidobacterium* and *Blautia*, among others, and decreases in *Shigella* and *Klebsiella* (in this case with low mean at baseline). Detailed data are shown in the Supplementary Materials, with Table S1 showing sequence distribution as well as sample metadata; Table S2 includes microbiome profiles at phylum, family, genus, and species levels, and Table S3 summarizes different populations at the genus level by Deseq2 analysis.

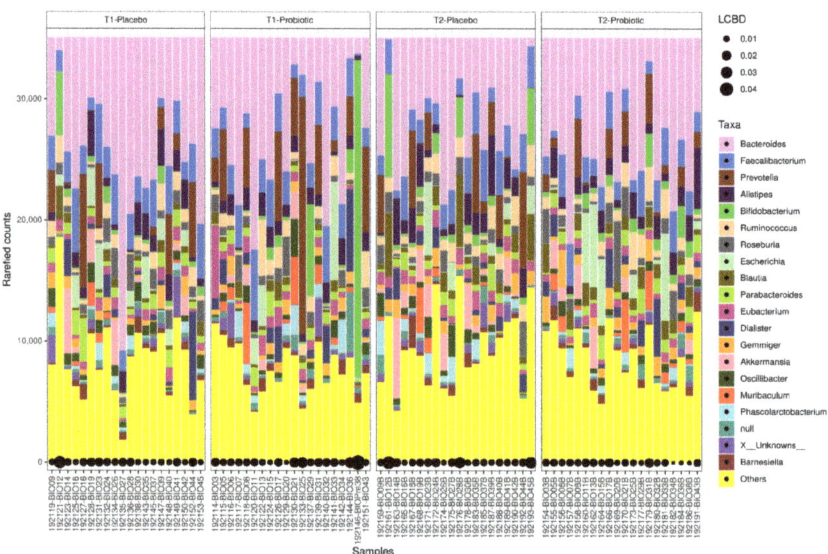

Figure 2. Local contributions to beta diversity (LCBD) analysis at family level (**right**) and genus level (**left**) from taxonomic identification of the samples sequenced (42 samples in the placebo group and 44 samples in the probiotic group; T1: before test #1, T2: at 6 weeks before test #2).

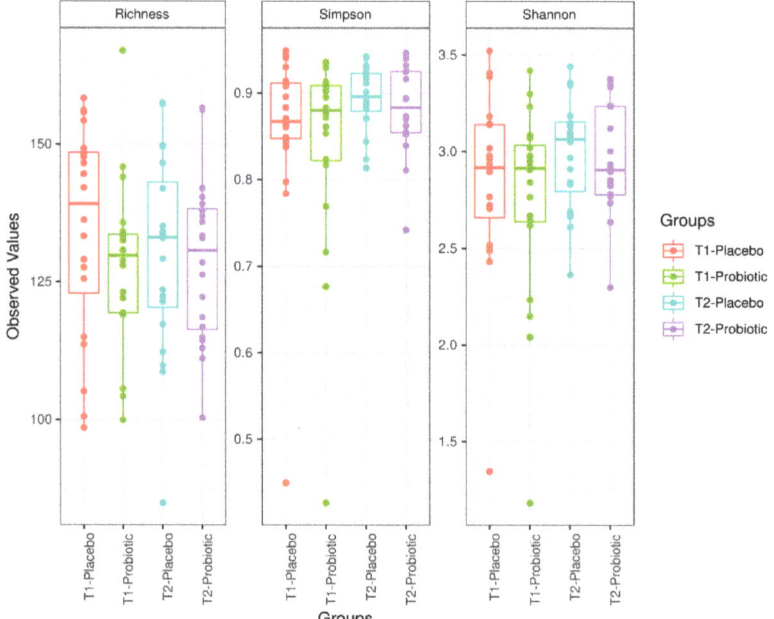

Figure 3. Richness, Simpson diversity index, and Shannon diversity index (from **left** to **right**) in the placebo and probiotic group at baseline (T1) and at 6 weeks (end of study) (T2).

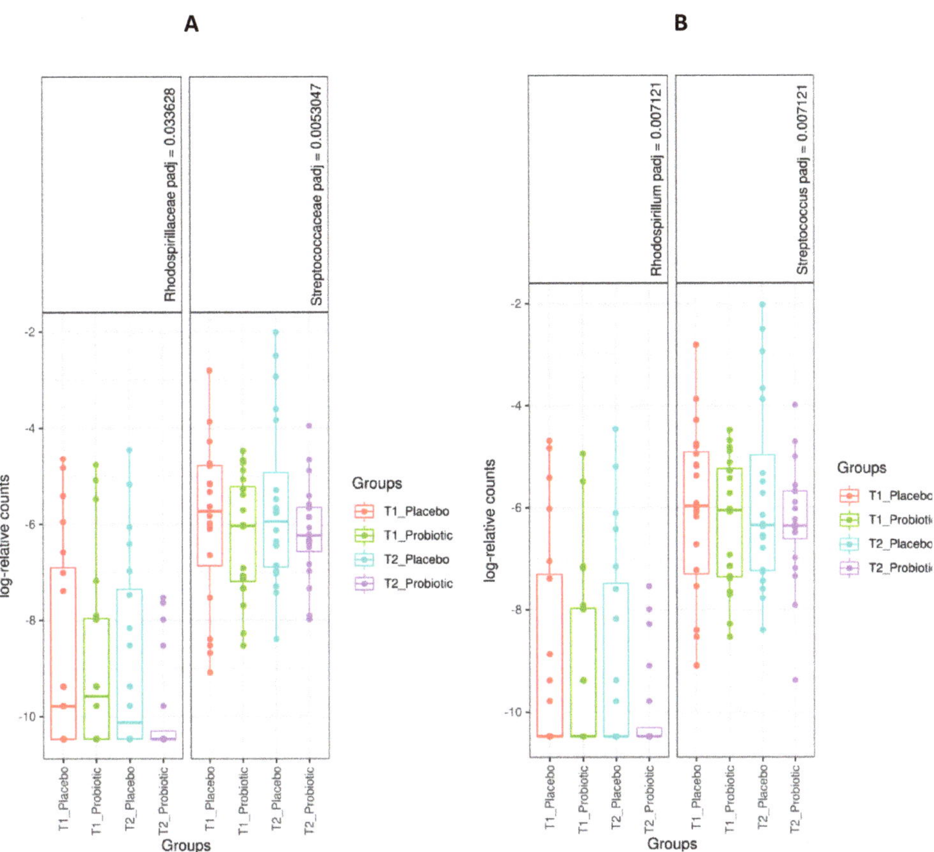

Figure 4. Differences between the placebo and probiotic groups at the end of the study (6 weeks) at the level of families (**A**) and genera (**B**).

The probiotic product was well tolerated, and no adverse effects were observed. Additionally, laboratory tests at the end of the study did not show any abnormalities.

4. Discussion

In an oxidative stress model of high-intensity and duration physical exercise in male cyclists, daily intake of a probiotic product based on a mixture of *B. longum*, *L. casei* and *L. rhamnosus* for 6 weeks was associated with a significant reduction of lipid-related oxidative stress biomarkers, such as serum MDA, serum Ox-LDL, and DNA-related oxidative stress biomarker, such as urinary 8-OHdG. Several studies have shown that high-intensity and duration physical exercise results in oxidative stress, due to ROS being generated excessively by enhanced oxygen consumption, as well as in changes in muscle antioxidant enzyme activity [24–27]. Additionally, physical exercise models in endurance-trained competitive and non-competitive athletes have been used to assess the benefits of different supplements with antioxidant capacity [28–32].

Probiotic supplements are nutraceuticals with wide applications in different aspects of human health and have recently gained increasing interest for their potential effects as antioxidants due to anti-oxidative enzyme upregulation, stimulation of the production of a variety of bioactive peptides, and gut flora re-establishment [33]. However, there is limited evidence of the influence of probiotic supplementation on oxidative markers in athletes, and as far as we are aware there are only four studies examining antioxidant potential of probiotics in athletes. In a randomized double-blind, placebo-controlled study, 22 elite athletes received *Lactobacillus helveticus* (n = 10) or placebo (n = 12) for 3 months, and it a significant decrease of MDA and advanced oxidation protein products (AOPP) was found, without modifications in antioxidant enzyme SOD activity [34]. In a comparative study of two groups of 12 athletes each, probiotic supplementation with a combination of *Lactobacillus rhamnosus* IMC 501 and *Lactobacillus paracasei* IMC 502 administered for 4 weeks vs. no supplementation (controls) was associated with an increase in plasma antioxidant levels, thus neutralizing ROS [35]. A randomized, double-blinded, placebo controlled trial conducted in 23 trained men who received multi-species probiotics (n = 11) or placebo (n = 12) over 14 weeks, was designed to evaluate changes of markers of intestinal barrier, oxidation, and inflammation associated with the use of probiotic supplementation at rest and after intense exercise [36]. Participants performed a 90-min intense cycle ergometry at baseline and after 14 weeks. In this study, supplementation had no effect on protein carbonyl and MDA but decreased zonulin in feces as a marker, indicating enhanced gut permeability [36]. Finally, in a study of marathon runners, *Lactobacillus rhamnosus* GG (probiotic group) or placebo drink (placebo group) were given during the 3-month training period, 6-day preparation period, and marathon run, but probiotics did not show any effect on serum total antioxidant potential Ox-LDL [37]. However, studies requiring larger samples of athletes are needed to assess the beneficial role of probiotic supplementation on markers of oxidative stress damage.

On the other hand, other studies have examined the association between gut microbiota and oxidative stress in diseases in which oxidative stress plays a well-known pathogenetic role, such as type 2 diabetes mellitus. In a systematic review and meta-analysis of 13 randomized clinical trials involving 840 subjects, probiotics intake resulted in significant improvement in serum levels of total antioxidant status, MDA, and total glutathione (GSH), but there was a modest effect on serum glucose levels and glycated hemoglobin (HbA1c) [38]. Wang et al. [15] reported an in-depth review of the antioxidant mechanisms of probiotics, summarizing their involvement in decreasing radical generation and improving the antioxidant system based on modulation of the redox status of the host via their metal ion chelating ability, regulation of signaling pathways, antioxidant systems, ROS-producing enzymes, and gut microbiota.

A diversity of exogenous and endogenous stimuli are involved in complex molecular and cellular changes, including oxidative DNA damage and participation in cancer development [39], and different studies have explored the potential of probiotics (*L. casei* and *L. rhamnosus*) as cell-free supernatants to inhibit colon cancer cell invasion [40], the antiproliferative and apoptotic effects driven by *L. casei* ATCC 393 against experimental colon cancer [41], or *Lactobacilli* strains as modulators of *Fiaf* gene expression in human epithelial intestinal cells [42].

8-hydroxy-2′deoxiguanosine (8OHdG) is usually measured as an index of oxidative DNA damage [43,44] with oxidative modification of DNA that causes mutations during replication [45]. In recent years, there has been an increasing interest in the impact of exercise on epigenetic events; in particular, ROS-mediated methylation patterns are being investigated. The understanding of the mechanisms leading to ROS-associated epigenetic modifications may contribute to a better knowledge of carcinogenesis and its progression, together with discovering of implicated biomarkers [46,47].

An interesting aspect of the present study was the assessment of changes in microbiome besides improvement of biomarkers of oxidative damage induced by a model of high-intensity and duration physical exercise in response to supplementation with the probiotic product. The microbiota can be considered as a true endocrine organ, and the interactions between exercise and its adaptations, probiotics, and the microbiota itself could help athletes by producing beneficial metabolic, antioxidant, or anti-inflammatory effects that improve training. *Methanobrevibacter*, *Holdemanella*, and *Blautia* increased in participants consuming probiotics, whereas *Lactobacillus* and *Lachnospira* were within the taxa that decreased at the final point. *M. smithii* is a prominent microbe with methanogenic properties. In a humanized gnotobiotic mouse model of host–archaeal–bacterial mutualism, it was shown that *M. smithii* removed H_2, which was related with more effective bacterial fermentation and subsequently more efficient short-chain fatty acids (SCFAs) production, increasing energy absorption [48,49]. *Holdemanella* is considered a butyrate producer. In a study of fecal microbiota collected from obese adults aimed to assess the effect of a pectin extracted from lemon and the probiotic strain *B. longum* BB-46, given in combination or alone, there was a positive correlation of *Holdemanella* with acetic and butyric acid, and a negative correlation with ammonium ions [50]. In an experimental high fat-induced oxidative stress, polyphenol supplementation affected different taxonomic levels of the gut microbiome by improving the proportion of *Blautia* (a butyrate producer) [51]. *Blautia* is one of the major taxonomic groups of the human gut microbiota (a genus in the *Lachnospiraceae* bacterial family, degrading complex polysaccharides to acetate, butyrate, and propionate (short chain fatty acids) that can be used by the host for energy and as a source of butyrate [52]. In a study of subjects who completed a 6-week endurance-based exercise intervention, there was an increase in butyrate concentrations induced by the exercise as a result of an increase in *Lachnospira* spp. [53]. This increase was independent of the BMI and decreased after return to sedentary activity. Surprisingly, *Lactobacillus* was found to be decreased at the end of the study, even being part of the probiotic. The reason for this finding is unknown. A point to be considered is how these strains could be able to resist the digestive system and arrive in sufficient amounts to detect enrichment of this genus. Conversely, both were detected by species-specific PCR in preliminary acute ingestion assays in feces (data not shown), although these results cannot be directly extrapolated to humans. Discussing a possible explanation for the functional effect of the formulation, even with a decrease in lactobacilli relative levels, is the potential capacity of extracellular metabolites of lactic acid bacteria to act as a prebiotic for key bacteria, influencing not only their growth and cell death, but also the expression of genes related to cell protection [54]. However, it seems that changes in microbiome do not directly correlate with the strains consumed, pointing that other mechanisms not necessary based on simple colonization might have a role on the results obtained.

The mechanisms by which the microbiome can impact upon oxidative stress and its effects are diverse. Among these, the production by the microbiota of toxic compounds can have a key impact on the health of the individual. Within this group, tryptophan catabolism by tryptophanase of certain bacterial groups produces indole, which is metabolized further to indoxyl-sulfate or indole-3 acetic acid. The latter toxins are secreted into the urine and are accumulated in the case of renal failure. These toxins decrease glutathione levels in renal tubular epithelial cells ren-dering them more vulnerable to oxidative stress [55]. Also, by activating ar-yl-hydrocarbon receptor (AhR) they can exert various deleterious effects [56,57].

Short-chain fatty acids, products of bacterial metabolism, have also been identified as an oxidative stress control mechanism. In a model of apoptosis in β-cells, butyrate and acetate attenuated the overproduction of ROS and NO and prevented cell apoptosis, and reduced viability and mitochondrial dysfunction [58]. Moreover, a bidirectional connection between mitochondrial genotype, ROS production, and gut microbiome has been recently established [59].

The present findings should be interpreted taking into account the limitations of the study, such as the small study population and the short duration of the intervention of only 6 weeks. Therefore, further studies with a larger sample size and duration of consumption of the probiotic product are warranted. It should be noted that in the present study, SOD and GPx were measured in serum samples, and significant differences between the study groups were not observed. However, it may be possible that significant differences could have been obtained by measurement of SOD and GPx in red blood cells.

5. Conclusions

Consumption of a probiotic product based on the three strains of *B. longum*, *L. casei*, and *L. rhamnosus* for 6 weeks in male amateur cyclists undergoing high-intensity and duration physical exercise was associated with a reduction of lipid-related oxidative stress biomarkers, without an increase in antioxidative enzymes. These findings suggest an antioxidant effect of the probiotic product on underlying interacting oxidative stress mechanisms and their modulation in healthy subjects.

Supplementary Materials: The following are available online at https://www.mdpi.com/2076-3921/10/2/323/s1, Table S1: Sequences_distribution; Table S2: Microbiome_profile; Table S3: DESeq2_differential-populationGenus.

Author Contributions: Conceptualization, F.J.L.-R. and V.Á.-G.; methodology, F.J.L.-R., A.J.L.-R., and V.Á.-G.; software, S.P.-P. and A.M.G.-M.; validation, D.V.-M., A.M.G.-M., and S.P.-P.; formal analysis, A.M.G.-M., M.S.M., E.C. (Eric Climent), S.G., D.R., and E.C. (Empar Chenoll); investigation, M.S.M., M.S.A.-R., D.V.-M., A.M.G.-M., and S.P.-P.; data curation, F.J.L.-R.; writing—original draft preparation, M.S.M., M.S.A.-R., and F.C.; writing—review and editing, M.S.M., M.S.A.-R., and F.C.; visualization, F.C.; supervision, V.Á.-G.; project administration, S.P.-P.; funding acquisition, F.J.L.-R. All authors have read and agreed to the published version of the manuscript.

Funding: We received funds from the CDTI agency of the Spanish Ministry of Economy and Competitiveness and European Regional Development Fund (ERDF), under the call of the Strategic Program of the Consortia of National Business Research (CIEN), project SMARTFOODS.

Institutional Review Board Statement: The study was conducted according to the guidelines of the Declaration of Helsinki, and approved by the Ethics Committee of UCAM (CE101703).

Informed Consent Statement: Informed consent was obtained from all subjects involved in the study.

Data Availability Statement: No new data were created or analyzed in this study. Data sharing is not applicable to this article.

Acknowledgments: The authors thank Marta Pulido for editing the manuscript and for editorial assistance.

Conflicts of Interest: Eric Climent, Salvador Genovés, Daniel Ramon and Empar Chenoll are employees of ADM-Biopolis. All other authors declare that they have no conflict of interest.

References

1. Pizzino, G.; Irrera, N.; Cucinotta, M.; Pallio, G.; Mannino, F.; Arcoraci, V.; Squadrito, F.; Altavilla, D.; Bitto, A. Oxidative stress: Harms and benefits for human health. *Oxid. Med. Cell Longev.* **2017**, *2017*, 8416763. [CrossRef]
2. Schieber, M.; Chandel, N.S. ROS function in redox signaling and oxidative stress. *Curr. Biol.* **2014**, *24*, R453–R462. [CrossRef]
3. Ray, P.D.; Huang, B.-W.; Tsuji, Y. Reactive oxygen species (ROS) homeostasis and redox regulation in cellular signaling. *Cell Signal.* **2012**, *24*, 981–990. [CrossRef]
4. Mittal, M.; Siddiqui, M.R.; Tran, K.; Reddy, S.P.; Malik, A.B. Reactive oxygen species in inflammation and tissue injury. *Antioxid. Redox. Signal.* **2014**, *20*, 1126–1167. [CrossRef]
5. Simpson, D.S.A.; Oliver, P.L. ROS generation in microglia: Understanding oxidative stress and inflammation in neurodegenerative disease. *Antioxidants* **2020**, *9*, 743. [CrossRef]
6. Dubois-Deruy, E.; Peugnet, V.; Turkieh, A.; Pinet, F. Oxidative stress in cardiovascular diseases. *Antioxidants* **2020**, *9*, 864. [CrossRef] [PubMed]
7. Yaribeygi, H.; Sathyapalan, T.; Atkin, S.L.; Sahebkar, A. Molecular Mechanisms Linking Oxidative Stress and Diabetes Mellitus. Available online: https://www.hindawi.com/journals/omcl/2020/8609213/ (accessed on 27 January 2021).
8. Romano, A.D.; Serviddio, G.; de Matthaeis, A.; Bellanti, F.; Vendemiale, G. Oxidative stress and aging. *J. Nephrol.* **2010**, *23* (Suppl. S15), S29–S36.
9. Vasquez, E.C.; Pereira, T.M.C.; Campos-Toimil, M.; Baldo, M.P.; Peotta, V.A. Gut microbiota, diet, and chronic diseases: The role played by oxidative stress. *Oxid. Med. Cell Longev.* **2019**, *2019*. [CrossRef]
10. Jones, R.M.; Mercante, J.W.; Neish, A.S. Reactive oxygen production induced by the gut microbiota: Pharmacotherapeutic implications. *Curr. Med. Chem.* **2012**, *19*, 1519–1529. [CrossRef]
11. Neish, A.S. Microbes in gastrointestinal health and disease. *Gastroenterology* **2009**, *136*, 65–80. [CrossRef]
12. Chung, H.; Kasper, D.L. Microbiota-stimulated immune mechanisms to maintain gut homeostasis. *Curr. Opin. Immunol.* **2010**, *22*, 455–460. [CrossRef]

13. Ismail, A.S.; Hooper, L.V. Epithelial cells and their neighbors. IV. bacterial contributions to intestinal epithelial barrier integrity. *Am. J. Physiol. Gastrointest. Liver Physiol.* **2005**, *289*, G779–G784. [CrossRef] [PubMed]

14. Prado, C.; Michels, M.; Ávila, P.; Burger, H.; Milioli, M.V.M.; Dal-Pizzol, F. The protective effects of fecal microbiota transplantation in an experimental model of necrotizing enterocolitis. *J. Pediatr. Surg.* **2019**, *54*, 1578–1583. [CrossRef]

15. Wang, Y.; Wu, Y.; Wang, Y.; Xu, H.; Mei, X.; Yu, D.; Wang, Y.; Li, W. Antioxidant properties of probiotic bacteria. *Nutrients* **2017**, *9*, 521. [CrossRef] [PubMed]

16. He, F.; Li, J.; Liu, Z.; Chuang, C.-C.; Yang, W.; Zuo, L. Redox mechanism of reactive oxygen species in exercise. *Front. Physiol.* **2016**, *7*. [CrossRef]

17. Radak, Z.; Zhao, Z.; Koltai, E.; Ohno, H.; Atalay, M. Oxygen consumption and usage during physical exercise: The balance between oxidative stress and ROS-dependent adaptive signaling. *Antioxid. Redox. Signal.* **2013**, *18*, 1208–1246. [CrossRef]

18. Kawamura, T.; Muraoka, I. Exercise-induced oxidative stress and the effects of antioxidant intake from a physiological viewpoint. *Antioxidants* **2018**, *7*, 119. [CrossRef] [PubMed]

19. Pingitore, A.; Lima, G.P.P.; Mastorci, F.; Quinones, A.; Iervasi, G.; Vassalle, C. Exercise and oxidative stress: Potential effects of antioxidant dietary strategies in sports. *Nutrition* **2015**, *31*, 916–922. [CrossRef]

20. Powers, S.K.; DeRuisseau, K.C.; Quindry, J.; Hamilton, K.L. Dietary antioxidants and exercise. *J. Sports Sci.* **2004**, *22*, 81–94. [CrossRef] [PubMed]

21. Genomic Sequence and Pre-Clinical Safety Assessment of Bifidobacterium Longum CECT 7347, a Probiotic Able to Reduce the Toxicity and Inflammatory Potential of Gliadin-Derived Peptides | Abstract. Available online: https://www.longdom.org/abstract/genomic-sequence-and-preclinical-safety-assessment-of-embifidobacterium-longumem-cect-7347-a-probiotic-able-to-reduce-th-33128.html (accessed on 27 January 2021).

22. Klindworth, A.; Pruesse, E.; Schweer, T.; Peplies, J.; Quast, C.; Horn, M.; Glöckner, F.O. Evaluation of general 16S ribosomal RNA gene PCR primers for classical and next-generation sequencing-based diversity studies. *Nucleic Acids Res.* **2013**, *41*, e1. [CrossRef] [PubMed]

23. Krotkiewski, M.; Brzezinska, Z. Lipid peroxides production after strenuous exercise and in relation to muscle morphology and capillarization. *Muscle Nerve* **1996**, *19*, 1530. [CrossRef]

24. Powers, S.K.; Jackson, M.J. Exercise-induced oxidative stress: Cellular mechanisms and impact on muscle force production. *Physiol. Rev.* **2008**, *88*, 1243–1276. [CrossRef]

25. Criswell, D.; Powers, S.; Dodd, S.; Lawler, J.; Edwards, W.; Renshler, K.; Grinton, S. High intensity training-induced changes in skeletal muscle antioxidant enzyme activity. *Med. Sci. Sports Exerc.* **1993**, *25*, 1135–1140. [CrossRef]

26. Hammeren, J.; Powers, S.; Criswell, D.; Martin, A.; Lowenthal, D.; Pollock, M. Exercise training-induced alterations in skeletal muscle oxidative and antioxidant enzyme activity in senescent rats. *Int. J. Sports Med.* **1992**, *13*, 412–416. [CrossRef]

27. Theofilidis, G.; Bogdanis, G.C.; Koutedakis, Y.; Karatzaferi, C. Monitoring exercise-induced muscle fatigue and adaptations: Making sense of popular or emerging indices and biomarkers. *Sports* **2018**, *6*, 153. [CrossRef]

28. De Salazar, L.; Torregrosa-García, A.; Luque-Rubia, A.J.; Ávila-Gandía, V.; Domingo, J.C.; López-Román, F.J. Oxidative stress in endurance cycling is reduced dose-dependently after one month of re-esterified DHA supplementation. *Antioxidants* **2020**, *9*, 1145. [CrossRef]

29. Torregrosa-García, A.; Ávila-Gandía, V.; Luque-Rubia, A.J.; Abellán-Ruiz, M.S.; Querol-Calderón, M.; López-Román, F.J. Pomegranate extract improves maximal performance of trained cyclists after an exhausting endurance trial: A randomised controlled trial. *Nutrients* **2019**, *11*, 721. [CrossRef] [PubMed]

30. Martínez-Sánchez, A.; Alacid, F.; Rubio-Arias, J.A.; Fernández-Lobato, B.; Ramos-Campo, D.J.; Aguayo, E. Consumption of watermelon juice enriched in L-citrulline and pomegranate ellagitannins enhanced metabolism during physical exercise. *J. Agric. Food Chem.* **2017**, *65*, 4395–4404. [CrossRef]

31. López-Román, F.J.; Ávila-Gandía, V.; Contreras-Fernández, C.J.; Luque-Rubia, A.J.; Villegas-García, J.A. Effect of docosahexaenoic acid supplementation on differences of endurance exercise performance in competitive and non-competitive male cyclists. *Gazz. Med. Ital. Arch. Sci. Med.* **2019**, *178*, 411–416. [CrossRef]

32. Ramos-Campo, D.J.; Ávila-Gandía, V.; López-Román, F.J.; Miñarro, J.; Contreras, C.; Soto-Méndez, F.; Domingo Pedrol, J.C.; Luque-Rubia, A.J. Supplementation of re-esterified docosahexaenoic and eicosapentaenoic acids reduce inflammatory and muscle damage markers after exercise in endurance athletes: A randomized, controlled crossover trial. *Nutrients* **2020**, *12*, 719. [CrossRef] [PubMed]

33. Mishra, V.; Shah, C.; Mokashe, N.; Chavan, R.; Yadav, H.; Prajapati, J. Probiotics as potential antioxidants: A systematic review. *J. Agric. Food Chem.* **2015**, *63*, 3615–3626. [CrossRef] [PubMed]

34. Michalickova, D.; Kotur-Stevuljevic, J.; Miljkovic, M.; Dikic, N.; Kostic-Vucicevic, M.; Andjelkovic, M.; Koricanac, V.; Djordjevic, B. Effects of probiotic supplementation on selected parameters of blood prooxidant-antioxidant balance in elite athletes: A double-blind randomized placebo-controlled study. *J. Hum. Kinet.* **2018**, *64*, 111–122. [CrossRef] [PubMed]

35. Martarelli, D.; Verdenelli, M.C.; Scuri, S.; Cocchioni, M.; Silvi, S.; Cecchini, C.; Pompei, P. Effect of a probiotic intake on oxidant and antioxidant parameters in plasma of athletes during intense exercise training. *Curr. Microbiol.* **2011**, *62*, 1689–1696. [CrossRef]

36. Lamprecht, M.; Bogner, S.; Schippinger, G.; Steinbauer, K.; Fankhauser, F.; Hallstroem, S.; Schuetz, B.; Greilberger, J.F. Probiotic supplementation affects markers of intestinal barrier, oxidation, and inflammation in trained men; a randomized, double-blinded, placebo-controlled trial. *J. Int. Soc. Sports Nutr.* **2012**, *9*, 45. [CrossRef] [PubMed]

37. Välimäki, I.A.; Vuorimaa, T.; Ahotupa, M.; Kekkonen, R.; Korpela, R.; Vasankari, T. Decreased training volume and increased carbohydrate intake increases oxidized LDL levels. *Int. J. Sports Med.* **2012**, *33*, 291–296. [CrossRef]
38. Ardeshirlarijani, E.; Tabatabaei-Malazy, O.; Mohseni, S.; Qorbani, M.; Larijani, B.; Baradar Jalili, R. Effect of probiotics supplementation on glucose and oxidative stress in type 2 diabetes mellitus: A meta-analysis of randomized trials. *Daru* **2019**, *27*, 827–837. [CrossRef]
39. Valko, M.; Izakovic, M.; Mazur, M.; Rhodes, C.J.; Telser, J. Role of oxygen radicals in DNA damage and cancer incidence. *Mol. Cell Biochem.* **2004**, *266*, 37–56. [CrossRef]
40. Escamilla, J.; Lane, M.A.; Maitin, V. Cell-free supernatants from probiotic lactobacillus casei and lactobacillus rhamnosus GG decrease colon cancer cell invasion in vitro. *Nutr. Cancer* **2012**, *64*, 871–878. [CrossRef] [PubMed]
41. Tiptiri-Kourpeti, A.; Spyridopoulou, K.; Santarmaki, V.; Aindelis, G.; Tompoulidou, E.; Lamprianidou, E.E.; Saxami, G.; Ypsilantis, P.; Lampri, E.S.; Simopoulos, C.; et al. Lactobacillus casei exerts anti-proliferative effects accompanied by apoptotic cell death and up-regulation of TRAIL in colon carcinoma cells. *PLoS ONE* **2016**, *11*, e0147960. [CrossRef]
42. Jacouton, E.; Mach, N.; Cadiou, J.; Lapaque, N.; Clément, K.; Doré, J.; van Hylckama Vlieg, J.E.T.; Smokvina, T.; Blottière, H.M. Lactobacillus rhamnosus CNCMI-4317 modulates fiaf/angptl4 in intestinal epithelial cells and circulating level in mice. *PLoS ONE* **2015**, *10*, e0138880. [CrossRef]
43. Black, C.N.; Bot, M.; Scheffer, P.G.; Cuijpers, P.; Penninx, B.W.J.H. Is depression associated with increased oxidative stress? A systematic review and meta-analysis. *Psychoneuroendocrinology* **2015**, *51*, 164–175. [CrossRef]
44. Collins, A.R.; Cadet, J.; Möller, L.; Poulsen, H.E.; Viña, J. Are we sure we know how to measure 8-Oxo-7,8-dihydroguanine in DNA from human cells? *Arch. Biochem. Biophys.* **2004**, *423*, 57–65. [CrossRef]
45. Dizdaroglu, M.; Jaruga, P.; Birincioglu, M.; Rodriguez, H. Free radical-induced damage to DNA: Mechanisms and measurement. *Free Radic. Biol. Med.* **2002**, *32*, 1102–1115. [CrossRef]
46. Mikhed, Y.; Görlach, A.; Knaus, U.G.; Daiber, A. Redox regulation of genome stability by effects on gene expression, epigenetic pathways and DNA damage/repair. *Redox. Biol.* **2015**, *5*, 275–289. [CrossRef] [PubMed]
47. Wu, Q.; Ni, X. ROS-mediated DNA methylation pattern alterations in carcinogenesis. *Curr. Drug Targets* **2015**, *16*, 13–19. [CrossRef]
48. Samuel, B.S.; Gordon, J.I. A humanized gnotobiotic mouse model of host-archaeal-bacterial mutualism. *Proc. Natl. Acad. Sci. USA* **2006**, *103*, 10011–10016. [CrossRef]
49. Samuel, B.S.; Hansen, E.E.; Manchester, J.K.; Coutinho, P.M.; Henrissat, B.; Fulton, R.; Latreille, P.; Kim, K.; Wilson, R.K.; Gordon, J.I. Genomic and metabolic adaptations of methanobrevibacter smithii to the human gut. *Proc. Natl. Acad. Sci. USA* **2007**, *104*, 10643–10648. [CrossRef] [PubMed]
50. Bianchi, F.; Larsen, N.; de Mello Tieghi, T.; Adorno, M.A.T.; Kot, W.; Saad, S.M.I.; Jespersen, L.; Sivieri, K. Modulation of gut microbiota from obese individuals by in vitro fermentation of citrus pectin in combination with bifidobacterium longum BB-46. *Appl. Microbiol. Biotechnol.* **2018**, *102*, 8827–8840. [CrossRef] [PubMed]
51. Yang, C.; Deng, Q.; Xu, J.; Wang, X.; Hu, C.; Tang, H.; Huang, F. Sinapic acid and resveratrol alleviate oxidative stress with modulation of gut microbiota in high-fat diet-fed rats. *Food Res. Int.* **2019**, *116*, 1202–1211. [CrossRef]
52. Eren, A.M.; Sogin, M.L.; Morrison, H.G.; Vineis, J.H.; Fisher, J.C.; Newton, R.J.; McLellan, S.L. A single genus in the gut microbiome reflects host preference and specificity. *ISME J.* **2015**, *9*, 90–100. [CrossRef]
53. Allen, J.M.; Mailing, L.J.; Niemiro, G.M.; Moore, R.; Cook, M.D.; White, B.A.; Holscher, H.D.; Woods, J.A. Exercise alters gut microbiota composition and function in lean and obese humans. *Med. Sci. Sports Exerc.* **2018**, *50*, 747–757. [CrossRef] [PubMed]
54. Lebas, M.; Garault, P.; Carrillo, D.; Codoñer, F.M.; Derrien, M. Metabolic response of *Faecalibacterium prausnitzii* to cell-free supernatants from lactic acid bacteria. *Microorganisms* **2020**, *8*, 1528. [CrossRef]
55. Edamatsu, T.; Fujieda, A.; Itoh, Y. Phenyl sulfate, indoxyl sulfate and p-cresyl sulfate decrease glutathione level to render cells vulnerable to oxidative stress in renal tubular cells. *PLoS ONE* **2018**, *13*, e0193342. [CrossRef]
56. Asai, H.; Hirata, J.; Hirano, A.; Hirai, K.; Seki, S.; Watanabe-Akanuma, M. Activation of aryl hydrocarbon receptor mediates suppression of hypoxia-inducible factor-dependent erythropoietin expression by indoxyl sulfate. *Am. J. Physiol. Cell Physiol.* **2016**, *310*, C142–C150. [CrossRef] [PubMed]
57. Eleftheriadis, T.; Pissas, G.; Antoniadi, G.; Liakopoulos, V.; Stefanidis, I. Kynurenine, by activating aryl hydrocarbon receptor, decreases erythropoietin and increases hepcidin production in HepG2 cells: A new mechanism for anemia of inflammation. *Exp. Hematol.* **2016**, *44*, 60–67. [CrossRef] [PubMed]
58. Hu, S.; Kuwabara, R.; de Haan, B.J.; Smink, A.M.; de Vos, P. Acetate and butyrate improve β-cell metabolism and mitochondrial respiration under oxidative stress. *Int. J. Mol. Sci.* **2020**, *21*, 1542. [CrossRef] [PubMed]
59. Yardeni, T.; Tanes, C.E.; Bittinger, K.; Mattei, L.M.; Schaefer, P.M.; Singh, L.N.; Wu, G.D.; Murdock, D.G.; Wallace, D.C. Host mitochondria influence gut microbiome diversity: A role for ROS. *Sci. Signal.* **2019**, *12*, 588. [CrossRef]

 *antioxidants*

Article

Effects of Two-Week Betaine Supplementation on Apoptosis, Oxidative Stress, and Aerobic Capacity after Exhaustive Endurance Exercise

Ming-Ta Yang [1], Xiu-Xin Lee [2], Bo-Huei Huang [3], Li-Hui Chien [4], Chia-Chi Wang [5] and Kuei-Hui Chan [4,]*

[1] Center for General Education, Taipei Medical University, Taipei 110301, Taiwan; yangrugby@tmu.edu.com
[2] Department of Primary Care Medicine, Taipei Medical University-Shuang Ho Hospital, New Taipei City 23561, Taiwan; b101103133@tmu.edu.tw
[3] Charles Perkins Centre, School of Health Sciences, Faculty of Medicine and Health, The University of Sydney, Camperdown 2006, Australia; 1040609@ntsu.edu.tw
[4] Graduate Institute of Athletics and Coaching Science, National Taiwan Sport University, Taoyuan 333325, Taiwan; chienlihui@gmail.com
[5] Office of Physical Education, National Taipei University of Business, Taipei 10051, Taiwan; sunnywango@gmail.com
* Correspondence: quenhuen@ntsu.edu.tw; Tel.: +88-63-3283-201 (ext. 2423)

Received: 28 October 2020; Accepted: 23 November 2020; Published: 27 November 2020

Abstract: This study evaluated the effects of 2 weeks of betaine supplementation on apoptosis, oxidative stress, and aerobic capacity after exhaustive endurance exercise (EEE). A double-blind, crossover, and counterbalanced design was adopted, with 10 healthy male participants asked to consume betaine (1.25 g of betaine mixed with 300 mL of sports beverage, twice per day for 2 weeks) or placebo (300 mL of sports beverage). All participants performed a graded exercise test on a treadmill to determine the maximal oxygen consumption (VO_{2max}) before supplementation and then performed the EEE test at an intensity of 80% VO_{2max} after 2 weeks of supplementation. The time to exhaustion, peak oxygen consumption, maximal heart rate, and average heart rate were recorded during the EEE test. Venous blood samples were drawn before, immediately after, and 3 h after the EEE test to assess apoptosis and the mitochondrial transmembrane potential (MTP) decline of lymphocytes as well as the concentrations of thiobarbituric acid reactive substance and protein carbonyl. The results indicated that lymphocyte apoptosis was significantly higher immediately after and 3 h after EEE than before exercise in participants in the placebo trial. However, lymphocyte apoptosis exhibited no significant differences among the three time points in participants in the betaine trial. Moreover, apoptosis in the betaine trial was significantly lower immediately after and 3 h after exercise compared with the placebo trial. No differences were noted for other variables. Thus, 2 weeks of betaine supplementation can effectively attenuate lymphocyte apoptosis, which is elevated by EEE. However, betaine supplementation exhibited no effects on MTP decline, oxidative stress, or aerobic capacity.

Keywords: lymphocytes; mitochondrial transmembrane potential decline; thiobarbituric acid reactive substance; protein carbonyl

---

1. Introduction

In a normal body state, the naturally occurring free radicals in humans have been reported to exert significant positive effects on immune system regulation [1] and to have a significant negative effect, namely peroxidation, on lipids, proteins, and DNA [2,3]. Exercise has been demonstrated to enhance muscular metabolism, and consequently, oxygen uptake, which further enhances the generation of free radicals and oxidative stress [4,5]. Apoptosis is a phenomenon that occurs when free radicals damage human DNA and cause accelerated programmed cell death [6]. One study indicated that the percentage of apoptotic cells increased significantly after running at an intensity of 80% maximal oxygen consumption (VO_{2max}) until exhaustion, whereas it remained unchanged after running at an intensity of 60% VO_{2max} for an identical running time [7]. Therefore, in athletes, the higher the exercise intensity is, the more apoptosis occurs. Adequate apoptosis is an essential mechanism in the human body [8], but a high percentage of apoptosis can induce alterations in the physiology and viability of circulating leucocytes, which have a causal relationship with exercise-induced immune distress [9]. Nutritional interventions to attenuate inflammation and apoptosis may directly or indirectly benefit muscular recovery and subsequent performance [10].

Betaine, a natural compound, is commercially obtained from sugar beet [11]. It was first discovered in *Beta vulgaris* in the 19th century [12] and has been noted to be present in microorganisms, plants, and animals [13], with wheat, shellfish, spinach, and beetroot containing high levels of the compound [14,15]. Betaine can not only be absorbed by the human body from diet but also be converted from choline. Choline, the precursor of betaine, can be oxidized to betaine aldehyde by choline dehydrogenase [12]. The betaine aldehyde can be oxidized to betaine by betaine aldehyde dehydrogenase in the presence of NAD^+ [16]. Therefore, the human body can also obtain betaine from foods rich in choline, such as eggs, meat, fish, and whole grains [17]. About 50% of choline in the intestine will be converted into betaine [18] and humans can obtain average 1 g of choline from daily diet [17]. As early as the 1990s, betaine was added to animal feed to evaluate its effects on growth performance [19,20] and disease prevention [21]. In the first study involving humans related to betaine supplementation and exercise performance, Armstrong et al. [22] observed that oxygen consumption during sprinting after acute betaine supplementation (5 g of betaine mixed in 1 L of carbohydrate–electrolyte fluid) was significantly higher than in those who consumed only carbohydrate–electrolyte fluid. However, betaine supplementation did not improve sprint performance in a hot environment. Furthermore, long-term betaine supplementation (1.25 g twice per day for 14 days) before an acute exercise session was noted to significantly increase the concentrations of growth hormone and insulin-like growth factor-1 as well as significantly decrease cortisol concentration [23]. Notably, a 14-day betaine supplementation was suggested to effectively promote protein synthesis. However, other studies have revealed that long-term betaine supplementation had no benefits on jump squat power, the number of bench press or squat repetitions [24], or the peak concentric or eccentric force outputs during isokinetic chest press [25]. Therefore, the effects of betaine supplementation on exercise performance remain unclear.

In addition to improving strength and power, betaine regulates organic osmolytes and protects the function of cells and mitochondria [12]. Therefore, some cell culture and animal studies have investigated the effects of betaine supplementation on apoptosis and oxidative stress in damaged cells [26–29]. A study by Veskovic et al. [26] observed that betaine decreased the liver's expression of proapoptotic mediator Bax and increased antiapoptotic Bcl-2 in nonalcoholic fatty liver disease induced by a methionine–choline-deficient diet in mice. In addition, betaine increased superoxide-dismutase, catalase, glutathione peroxidase, and paraoxonase activities. Studies have suggested that betaine can effectively attenuate apoptosis and improve antioxidative defense. In addition, other studies have determined that betaine exerts antiapoptotic effects in human corneal epithelial cells [27] and antioxidative stress effects in the liver of rats [28,29]. Furthermore, our laboratory data revealed that a single dose of betaine supplement (1.25 g of betaine mixed in 300 mL of sports beverage) 1 h before an exhaustive endurance exercise (EEE) significantly decreased lymphocyte apoptosis but had no effects on mitochondrial transmembrane potential (MTP) decline [30].

Based on these aforementioned results, we hypothesized that long-term betaine supplementation attenuates apoptosis and oxidative stress induced by exercise and enhances aerobic capacity. Therefore, this study evaluated the effects of 2 weeks of betaine supplementation on apoptosis, oxidative stress, and aerobic capacity after EEE.

2. Materials and Methods

2.1. Participants

Ten healthy male participants were recruited from National Taiwan Sport University, Taoyuan City, Taiwan. Individuals with diabetes and cardiovascular, renal, liver, or autoimmune diseases were excluded. All participants were requested to maintain their regular eating habits, avoid alcohol, and avoid consuming other nutritional supplements during the experimental period. The participants were informed of the requirements, benefits, and risks of the study before obtaining written informed consent. The study was approved by the Institutional Review Board of Fu Jen Catholic University, New Taipei City, Taiwan, with the IRB number C102053. The anthropometric data of participants were as follows: age: 24.60 ± 3.06 years; weight: 76.45 ± 9.58 kg; height: 177.00 ± 8.26 cm. Moreover, the performance in endurance exercise capacities of VO_{2max}, 80% VO_{2max}, and mean relative running velocity at 80% VO_{2max} were 50.21 ± 8.38 mL/kg/min, 40.20 ± 6.71 mL/kg/min, and 3.07 ± 0.26 m/s, respectively.

2.2. Experimental Design

Regarding design, this was a double-blind, crossover, and counterbalanced study to evaluate the influence of 2 weeks of betaine supplementation on apoptosis, oxidative stress, and aerobic capacity after EEE in a healthy male population. The participants were asked to consume betaine or placebo, with at least 3 weeks of washout period between the trials. All participants performed a graded exercise test (GXT) until exhaustion on the treadmill to determine VO_{2max} before supplementation [31], and the speed equivalent to 80% VO_{2max} was implemented for an EEE test conducted on a treadmill for 30 min, after which speed was increased by 0.2 m/s every 1 min until exhaustion. The participants were instructed to consume a standard breakfast consisting of 648 kcal of total energy (66% carbohydrate, 11% protein, and 23% fat) 1 h before exercise. Blood samples were drawn before exercise (Pre), immediately after (Post-0), and 3 h after (Post-3) the EEE to determine the biomarkers of apoptosis and oxidative stress. Furthermore, plasma concentrations of betaine and choline before supplementation and at the Pre time point of EEE (after two weeks of supplementation) were also analyzed. In addition, the time to exhaustion, peak oxygen consumption, maximal heart rate (HR_{max}), and average heart rate (AHR) were recorded during the exercise period. Figure 1 illustrates the scheme of the study.

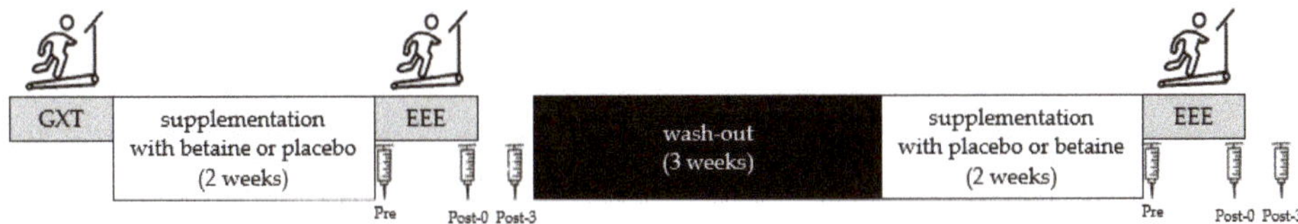

Figure 1. Experimental scheme. GXT, graded exercise test; EEE, exhaustive endurance exercise; Pre, before exercise; Post-0, immediately after exercise; Post-3, 3 h after exercise.

2.3. Supplementation Protocol

The betaine supplementation strategy in the present study was based on the one suggested by Apicella et al. [23], wherein all participants consumed either 1.25 g of betaine (betaine powder; Twinlab, CO, USA) mixed with 300 mL sports beverage (Pocari Sweat, Otsuka, Taipei, Taiwan) or placebo (only sports beverage) twice per day for 2 weeks. The betaine for the supplementation was extracted from natural sugar beet and subsequently purified (99% pure). The sports beverage contains carbohydrates (66 g/L) and electrolytes of sodium (21 mEq/L), chloride (16 mEq/L), and potassium (4.9 mEq/L). The supplements for both trials had the same color and taste. The participants consumed supplements after breakfast and dinner during the experiment period.

2.4. Graded Exercise Test

The participants performed a GXT until exhaustion to determine the exercise intensity (80% VO_{2max}). The treadmill (pulsar; h/p/cosmos, Nussdorf, Germany) GXT protocol was based on a previous study [31], wherein the treadmill speed started at 2.0 m/s and increased by 0.5 m/s every 4 min for the first three stages, after which the intensity was further elevated in increments of 0.5 m/s every 2 min until the participant reached exhaustion. Expired gas, VO_2, and VCO_2 were analyzed using gas analysis (Vmax Spectra 29c; SensorMedics, Yorba Linda, CA, USA), and HR was monitored (Polar S610; Kempele, Finland) at the same time points. Individual VO_{2max} was assumed to have been achieved when two of the following criteria were met: (1) respiratory exchange ratio greater than 1.1, (2) rating of perceived exertion greater than 18, and (3) HR within 15 beats/min of individual predicted HR_{max}.

2.5. Blood Sampling and Analysis

Blood samples were drawn from the participants through antecubital vein suction into ethylenediaminetetraacetic acid tubes. A small portion of whole blood (400 µL) was then immediately analyzed for lymphocyte apoptosis and MTP decline in lymphocytes. Other blood samples were centrifuged at 2500× g for 20 min at 4 °C in an Eppendorf centrifuge 5804 R (Eppendorf AG, Hamburg, Germany). The supernatant was collected and stored at −20 °C before being analyzed for the concentrations of betaine, choline, thiobarbituric acid reactive substance (TBARS), and protein carbonyl (PC).

2.5.1. Lymphocyte Apoptosis

The Annexin V-FITC apoptosis detection kit (BioVision, Milpitas, CA, USA) was applied to detect the lymphocyte apoptosis. After being gently mixed, 200 µL of the whole blood sample was transferred to a conical polypropylene test tube, and 4 mL of lysing buffer (Thermo Fisher Scientific, Waltham, MA, USA) was added. The solution was incubated at room temperature for 7 min before being centrifuged at 1500× g for 5 min at 20 °C, and the supernatant was then discarded. Subsequently, 3 mL of phosphate buffered saline (PBS) (Corning®, Corning, NY, USA) was added to wash the cells, the solution was centrifuged for 5 min, and the supernatant was then discarded. Lymphocytes were suspended with 300 µL of binding buffer. Thereafter, 3 µL of Annexin V-FETC and propidium iodide were added, and the solution was incubated in the dark at room temperature for 5 min. Finally, samples were gently drawn and then the subject to analysis by the BD FACSCalibur™ flow cytometry (Becton Dickson, San Jose, CA, USA).

2.5.2. MTP Decline in Lymphocyte

The MitoProbe™ JC-1 Assay Kit (Molecular Probes, Eugene, OR, USA) was applied to detect the MTP decline in lymphocyte. After being gently mixed, 200 µL of whole blood sample was transferred to a conical polypropylene test tube, and 4 mL of lysing buffer (Thermo Fisher Scientific, Waltham, MA, USA) was added. The solution was incubated at room temperature for 7 min before being centrifuged at 1500× g for 5 min at 20 °C, and the supernatant was then discarded. Subsequently, 3 mL of PBS (Corning®, Corning, NY, USA) was added to wash the cells, the solution was centrifuged for 5 min, and the supernatant was then discarded. Lymphocytes were suspended with 500 µL of PBS by using a MitoProbe™ JC-1 assay kit, and then subjected to a dry bath in an incubator (MD-01N, Major Science, Taoyuan, Taiwan) at 37 °C for 15 min. Finally, samples were gently drawn and subjected to analysis by the BD FACSCalibur™ flow cytometry (Becton Dickson, San Jose, CA, USA).

2.5.3. Concentrations of Plasma Betaine and Choline

To determine the blood concentrations of betaine and choline, d_{11}-betaine and methyl-d_9-choline were used as internal standards. Betaine hydrochloride, the standard of betaine, and d_{11}-betaine were obtained from Chem Service (West Chester, PA, USA) and Cambridge Isotope Laboratories

(Tewksbury, MA, USA), respectively. Choline chloride, the standard of choline, and methyl-d_9-choline were purchased from Sigma-Aldrich (St. Louis, MO, USA). Initially, 30 µL of plasma was mixed with 90 µL of internal standard solution (10 ng/mL d_{11}-betaine and d_9-choline in methanol) in a microcentrifuge tube before being vortexed for 5 min and subsequently centrifuged at 5350× g for 10 min at 4 °C. The supernatants (60 µL) were placed in an LC-MS vial, and then analyzed using an AB SCIEX API 2000 liquid chromatography tandem-mass spectrometry system (Sciex Division of MDS, Toronto, ON, Canada). The analysis was performed using an Agilent 1260 Infinity Binary high-performance liquid chromatography system (Agilent Technologies, Santa Clara, CA, USA) equipped with an integrated degasser (G1322A), a pump (G1312C), an autosampler (G1329B), and a thermostat (G1330B). Analytes were chromatographically separated using an Waters XBridge BEH Amide column (100 mm × 2.1 mm, 3.5 µm) with a gradient mobile phase comprising (A) 0.1% formic acid (Sigma-Aldrich, St. Louis, MO, USA) in H_2O, and (B) 0.1% formic acid in acetonitrile (Duksan, Seoul, Korea) under linear-gradient conditions (Period, A:B-%, *v/v*: 0–0.5 min, 30:70; 0.5–4 min, 30:70; 4–4.1 min, 50:50; 4.1–8 min, 30:70) at a flow rate of 400 µL/min; this procedure was a modification of that described by Bruce et al. [32]. The column was maintained at 28 °C. The monitoring conditions for multiple reactions are presented in Table 1, and Figure 2A,B present the LC MS/MS chromatograms of the standards and one blood sample.

Table 1. Multiple reaction monitoring conditions of betaine and choline analysis.

Compounds	Parent Ion (m/z)	Molecular Ion (m/z)	DP (V)	FP (V)	EP (V)	CE (V)	CXP (V)
choline	104	60	28	320	8	26	7
d_9-choline	113	69	35	367	8.8	26	8
betaine	118	58	35	350	10	38	6
d_{11}-betaine	129	66	38	350	8	43	7

DP, declustering potential; FP, focusing potential; EP, entrance potential; CE, collision energy; CXP, collision cell exit potential.

2.5.4. TBARS Concentration

TBARS concentration was analyzed using colorimetry, as described in a previous study [33]. Briefly, 200 µL of plasma was mixed with 200 µL of trichloroacetic acid (TCA; JT Baker, Phillipsburg, NJ, USA), 200 µL of Tris-HCl (Serva Electrophoresis, Heidelberg, Germany) and incubated at room temperature for 10 min. Thereafter, 400 µL of TBARS reagent, including 55 mM thiobarbituric acid (TBA) (from Merck, Darmstadt, Gemany) and 2 M sodium persulfate (Sigma-Aldrich Chemie GmbH, Steinheim, Germany) were added into the tube. Samples were kept to a dry bath in an incubator at 95 °C for 45 min and put on ice for 5 min. Subsequently, 400 µL of 75% TCA was added to the tube and mixed well, and the solution was then centrifuged for 5 min at 10,000× g. Thereafter, 100 µL of supernatant was loaded onto 96-well plates, and various concentrations of 1,1,3,3-tetraethoxypropane (Aldrich, St. Louis, MO, USA) were used to construct a standard curve, per the procedure described by Boadi et al. [34]. The absorbance was read at 530 nm using a Tecan Infinite M200 microplate reader (Tecan Austria GmbH, Grödig, Austria), and the result was applied to the regression formula to obtain the TBARS concentration.

(A)

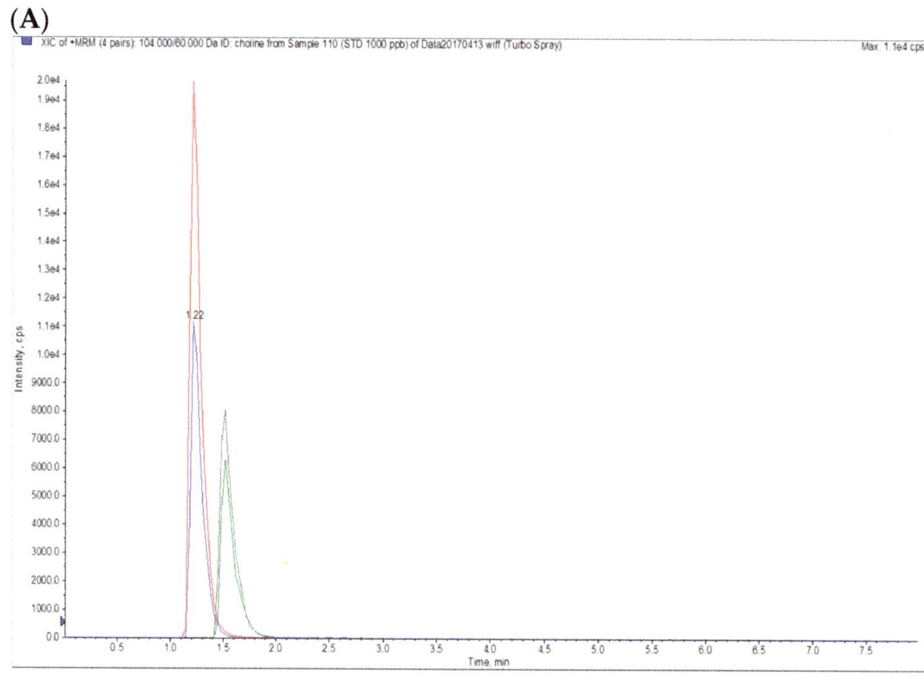

(B)

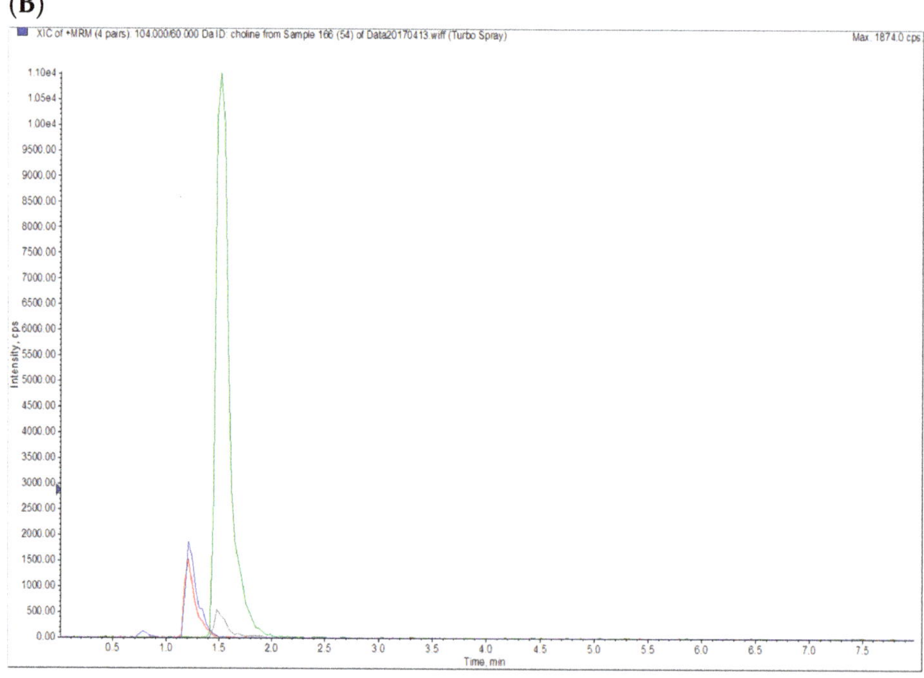

Figure 2. LC MS/MS Chromatograms of (**A**) betaine and choline standards (10 ng/mL each) and (**B**) a plasma sample. Blue: choline; Red: d_9-choline; Green: betaine; Gray: d_{11}-betaine.

2.5.5. PC Concentration

PC concentration was analyzed using an enzyme-linked immunosorbent assay with a commercial assay kit (BioVision, Milpitas, CA, USA). Briefly, 100 µL of plasma was mixed with 10 µL of streptozocin solution, and the mixture was incubated at room temperature for 15 min before being centrifuged for 5 min; the supernatant was then transferred to a new tube. Subsequently, 100 µL of DNPH was added to each sample, which was then vortexed and incubated at room temperature for 10 min. Thereafter, 30 µL of TCA was added to each sample, which was then put on ice for 5 min, and centrifuged for 2 min. Then, the supernatant was removed, 500 µL of cold acetone was added to each tube, and the solution was placed in a sonicating bath for 30 s. Subsequently, samples were placed at $-20\ °C$ for 5 min before being centrifuged for 2 min, and the acetone was then carefully removed. These steps were repeated twice with 500 µL of cold acetone added. Finally, 200 µL of guanidine solution was added, and the solution was sonicated and incubated at $60\ °C$ for 15 min. The 100 µL sample was loaded onto 96-well plates and read by the Tecan Infinite M200 microplate reader (Tecan Austria GmbH, Grödig, Austria) at an absorbance of 375 nm. The result was applied to the regression formula to obtain the PC concentration.

2.6. Statistical Analysis

Statistical analysis was performed using SPSS version 22.0 (IBM, Armonk, NY, USA). Data are expressed as mean ± standard deviation (SD). A paired *t* test was used to compare the variables of time to exhaustion, peak oxygen consumption, HR_{max}, and AHR between the trials. A 2 (trials: betaine and placebo) × 3 (time points: Pre, Post-0, and Post-3) two-way repeated-measures analysis of variance was used to compare the variables of apoptosis, MTP decline, TBARS, and PC. When an interaction was noted, least significant difference post hoc tests were performed to determine where the difference occurred. The significance level was set at $p < 0.05$.

3. Results

3.1. Blood Concentrations of Betaine and Coline

Figure 3A illustrates the results of betaine concentrations in plasma before and after 2 weeks of betaine or placebo supplementation. A two-way repeated ANOVA with treatment and time indicated the interaction between two factors was significant ($p < 0.05$). The plasma betaine concentration in subjects in the placebo trial exhibited no significant differences before and after the supplementation. However, the plasma betaine concentration in the participants of betaine trial after 2 weeks of supplementation was significantly higher than before supplementation (6.53 ± 3.71 µg/mL vs. 2.04 ± 0.61 µg/mL, $p < 0.05$). Moreover, after 2 weeks of supplementation, the plasma betaine concentration was significantly higher for the betaine trial than the placebo trial (6.53 ± 3.71 µg/mL vs. 1.90 ± 0.23 µg/mL, $p < 0.05$). Figure 3B presents of choline concentrations in plasma before and after 2 weeks of betaine or placebo supplementation. A two-way repeated ANOVA with treatment and time indicated there is no interaction between two factors. The result shows that the supplementation strategy of present study can effectively enhance the betaine concentrations in plasma, rather than choline metabolism.

3.2. Effects of Betaine Supplementation on Lymphocyte Apoptosis and MTP Decline

Figure 4A presents the results regarding lymphocyte apoptosis after 2 weeks of betaine or placebo supplementation. A two-way repeated ANOVA with treatment and time indicated the interaction between two factors was significant ($p < 0.05$). Lymphocyte apoptosis in participants undergoing 2 weeks of placebo supplementation was significantly higher at the Post-0 and Post-3 time points than at the Pre time point (22.61 ± 13.13%, 24.32 ± 11.49% vs. 6.77 ± 2.34%, respectively, $p < 0.05$). However, lymphocyte apoptosis exhibited no significant differences among the three time points in the betaine trial. Moreover, apoptosis in the betaine trial at Post-0 and Post-3 time points was significantly lower than that in the placebo trial. Figure 4B presents the results of the phenomenon of MTP decline after 2 weeks of betaine or placebo supplementation. No significant difference was observed in the phenomenon of MTP decline between the trials and time points ($p > 0.05$). The results indicated that twice per day supplementation of 1.25 g of betaine mixed in 300 mL of sports beverage for 2 weeks attenuated lymphocyte apoptosis immediately after EEE but can't benefit on MTP decline.

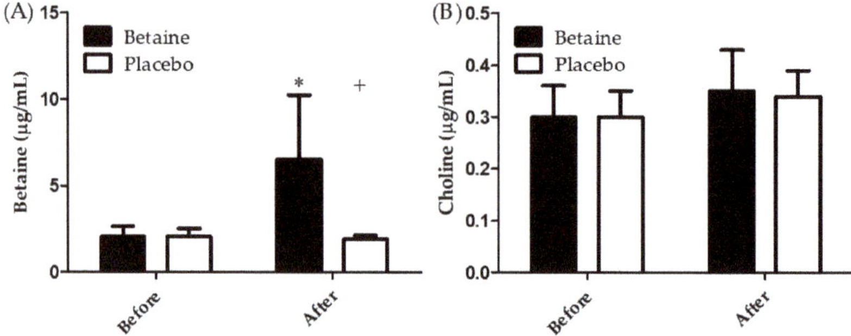

Figure 3. Blood (**A**) betaine and (**B**) choline concentrations before and after 2 weeks of supplementation. * Significant ($p < 0.05$) difference from before. + Significant ($p < 0.05$) difference from betaine trial.

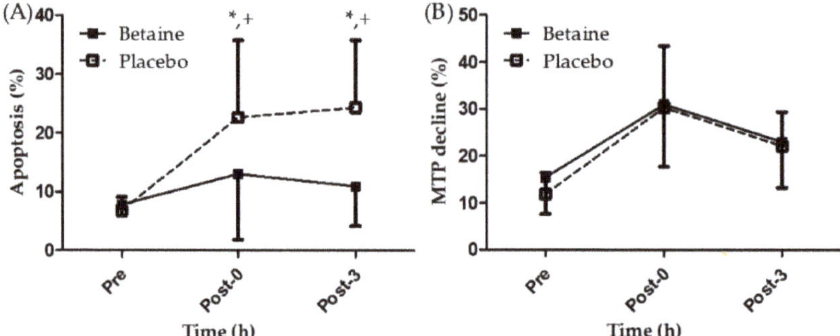

Figure 4. Changes of (**A**) lymphocyte apoptosis and (**B**) mitochondrial transmembrane potential decline in the exhaustive exercise after 2 weeks of betaine supplementation. * Significant ($p < 0.05$) difference from Pre. + Significant ($p < 0.05$) difference from betaine. Pre, before exercise; Post-0, immediately after exercise; Post-3, 3 h after exercise.

3.3. Effect of Betaine Supplementation on Oxidative Stress

Figure 5A,B present the concentrations of TBARS and PC after 2 weeks of betaine or placebo supplementation. A two-way repeated ANOVA with treatment and time indicated there is no interaction between two factors in TBARS and PC. Furthermore, no significant differences were observed in the concentrations of TBARS and PC between the trials and time points ($p > 0.05$). The results showed that two weeks of betaine supplementation had no effects on biomarkers of exercise induced oxidative stress (TBARS and PC).

3.4. Effect of Betaine Supplementation on Aerobic Capacity

All participants completed EEE, which involved running at an intensity of 80% VO_{2max} for 30 min on a treadmill, followed by a 0.2 m/s increase in speed every min until exhaustion. Table 2 indicates no significant differences between the trials regarding the time to exhaustion, peak oxygen consumption, HR_{max}, or AHR ($p > 0.05$). The result shows that two weeks of betaine supplementation had no effects on the indicators of aerobic capacity (time to exhaustion, VO_{2max}, HRmax, and AHR).

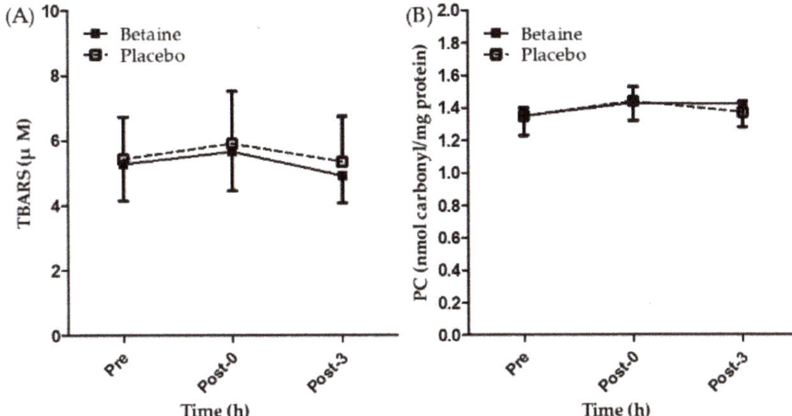

Figure 5. Changes of (**A**) TBARS and (**B**) PC in the exhaustive exercise after 2 weeks of betaine supplementation. Pre, before exercise; Post-0, immediately after exercise; Post-3, 3 h after exercise.

Table 2. Changes in aerobic capacity after 2 weeks of betaine supplementation.

Trial	TTE (min)	HR_{max} (beats/min)	AHR (beats/min)	VO_{2peak} (mL/kg/min)
Betaine	33.61 ± 2.33	195.70 ± 7.06	173.33 ± 10.05	50.58 ± 6.50
Placebo	33.32 ± 1.62	195.70 ± 7.32	173.26 ± 8.99	47.40 ± 4.10

The data are presented as mean ± SD ($n = 10$). TTE, time to exhaustion; HR_{max}, maximal heart rate; AHR, average heart rate; VO_{2peak}, peak oxygen intake.

4. Discussion

This study examined the effects of 2 weeks of betaine supplementation on lymphocyte apoptosis, MTP decline, oxidative stress (TBARS and PC), and aerobic capacity (time to exhaustion, VO_{2max}, HR_{max}, and AHR) after EEE. The findings revealed that twice per day supplementation of 1.25 g of betaine mixed in 300 mL of sports beverage for 2 weeks attenuated lymphocyte apoptosis immediately after EEE. However, a 14-day betaine supplementation had no effects on MTP decline, oxidative stress, or aerobic capacity. In addition, the results revealed that the concentration of betaine in participants' plasma increased significantly with 2 weeks of betaine supplementation. This result was consistent with a previous study that indicated 2 weeks of betaine supplementation (1.25 g of betaine mixed in 300 mL of sports beverage twice per day) can effectively enhance the concentration of betaine in plasma [24]. Moreover, no significant differences were noted in the concentration of choline in plasma between trials (Figure 3B). Therefore, the reason for the elevated betaine concentration in plasma with betaine supplementation was due to of the supplementation rather than choline metabolism. Therefore, the beneficial effects of attenuation of lymphocyte apoptosis after EEE can be attributed to the elevation in plasma betaine concentration. This study also revealed that 2 weeks of betaine supplementation could not reduce the production of TBARS and PC after EEE. The reactive oxygen species production is considered essential for exercise adaptations to occur [35], our result indicated the effectiveness of the betaine supplementation to reduce lymphocyte apoptosis is not via the redox processes.

The MTP is negative inside and positive outside under the normal status; however, when apoptosis is initiated through the depolarization of MTP, which opens the permeability transition pore of mitochondria, the increased membrane permeability of mitochondria results in MTP decline [36]. Therefore, mitochondria are believed to play a pivotal role in the intrinsic apoptosis pathway [37]. In addition, the extrinsic pathway is triggered by pro-inflammatory marker like tumor necrosis factor-α or by death receptors and ligands (Fas/FasL complex) [38]. Therefore, apoptosis is induced by two distinct cell death pathways, either the intrinsic or extrinsic pathway [39,40]. This study showed that betaine supplementation could not reduce the production of TBARS and PC after EEE (Figure 4A,B), potentially because betaine supplementation is unable to effectively inhibit MTP decline after EEE. Craig [12] indicated that active coupled sodium and chloride ions and passive sodium ion independent transport systems promote the cellular absorption of betaine. Based on this finding, most studies regarding betaine supplementation have used betaine mixed in a sports beverage, with the sports beverage employed as the placebo [23,24,41–43]. Notably, sodium and chloride ions are essential electrolytes present in the blood and extracellular fluids. Moreover, sodium plays a crucial role in muscle contractions, and its depletion may affect the whole body [44]. Therefore, sodium and chloride ions, the essential ingredients of a sports drink, may have had an ergogenic effect on MTP, causing the consistent changes in MTP decline in both betaine and placebo trials. Hence, further studies should consider only betaine intake to evaluate its effects on MTP decline after EEE. In addition, this study revealed that 2 weeks of betaine supplementation can effectively attenuate lymphocyte apoptosis after EEE (Figure 3A). This result is consistent with in vitro studies, which have indicated that betaine can effectively decrease lymphocytes apoptosis in cell cultures [45,46]. Moreover, this result is consistent with a previous study in our laboratory that determined a single dose of betaine supplementation can effectively attenuate lymphocyte apoptosis immediately after EEE [30]. However, the present study revealed that 2 weeks of betaine supplementation can significantly attenuate lymphocyte apoptosis not only immediately after EEE but also up to 3 h after EEE. Therefore, the present study demonstrated that the ergogenic effect of 2 weeks of betaine supplementation is better than that of acute betaine supplementation. Lymphocyte apoptosis represents an extrinsic apoptosis pathway, which is initiated when the death ligands, such as tumor necrosis factor-α, bind to the death receptor [47]. Nevertheless, future studies might evaluate the inflammatory response after betaine supplementation to clarify the mechanism of lymphocyte apoptosis attenuation.

Animal studies have indicated that betaine can decrease oxidative stress [48,49]. Therefore, the present study hypothesized that betaine supplementation effectively attenuates oxidative stress caused by exhaustive exercise and enhances aerobic capacity. The present study did not observe an attenuation in lipid and protein peroxidation with 2 weeks of betaine supplementation. This result differs from that of previous studies involving cultured cells and animal tissue, which indicated that betaine can attenuate oxidative stress [48–53]. Halliwell [54] reviewed numerous studies with antioxidant paradox and concluded that the levels of oxidative damage measured in laboratory animals seem more responsive to being decreased by dietary antioxidants than they are in humans. This may one of the reasons with the inconsistent results. Moreover, the validity of measuring TBARS via colorimetry to be the indicator of lipid peroxidation is questionable [55]. This may the limitation of this study. Furthermore, some studies reported positive effects of betaine supplementation related to oxidative stress were conducted over an experimental period lasting more than 3 weeks [52,53,56]. More human studies with longer supplementation periods should be conducted to clarify the effects of betaine supplementation on oxidative stress.

The present study revealed that 1.25 g of betaine mixed in 300 mL of sports beverage twice per day for 2 weeks had no effects on the time to exhaustion, peak oxygen consumption, HR_{max}, or AHR (Table 2). This finding is consistent with that of the previous study in our laboratory, which determined that acute betaine supplementation cannot enhance aerobic capacity [30]. Mitochondrial density can affect the consumption of oxygen during oxidative phosphorylation [57]. The fact that MTP decline is not suppressed by betaine supplementation may be one of the reasons for the lack of aerobic capacity enhancement. Furthermore, one study indicated that betaine supplementation effectively increases the force and power in participants who had a minimum of 3 months of prior resistance training [24]. And betaine supplementation had no effects on muscular strength in untrained participants [41]. The participants in the present study were healthy untrained men, which may explain why 2 weeks of betaine supplementation did not enhance aerobic capacity. Study of Schwab et al. [58] indicated that consuming 6 g per day of betaine for at least 4 weeks elevates plasma betaine concentration. However, other studies have revealed that consuming 6 g of betaine per day for 6 weeks and 12 weeks can increase risks to human health by elevating concentrations of low-density lipoprotein cholesterol and triglycerides as a consequence of increasing the concentration of homocysteine in blood [58,59]. Future studies should attempt to evaluate how various periods and dosages of betaine supplementation affect blood lipids to understand the effects of betaine supplementation on exercise.

5. Conclusions

Our study indicated that a 2 weeks supplementation of 1.25 g of betaine mixed with 300 mL of sports beverage twice per day significantly increased the plasma betaine concentration and effectively attenuated lymphocyte apoptosis after EEE. However, betaine supplementation had no effects on MTP decline, oxidative stress (TBARS and PC), or aerobic capacity (time to exhaustion, VO_{2max}, HR_{max}, and AHR). This study provides a nutritional supplementation strategy to attenuate exercise-induced lymphocyte apoptosis for intensive training athletes. Athletes may have better physiological status to conduct the training and shorten the recovery time during long term training periods. We recommend future studies use a higher dosage (more than 3 g, but less than 6 g per day) of betaine or combine it with other nutritional supplements to evaluate the effects of betaine supplementation on other athletic performance parameters or other related biomarkers of apoptosis.

Author Contributions: Formal analysis, M.-T.Y. and K.-H.C.; Investigation, M.-T.Y., X.-X.L., B.-H.H., L.-H.C., and C.-C.W.; Methodology, M.-T.Y. and K.-H.C.; Project administration, K.-H.C.; Writing-original draft, M.-T.Y., X.-X.L. and K.-H.C.; Writing-review and editing, M.-T.Y. and K.-H.C. All authors have read and agreed to the published version of the manuscript.

Funding: This study was supported by the Ministry of Science and Technology (MOST) of the Executive Yuan, Taiwan, under grant No. NSC 102-2410-H-179-002.

Acknowledgments: The authors are grateful to all participants who completed the study and the anonymous reviewers who aided in manuscript preparation. The authors also wish to thank Kuan-Wei Lee for the assistance with biochemical analysis and the Core Facility Center of Taipei Medical University for performing the liquid chromatography–mass spectrometry analysis.

Conflicts of Interest: The authors declare no conflict of interests.

References

1. Aslani, B.A.; Ghobadi, S. Studies on oxidants and antioxidants with a brief glance at their relevance to the immune system. *Life Sci.* **2016**, *146*, 163–173. [CrossRef] [PubMed]
2. Jornot, L.; Petersen, H.; Junod, A.F. Hydrogen peroxide-induced DNA damage is independent of nuclear calcium but dependent on redox-active ions. *Biochem. J.* **1998**, *335*, 85–94. [CrossRef] [PubMed]
3. Kono, Y.; Fridovich, I. Superoxide radical inhibits catalase. *J. Biol. Chem.* **1982**, *257*, 5751–5754. [PubMed]
4. Clarkson, P.M.; Thompson, H.S. Antioxidants: What role do they play in physical activity and health? *Am. J. Clin. Nutr.* **2000**, *72*, 637S–646S. [CrossRef] [PubMed]
5. Finaud, J.; Lac, G.; Filaire, E. Oxidative stress: Relationship with exercise and training. *Sports Med.* **2006**, *36*, 327–358. [CrossRef]
6. Phaneuf, S.; Leeuwenburgh, C. Apoptosis and exercise. *Med. Sci. Sports Exerc.* **2001**, *33*, 393–396. [CrossRef]
7. Mooren, F.C.; Blöming, D.; Lechtermann, A.; Lerch, M.M.; Völker, K. Lymphocyte apoptosis after exhaustive and moderate exercise. *J. Appl. Physiol.* **2002**, *93*, 147–153. [CrossRef]
8. Ådén, J.; Mushtaq, A.U.; Dingeldein, A.; Wallgren, M.; Gröbner, G. A novel recombinant expression and purification approach for the full-length anti-apoptotic membrane protein Bcl-2. *Protein Expr. Purif.* **2020**, *172*, 105628. [CrossRef]
9. Tuan, T.C.; Hsu, T.G.; Fong, M.C.; Hsu, C.F.; Tsai, K.K.; Lee, C.Y.; Kong, C.W. Deleterious effects of short-term, high-intensity exercise on immune function: Evidence from leucocyte mitochondrial alterations and apoptosis. *Br. J. Sports Med.* **2008**, *42*, 11–15. [CrossRef]
10. Townsend, J.R.; Stout, J.R.; Jajtner, A.R.; Church, D.D.; Beyer, K.S.; Riffe, J.J.; Muddle, T.W.D.; Herrlinger, K.L.; Fukuda, D.H.; Hoffman, J.R. Polyphenol supplementation alters intramuscular apoptotic signaling following acute resistance exercise. *Physiol. Rep.* **2018**, *6*, e13552. [CrossRef]

11. Shakeri, M.; Cottrell, J.J.; Wilkinson, S.; Le, H.H.; Suleria, H.A.R.; Warner, R.D.; Dunshea, F.R. Growth performance and characterization of meat quality of broiler chickens supplemented with betaine and antioxidants under cyclic heat stress. *Antioxidants* **2019**, *8*, 336. [CrossRef] [PubMed]
12. Craig, S.A. Betaine in human nutrition. *Am. J. Clin. Nutr.* **2004**, *80*, 539–549. [CrossRef] [PubMed]
13. Zhao, G.; He, F.; Wu, C.; Li, P.; Li, N.; Deng, J.; Zhu, G.; Ren, W.; Peng, Y. Betaine in inflammation: Mechanistic aspects and applications. *Front. Immunol.* **2018**, *9*, 1070. [CrossRef] [PubMed]
14. Huang, F.; Chen, X.; Jiang, X.; Niu, J.; Cui, C.; Chen, Z.; Sun, J. Betaine ameliorates prenatal valproic-acid-induced autism-like behavioral abnormalities in mice by promoting homocysteine metabolism. *Psychiatry Clin. Neurosci.* **2019**, *73*, 317–322. [CrossRef]
15. Saeed, M.; Babazadeh, D.; Naveed, M.; Arain, M.A.; Hassan, F.U.; Chao, S. Reconsidering betaine as a natural anti-heat stress agent in poultry industry: A review. *Trop. Anim. Health Prod.* **2017**, *49*, 1329–1338. [CrossRef]
16. Dragolovich, J. Dealing with salt stress in animal cells: The role and regulation of glycine betaine concentrations. *J. Exp. Zool.* **1994**, *268*, 139–144. [CrossRef]
17. Zeisel, S.H.; Growdon, J.H.; Wurtman, R.J.; Magil, S.G.; Logue, M. Normal plasma choline responses to ingested lecithin. *Neurology* **1980**, *30*, 1226–1229. [CrossRef]
18. Flower, R.J.; Pollitt, R.J.; Sanford, P.A.; Smyth, D.H. Metabolism and transfer of choline in hamster small intestine. *J. Physiol.* **1972**, *226*, 473–489. [CrossRef]
19. Fernández-Fígares, I.; Wray-Cahen, D.; Steele, N.C.; Campbell, R.G.; Hall, D.D.; Virtanen, E.; Caperna, T.J. Effect of dietary betaine on nutrient utilization and partitioning in the young growing feed-restricted pig. *J. Anim. Sci.* **2002**, *80*, 421–428. [CrossRef]
20. Matthews, J.O.; Southern, L.L.; Pontif, J.E.; Higbie, A.D.; Bidner, T.D. Interactive effects of betaine, crude protein, and net energy in finishing pigs. *J. Anim. Sci.* **1998**, *76*, 2444–2455. [CrossRef]
21. Waldenstedt, L.; Elwinger, K.; Thebo, P.; Uggla, A. Effect of betaine supplement on broiler performance during an experimental coccidial infection. *Poult. Sci.* **1999**, *78*, 182–189. [CrossRef] [PubMed]
22. Armstrong, L.E.; Casa, D.J.; Roti, M.W.; Lee, E.C.; Craig, S.A.; Sutherland, J.W.; Fiala, K.A.; Maresh, C.M. Influence of betaine consumption on strenuous running and sprinting in a hot environment. *J. Strength Cond. Res.* **2008**, *22*, 851–860. [CrossRef] [PubMed]
23. Apicella, J.M.; Lee, E.C.; Bailey, B.L.; Saenz, C.; Anderson, J.M.; Craig, S.A.; Kraemer, W.J.; Volek, J.S.; Maresh, C.M. Betaine supplementation enhances anabolic endocrine and Akt signaling in response to acute bouts of exercise. *Eur. J. Appl. Physiol.* **2013**, *113*, 793–802. [CrossRef] [PubMed]
24. Lee, E.C.; Maresh, C.M.; Kraemer, W.J.; Yamamoto, L.M.; Hatfield, D.L.; Bailey, B.L.; Armstrong, L.E.; Volek, J.S.; McDermott, B.P.; Craig, S.A. Ergogenic effects of betaine supplementation on strength and power performance. *J. Int. Soc. Sports Nutr.* **2010**, *7*, 27. [CrossRef]
25. Hoffman, J.R.; Ratamess, N.A.; Kang, J.; Gonzalez, A.M.; Beller, N.A.; Craig, S.A. Effect of 15 days of betaine ingestion on concentric and eccentric force outputs during isokinetic exercise. *J. Strength Cond. Res.* **2011**, *25*, 2235–2241. [CrossRef]
26. Veskovic, M.; Mladenovic, D.; Milenkovic, M.; Tosic, J.; Borozan, S.; Gopcevic, K.; Labudovic-Borovic, M.; Dragutinovic, V.; Vucevic, D.; Jorgacevic, B.; et al. Betaine modulates oxidative stress, inflammation, apoptosis, autophagy, and Akt/mTOR signaling in methionine-choline deficiency-induced fatty liver disease. *Eur. J. Pharmacol.* **2019**, *848*, 39–48. [CrossRef]
27. Garrett, Q.; Khandekar, N.; Shih, S.; Flanagan, J.L.; Simmons, P.; Vehige, J.; Willcox, M.D. Betaine stabilizes cell volume and protects against apoptosis in human corneal epithelial cells under hyperosmotic stress. *Exp. Eye Res.* **2013**, *108*, 33–41. [CrossRef]
28. Giris, M.; Dogru-Abbasoglu, S.; Soluk-Tekkesin, M.; Olgac, V.; Uysal, M. Effect of betaine treatment on the regression of existing hepatic triglyceride accumulation and oxidative stress in rats fed on high fructose diet. *Gen. Physiol. Biophys.* **2018**, *37*, 563–570. [CrossRef]
29. Heidari, R.; Niknahad, H.; Sadeghi, A.; Mohammadi, H.; Ghanbarinejad, V.; Ommati, M.M.; Hosseini, A.; Azarpira, N.; Khodaei, F.; Farshad, O.; et al. Betaine treatment protects liver through regulating mitochondrial function and counteracting oxidative stress in acute and chronic animal models of hepatic injury. *Biomed. Pharmacother.* **2018**, *103*, 75–86. [CrossRef]
30. Yang, M.T.; Lin, S.C.; Chien, L.H.; Chan, K.H. Effects of acute betaine supplementation on apoptosis, oxidative stress and aerobic capacity after exhaustive endurance exercise. *Sports Exerc. Res.* **2018**, *20*, 373–383. [CrossRef]

31. Ho, C.F.; Shih, C.Y.; Chan, K.H.; Wang, T.Y. Effect of uphill high-intensity interval training on aerobic capacity and power of lower extremity in basketball players. *Sports Exerc. Res.* **2012**, *14*, 476–482. [CrossRef]

32. Bruce, S.J.; Guy, P.A.; Rezzi, S.; Ross, A.B. Quantitative measurement of betaine and free choline in plasma, cereals and cereal products by isotope dilution LC-MS/MS. *J. Agric. Food Chem.* **2010**, *58*, 2055–2061. [CrossRef] [PubMed]

33. Keles, M.S.; Taysi, S.; Sen, N.; Aksoy, H.; Akcay, F. Effect of corticosteroid therapy on serum and CSF malondialdehyde and antioxidant proteins in multiple sclerosis. *Can. J. Neurol. Sci.* **2001**, *28*, 141–143. [CrossRef] [PubMed]

34. Boadi, W.Y.; Iyere, P.A.; Adunyah, S.E. Effect of Quercetin and Genistein on Copper- And Iron-Induced Lipid Peroxidation in Methyl Linolenate. *J. Appl. Toxicol.* **2003**, *23*, 363–369. [CrossRef] [PubMed]

35. Margaritelis, N.V.; Theodorou, A.A.; Paschalis, V.; Veskoukis, A.S.; Dipla, K.; Zafeiridis, A.; Panayiotou, G.; Vrabas, I.S.; Kyparos, A.; Nikolaidis, M.G. Adaptations to endurance training depend on exercise-induced oxidative stress: Exploiting redox interindividual variability. *Acta Physiol.* **2018**, *222*. [CrossRef] [PubMed]

36. Moser, C.C.; Farid, T.A.; Chobot, S.E.; Dutton, P.L. Electron tunneling chains of mitochondria. *Biochim. Biophys. Acta* **2006**, *1757*, 1096–1109. [CrossRef]

37. Nilsson, M.I.; Tarnopolsky, M.A. Mitochondria and aging-the role of exercise as a countermeasure. *Biology* **2019**, *8*, 40. [CrossRef]

38. Andreotti, D.Z.; Silva, J.D.N.; Matumoto, A.M.; Orellana, A.M.; de Mello, P.S.; Kawamoto, E.M. Effects of physical exercise on autophagy and apoptosis in aged brain: Human and animal studies. *Front. Nutr.* **2020**, *7*, 94. [CrossRef]

39. Bratton, S.B.; MacFarlane, M.; Cain, K.; Cohen, G.M. Protein complexes activate distinct caspase cascades in death receptor and stress-induced apoptosis. *Exp. Cell Res.* **2000**, *256*, 27–33. [CrossRef]

40. van Loo, G.; Saelens, X.; van Gurp, M.; MacFarlane, M.; Martin, S.J.; Vandenabeele, P. The role of mitochondrial factors in apoptosis: A Russian roulette with more than one bullet. *Cell Death Differ.* **2002**, *9*, 1031–1042. [CrossRef]

41. del Favero, S.; Roschel, H.; Artioli, G.; Ugrinowitsch, C.; Tricoli, V.; Costa, A.; Barroso, R.; Negrelli, A.L.; Otaduy, M.C.; da Costa Leite, C.; et al. Creatine but not betaine supplementation increases muscle phosphorylcreatine content and strength performance. *Amino Acids* **2012**, *42*, 2299–2305. [CrossRef] [PubMed]

42. Pryor, J.L.; Craig, S.A.; Swensen, T. Effect of betaine supplementation on cycling sprint performance. *J. Int. Soc. Sports Nutr.* **2012**, *9*, 12. [CrossRef] [PubMed]

43. Trepanowski, J.F.; Farney, T.M.; McCarthy, C.G.; Schilling, B.K.; Craig, S.A.; Bloomer, R.J. The effects of chronic betaine supplementation on exercise performance, skeletal muscle oxygen saturation and associated biochemical parameters in resistance trained men. *J. Strength Cond. Res.* **2011**, *25*, 3461–3471. [CrossRef] [PubMed]

44. Gupta, N.; Jani, K.K.; Gupta, N. Hypertension: Salt restriction, sodium homeostasis, and other ions. *Indian J. Med. Sci.* **2011**, *65*, 121–132. [CrossRef]

45. Alfieri, R.R.; Cavazzoni, A.; Petronini, P.G.; Bonelli, M.A.; Caccamo, A.E.; Borghetti, A.F.; Wheeler, K.P. Compatible osmolytes modulate the response of porcine endothelial cells to hypertonicity and protect them from apoptosis. *J. Physiol.* **2002**, *540*, 499–508. [CrossRef]

46. Graf, D.; Kurz, A.K.; Reinehr, R.; Fischer, R.; Kircheis, G.; Häussinger, D. Prevention of bile acid-induced apoptosis by betaine in rat liver. *Hepatology* **2002**, *36*, 829–839. [CrossRef]

47. Teocchi, M.A.; D'Souza-Li, L. Apoptosis through death receptors in temporal lobe epilepsy-associated hippocampal sclerosis. *Mediators Inflamm.* **2016**, *2016*, 8290562. [CrossRef]

48. Bingül, İ.; Başaran-Küçükgergin, C.; Aydın, A.F.; Çoban, J.; Doğan-Ekici, I.; Doğru-Abbasoğlu, S.; Uysal, M. Betaine treatment decreased oxidative stress, inflammation, and stellate cell activation in rats with alcoholic liver fibrosis. *Environ. Toxicol. Pharmacol.* **2016**, *45*, 170–178. [CrossRef]

49. Hagar, H.; Medany, A.E.; Salam, R.; Medany, G.E.; Nayal, O.A. Betaine supplementation mitigates cisplatin-induced nephrotoxicity by abrogation of oxidative/nitrosative stress and suppression of inflammation and apoptosis in rats. *Exp. Toxicol. Pathol.* **2015**, *67*, 133–141. [CrossRef]

50. Alirezaei, M. Betaine protects cerebellum from oxidative stress following levodopa and benserazide administration in rats. *Iran. J. Basic Med. Sci.* **2015**, *18*, 950–957.

51. Guo, Y.; Xu, L.S.; Zhang, D.; Liao, Y.P.; Wang, H.P.; Lan, Z.H.; Guan, W.J.; Liu, C.Q. Betaine Effects on Morphology, Proliferation, and p53-induced Apoptosis of HeLa Cervical Carcinoma Cells in Vitro. *Asian Pac. J. Cancer Prev.* **2015**, *16*, 3195–3201. [CrossRef] [PubMed]

52. Harisa, G.I. Oxidative stress and paraoxonase activity in experimental selenosis: Effects of betaine administration. *Biol. Trace Elem. Res.* **2013**, *152*, 258–266. [CrossRef] [PubMed]

53. Kwon, D.Y.; Jung, Y.S.; Kim, S.J.; Park, H.K.; Park, J.H.; Kim, Y.C. Impaired sulfur-amino acid metabolism and oxidative stress in nonalcoholic fatty liver are alleviated by betaine supplementation in rats. *J. Nutr.* **2009**, *139*, 63–68. [CrossRef] [PubMed]

54. Halliwell, B. The antioxidant paradox: Less paradoxical now? *Br. J. Clin. Pharmacol.* **2013**, *75*, 637–644. [CrossRef] [PubMed]

55. Halliwell, B.; Whiteman, M. Measuring reactive species and oxidative damage in vivo and in cell culture: How should you do it and what do the results mean? *Br. J. Pharmacol.* **2004**, *142*, 231–255. [CrossRef]

56. Ganesan, B.; Anandan, R.; Lakshmanan, P.T. Studies on the protective effects of betaine against oxidative damage during experimentally induced restraint stress in Wistar albino rats. *Cell Stress Chaperones* **2011**, *16*, 641–652. [CrossRef]

57. Barstow, T.J. Characterization of VO_2 kinetics during heavy exercise. *Med. Sci. Sports Exerc.* **1994**, *26*, 1327–1334. [CrossRef]

58. Schwab, U.; Törrönen, A.; Toppinen, L.; Alfthan, G.; Saarinen, M.; Aro, A.; Uusitupa, M. Betaine supplementation decreases plasma homocysteine concentrations but does not affect body weight, body composition, or resting energy expenditure in human subjects. *Am. J. Clin. Nutr.* **2002**, *76*, 961–967. [CrossRef]

59. Olthof, M.R.; Verhoef, P. Effects of betaine intake on plasma homocysteine concentrations and consequences for health. *Curr. Drug Metab.* **2005**, *6*, 15–22. [CrossRef]

Publisher's Note: MDPI stays neutral with regard to jurisdictional claims in published maps and institutional affiliations.

Article

Diazoxide and Exercise Enhance Muscle Contraction during Obesity by Decreasing ROS Levels, Lipid Peroxidation, and Improving Glutathione Redox Status

Mariana Gómez-Barroso [1], Koré M. Moreno-Calderón [1], Elizabeth Sánchez-Duarte [2], Christian Cortés-Rojo [1], Alfredo Saavedra-Molina [1], Alain R. Rodríguez-Orozco [3] and Rocío Montoya-Pérez [1,*]

[1] Instituto de Investigaciones Químico-Biológicas, Universidad Michoacana de San Nicolás de Hidalgo, Francisco J. Múgica S/N, Col. Felicitas del Río, Morelia, Michoacán 58030, Mexico; 0939531k@umich.mx (M.G.-B.); 0935000j@umich.mx (K.M.M.-C.); christian.cortes@umich.mx (C.C.-R.); saavedra@umich.mx (A.S.-M.)

[2] Departamento de Ciencias Aplicadas al Trabajo, Universidad de Guanajuato Campus León, Eugenio Garza Sada 572, Lomas del Campestre Sección 2, León, Guanajuato 37150, Mexico; elizabeth.sanchez@ugto.mx

[3] Facultad de Ciencias Médicas y Biológicas "Dr. Ignacio Chávez", Universidad Michoacana de San Nicolás de Hidalgo Av. Dr. Rafael Carrillo S/N, Esq. Dr. Salvador González Herrejón, Bosque Cuauhtémoc, Morelia, Michoacán 58020, Mexico; alain.rodriguez@umich.mx

* Correspondence: rmontoya@umich.mx

Received: 19 October 2020; Accepted: 2 December 2020; Published: 4 December 2020

Abstract: Obesity causes insulin resistance and hyperinsulinemia which causes skeletal muscle dysfunction resulting in a decrease in contraction force and a reduced capacity to avoid fatigue, which overall, causes an increase in oxidative stress. K_{ATP} channel openers such as diazoxide and the implementation of exercise protocols have been reported to be actively involved in protecting skeletal muscle against metabolic stress; however, the effects of diazoxide and exercise on muscle contraction and oxidative stress during obesity have not been explored. This study aimed to determine the effect of diazoxide in the contraction of skeletal muscle of obese male Wistar rats (35 mg/kg), and with an exercise protocol (five weeks) and the combination from both. Results showed that the treatment with diazoxide and exercise improved muscular contraction, showing an increase in maximum tension and total tension due to decreased ROS and lipid peroxidation levels and improved glutathione redox state. Therefore, these results suggest that diazoxide and exercise improve muscle function during obesity, possibly through its effects as K_{ATP} channel openers.

Keywords: skeletal muscle; obesity; fatigue; oxidative stress; diazoxide; exercise

---

1. Introduction

Obesity is a chronic disease of preventable multifactorial origin; it is the fifth main risk factor for human death globally. It is a pathological state that impairs skeletal muscle [1,2], characterized by a decrease in the force of contraction, a reduced capacity to withstand fatigue, and cell damage [3].

Fat accumulation alters carbohydrates and lipids metabolism, which affects the normal contractile function [4], decrease in glucose uptake, and deterioration of the insulin signaling pathway, causing insulin resistance [5]. Together, it favors the development of fatigue [6], decreased mitochondrial respiration and ATP production, as well as an increase in mitochondrial production of reactive oxygen species (ROS) [7].

Opening of the ATP-sensitive potassium channels (K_{ATP} channels) has been identified as a defense mechanism to counter muscle fatigue and metabolic stress; in the cell membrane and the inner membrane of the mitochondria [8], and play an important role as sensors of intracellular ATP and ADP ratio [9], closing when ATP levels are high and opening when ADP levels increase [10]. Several studies have shown that K_{ATP} channels play an essential role in tissue protection [11,12], where the channel becomes crucial in preventing contractile dysfunction and fiber damage caused by oxidative stress [13].

K_{ATP} channels can be activated pharmacologically, making them an important target for myoprotection [14,15]. Diazoxide is a vasodilator and an inhibitor of insulin secretion. It is considered a drug that is beneficial for some pathologies, such as obesity, since it reduces intake of food and weight gain, prevents hyperinsulinemia, improves insulin sensitivity, and improves blood glucose and lipid profile, which together counteracts mitochondrial dysfunction [16,17].

Diazoxide, reported as a selective opener for mitoK_{ATP} channels, delays fatigue in mammalian fast skeletal muscle fibers [18]. Similarly, increased post-fatigue force in the slow skeletal muscle has been observed by diazoxide and nicorandil, another K_{ATP} channel opener, [19,20] and both, protects skeletal muscle against ischemia-reperfusion injury [15].

Exercise is a non-pharmacological treatment for metabolic disorders associated with obesity. It has been reported that it increases the expression of K_{ATP} channels [13] and promotes various metabolic adaptations in skeletal muscle [21]. Exercise decreases the fat stored in the muscle which improves muscle contraction, increases insulin sensitivity which prevents hyperinsulinemia and stimulates the transport of glucose into the cell [22], and increases antioxidant defenses through a hormonal mechanism which makes the muscle more resistant to oxidative stress [23]. Together these mechanisms protect against mitochondrial dysfunction through lower ROS levels by improving respiration and promoting beta-oxidation of fatty acids [7,24].

Therefore, we hypothesized that diazoxide and exercise improve muscle function in obesity by decreasing oxidative stress. Obesity and its effect on skeletal muscle have been extensively studied, as has exercise to counteract its adverse effects, while the opening of the K_{ATP} channels has been studied as a mechanism of muscular protection. However, this is the first study that shows how diazoxide, exercise, and the combination of both, improve the contraction and function of muscle fibers in obesity by reducing oxidative stress.

2. Materials and Methods

2.1. Experimental Animals and Groups

Male Wistar rats between 300 and 350 g were used. The animals were kept in acrylic cages under bioterium conditions at room temperature for a period of 12 h of light/12 h of darkness, with free access to food and water. Animals were randomly assigned to eight groups (See Table 1).

Table 1. Experimental groups.

Groups	Diet	Diazoxide 35 mg/kg	Exercise Moderate Intensity
Control (C)	Standard rodent chow®	no	no
Diazoxide (D)	Standard rodent chow®	yes	no
Exercise (E)	Standard rodent chow®	no	yes
Obese (O)	High fat diet	no	no
Obese diazoxide (OD)	High fat diet	yes	no
Obese exercise (OE)	High fat diet	no	yes
Obese exercise diazoxide (OED)	High fat diet	yes	yes
Exercise diazoxide (ED)	Standard rodent chow®	yes	yes

The diets are applied for eight weeks, while the protocol of moderate-intensity exercise is applied for five weeks. Diazoxide was administered for 14 days intraperitoneally at a dose of 35 mg/kg. Finally, the respective combinations were combined.

Diets were administered for eight weeks; the standard rodent chow® diet showed a caloric content of 336 cal/100 g with a proportion of 28.507% protein, 13.496% fat, and 57.996% carbohydrates. The high-fat diet contained 50% standard rodent chow® and 50% fat [2], showed a caloric content of 649.25 cal/100 g with a proportion of 14.05% protein, 69.5% fat, and 21.4% of carbohydrates (Mexican equivalent food system). The exercise protocol was applied for five weeks at moderate intensity (see Table 2) and diazoxide was applied at a dose of 35 mg/kg intraperitoneally for 14 days in the obese group, diazoxide was applied at the end of the obesity induction period, while in those subjected to the exercise protocol, the drug was applied during weeks four and five.

Table 2. Exercise protocol. Exercise protocol per week for groups: E; exercise, ED; diazoxide exercise, OE; obese exercise, OED; obese diazoxide exercise. The speed is displayed in meters per minute and the time for which this speed was applied is shown in parentheses.

Groups	E, ED	OE, OED
Week 1	10 m/min (10 min)	10 m/min (10 min)
Week 2	10 m/min (10 min) 16 m/min (5 min)	10 m/min (10 min) 16 m/min (5 min)
Week 3	10 m/min (5 min) 16 m/min (5 min) 22 m/min (5 min)	10 m/min (5 min) 16 m/min (10 min)
Week 4	10 m/min (5 min) 16 m/min (5 min) 22 m/min (5 min)	10 m/min (5 min) 16 m/min (10 min)
Week 5	10 m/min (5 min) 16 m/min (5 min) 22 m/min (10 min)	10 m/min (5 min) 16 m/min (15 min)

All procedures with animals were carried out by the Federal Regulations for the Use and Care of Animals (NOM-062-ZOO-1999) issued by the Ministry of Agriculture of Mexico.

2.2. Muscles Dissection

At the end of the protocols, the rats fasted for 12 h, the weight and glucose parameters of each group were measured to compare these values concerning each treatment. Subsequently, they were sacrificed for cervical dislocation; dissection was performed to obtain soleus and digitorum extensor longus (EDL) muscles of the two posterior extremities. The muscles of one of the extremities were maintained with Krebs-Ringer solution, to be later taken to isometric tension measurements. Simultaneously, the other extremity muscles were stored at −80 °C to be subsequently homogenized for biochemical tests. The Biuret method obtained the protein concentration of the homogenates [25].

2.3. Isometric Tension Measurements

The soleus and EDL muscles were placed in a Petri dish covered with a transparent resin bottom (Sylgard, World Precision Instruments, Sarasota, FL. USA) where they were fixed with the help of entomological pins immersed in Krebs-Ringer solution (118 mM NaCl, 4.75 mM KCl, 1.18 mM $MgSO_4$, 24.8 mM $NaHCO_3$, 1.18 mM KH_2PO_4, 10 mM glucose, and 2.54 mM $CaCl_2$) and carbogen gas (95% O_2 and 5% CO_2) was supplied. Excess connective and fatty tissue were removed under a stereoscopic microscope.

The muscle was mounted into a chamber for isometric tension measurements, with its proximal end attached to the bottom of the chamber and the distal end to the hook of an optical transducer, which was connected to an amplifier and this in turn to an analog–digital interface (World Precision Instruments, Sarasota, FL. USA) that allowed acquiring the tension generated by the muscle in a computer, using the MDAC software (World precision instruments, Sarasota, FL. USA). Two platinum electrodes were placed inside the recording chamber, which was connected to a stimulus isolation unit and an electric stimulator (Grass) in order to apply the protocol to induce fatigue, which consisted of 100 V pulses, 300 ms of duration, and a frequency of 45 Hz for soleus muscle and 50 Hz for EDL muscle. The stimulation was stopped once fatigue was presented. The fatigued muscles were stored at -80 °C to be subsequently homogenized for biochemical tests' performance (measurement of total reactive oxygen species, lipid peroxidation, and glutathione levels).

The muscle was stretched 1.3 times its resting length and left to perfuse in the physiological solution for 10 min before recording the isometric tension; the experiment was performed at a temperature of 25 °C.

2.4. Reactive Oxygen Species Analysis

ROS levels were determined by evaluating the oxidation of the 2′, 7′-dichlorodihydrofluorescein diacetate fluorescent probe (H_2DCFDA). A total of 0.5 mg/mL of homogenate from each muscle was placed in test tubes and incubated at 4 °C with constant shaking for 20 min in a buffer with 10 mM HEPES, 100 mM KCl, 3 mM $MgCl_2$, 3 mM KH_2PO_4 (pH 7.4), and 1.25 mM of H_2DCFDA in a total volume of 2 mL. This suspension was placed in a quartz cell, and the basal fluorescence was determined. One minute later, 10 mM glutamate/malate was added as substrate, and the changes in fluorescence were determined for an additional 20 min [26].

Fluorescence changes were measured on a Shimadzu RF-5301PC spectrofluorometer (λex 485 nm; λem 520 nm). The data were expressed as the difference in fluorescence obtained by subtracting the fluorescence units obtained at the end of the 20 min with the substrate minus the fluorescence obtained before the substrate's addition. The result was expressed as arbitrary fluorescence units/min.

2.5. Lipid Peroxidation Measurement

Lipid peroxidation was assessed using the levels of thiobarbituric acid reactive substances (TBARS). First, 0.5 mg/mL of the homogenate was resuspended in 1 mL of phosphate buffer (50 mM KH_2PO_4, pH 7.6) and incubated with 50 μM $FeSO_4$ for 30 min at 4 °C to induce lipid peroxidation [27].

At the end of the incubation time, 2 mL of acid solution (trichloroacetic acid, thiobarbituric acid, and hydrochloric acid) were added to each sample, and they were incubated in boiling water for 30 min. Subsequently, the tubes were placed on ice for 5 min and centrifuged at 7500 rpm for 5 min. The absorbance of each sample was determined at 532 nm on a Shimadzu UV-2550 spectrophotometer. Data were expressed as TBARS/mg protein numbers.

2.6. Glutathione Redox State Measurement

To 0.5 mg/mL of muscle homogenate, 5% (*v/v*) sulfosalicylic acid was added, and it underwent two cycles of freezing and thawing. Subsequently, it was centrifuged at 10,000 rpm for 5 min and the supernatant was extracted.

Total glutathione (GSH + GSSG) and oxidized glutathione (GSSG) were determined by an enzymatic method. GSH + GSSG levels were determined using 90 μL of the supernatant, resuspended in phosphate buffer (K_2HPO_4, 0.1 M, pH 7.5), and mixed with 3 mM 5,5′-dithiobis 2-nitrobenzoic acid (DTNB) and 0.115 μ/mL glutathione reductase in a final volume of 1 m. After 5 min incubation at room temperature, 2 mM NADPH was added and the reaction kinetics were determined for 15 min. The increase in absorbance at 412 nm measured on a Shimadzu UV-2550 spectrophotometer was converted to GSH concentration using a standard curve with known GSH values [28,29].

For GSSG determination, the same DNTB recycling assay was applied after incubating for 1 h at room temperature with 4% vinylpyridine (*v/v*) to derivatize the reduced GSH [30]. Reduced glutathione (GSH) was determined by subtracting the concentration of GSH + GSSG minus that of GSSG. Data were expressed as μMoles/mg protein.

2.7. Data Analysis

Results were expressed as the mean ± standard error of n = 8 independent experiments using samples from different animals. Statistical differences between groups were determined by one-way and two-way analysis of variance (ANOVA) and Tukey's post-hoc test. A $p \leq 0.05$ was established. The analysis was performed with GraphPad Prism software version 6.0.

3. Results

3.1. Effect of Diazoxide and Exercise on the Physiological Parameters of Obese Rats

The effect of diazoxide, exercise, and their combination on bodyweight and fasting serum glucose levels was evaluated at the end of each treatment. Table 3 shows these values. The obese group increased bodyweight by 70.6% compared to the control group; however, in the groups of obese rats treated with diazoxide, exercise, and the combination of both, a reduction in bodyweight of 18.35% was observed, 17.62% and 22.69%, respectively, to the group of obese rats without treatments. In blood glucose levels, it was observed how obesity increased blood glucose levels by 28.08% concerning the control group; however, a reduction in said levels of 13.95% was observed in the group of obese rats treated with diazoxide, 21.49% in the exercise group, and 26.12% in the exercise with diazoxide group.

Table 3. Weight and glucose at the end of the treatments; C: control; D: diazoxide; E: exercise; O: obese; OD: obese diazoxide; OE: obese exercise; OED: obese exercise diazoxide; ED: exercise diazoxide. The data are represented as the mean ± standard error. Different letters indicate statistically significant differences between groups ($p < 0.05$) one-way ANOVA, Tukey post-hoc test, $n = 8$.

Groups	Bodyweight (g)	Glucose
C	323.80 ± 2.80 g [a]	77 ± 1.13 mg/dL [a]
D	322.12 ± 2.27 g [a]	85 ± 1.56 mg/dL [b]
E	324.28 ± 5.37 g [a]	70 ± 1.47 mg/dL [c]
O	552.57 ± 3.61 g [b]	98 ± 1.25 mg/dL [d]
OD	451.16 ± 19.34 g [c]	84 ± 1.65 mg/dL [b]
OE	455.19 ± 21.66 g [c]	77 ± 1.51 mg/dL [a]
OED	427.16 ± 14.89 g [c]	72 ± 1.84 mg/dL [ac]
ED	354.33 ± 2.12 g [a]	75 ± 2.80 mg/dL [ac]

3.2. Effect of Diazoxide and Exercise on Maximum and Total Tension and Time of Resistance to Fatigue of Slow and Fast Skeletal Muscle of Obese Rats

To explore the effect of treatment with diazoxide, exercise, and the combination of both, maximum and total tension and fatigue resistance was assayed, a record of tension in the soleus muscle and EDL of the different groups was performed. Figure 1 shows the maximum and total tension and the resistance time to fatigue of the soleus muscle (A,C) and EDL (B,D). In the soleus muscle (Figure 1A) of obese rats, a 41.22% decrease in maximum tension, and a 50.18% decrease in total tension concerning the control group can be observed. However, an increase was observed in both tensions with each treatment, diazoxide and exercise, and even a better effect with the combination of both, since a 108.58% increment for maximum tension can be observed, and 115.03% in total tension. In comparison, the group of obese rats exercised presented an increase in maximum tension of 85.94% and 82.64% in total tension.

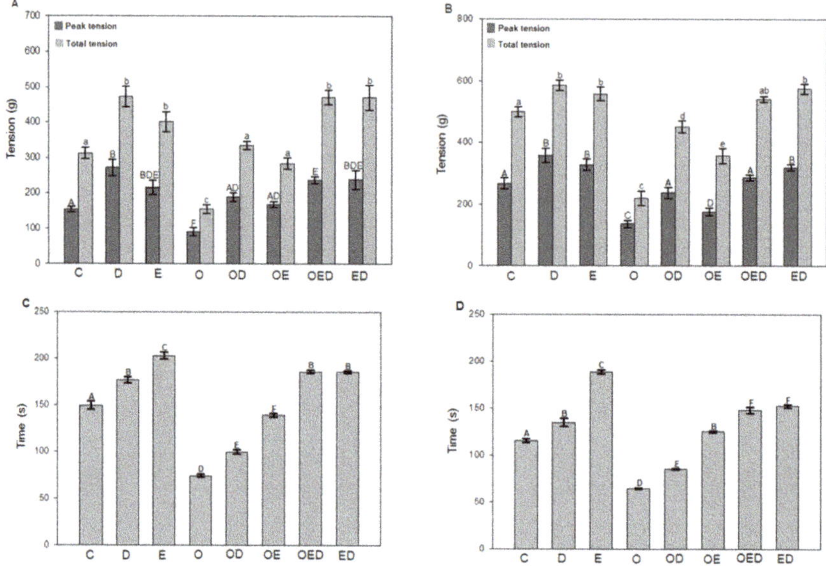

Figure 1. Effect of diazoxide and exercise on maximum, total tension, and time of resistance to fatigue of slow and fast skeletal muscle of obese rats. (**A**) Maximum tension and total soleus muscle tension, (**B**) maximum and total tension of the EDL muscle, (**C**) fatigue resistance time for soleus muscle, and (**D**) fatigue resistance time for EDL muscle. C: control; D: diazoxide; E: exercise; O: obese; OD: obese diazoxide; OE: obese exercise; OED: obese diazoxide exercise; ED: diazoxide exercise. Data are represented as the mean ± standard error. Different letters indicate statistically significant differences between groups, in graphs A and B, capital letters compare the maximum tension, lowercase letters compare the total tension ($p < 0.05$) two-way ANOVA, Tukey post-hoc test, $n = 8$.

Finally, in obese rats exercised with diazoxide, an increase of 162.12% was observed for maximum tension and 102.77% for total tension. EDL muscle (Figure 1B) of obese rats showed a decrease of 49.15% in the maximum tension and 56.17% for the total tension. However, there was an increase of both tensions with each treatment, and a more significant increase with the combination of both, since the obese group treated with diazoxide had an increase of 74.97% for maximum tension and 105.70% in total tension. In contrast, in the group of obese rats exercised there was an increase in the maximum tension of 29.28% and 63.28% in total tension was shown.

3.3. Diazoxide and Exercise Decrease ROS Levels and Lipid Peroxidation in Slow and Fast Skeletal Muscle of Obese Rats

To analyze the effect of diazoxide treatment and the combination of both on the oxidative stress of muscle tissue, ROS levels, and lipid peroxidation of the different groups' soleus and EDL muscles were evaluated. Figure 2 shows ROS and lipid peroxidation levels of the soleus muscle (Figure 2A,C) and EDL (Figure 2B,D) before and after fatigue. In the evaluation of ROS in the soleus muscle (Figure 2A) and the EDL muscle (Figure 2B), there was an increase in ROS levels of 51.96% and 35.89% before fatigue and 40.81% and 81.41% after fatigue, respectively, in the group of obese rats compared to the control group. However, a decrease in ROS levels with each of the treatments was observed; in the obese group treated with diazoxide, there was a decrease in ROS levels of 34.80% for soleus muscle and 45.46% for EDL muscle before fatigue, and 26.67% for soleus muscle and 29.26% after fatigue. The exercised obese group showed a decrease of 38.69% in soleus muscle and 28.59% for EDL muscle before fatigue and 58.57% in soleus muscle and 71.73% for EDL muscle after fatigue and finally in the group of obese rats exercised with diazoxide, a decrease of 39.95% for soleus muscle and 37.25% for EDL muscle before fatigue and 63.49% for soleus muscle and 69.38% for EDL muscle after fatigue, all compared to the group of obese rats without any treatment.

In lipid peroxidation, the results of the soleus muscle (Figure 2C) and EDL muscle (Figure 2D) showed an increase in lipid peroxidation levels of 70.91% before fatigue and 70.19% after fatigue and 43.48% before fatigue and 51.19% and after fatigue, respectively, in the group of obese rats compared to the control group. However, a decrease in lipid peroxidation levels was observed with each of the treatments separately and a synergistic effect with both. In the group of obese rats treated with diazoxide, a decrease of 26.91% was observed for soleus muscle and 19.70% for EDL muscle before fatigue, and 17.24% for soleus muscle and 23.91% for EDL muscle after fatigue. The group of exercised obese rats showed a decrease of 22.84% for the soleus muscle and 21.07% for the EDL muscle before fatigue, and 38.36% for the soleus muscle and 35.87% for the EDL muscle. Finally, in the group of obese rats exercised with diazoxide, a decrease of 39.96% was observed for soleus muscle and of 31.46% for EDL muscle before fatigue and 44.64% for soleus muscle and 43.77% for EDL muscle after fatigue, concerning the obese group without treatment.

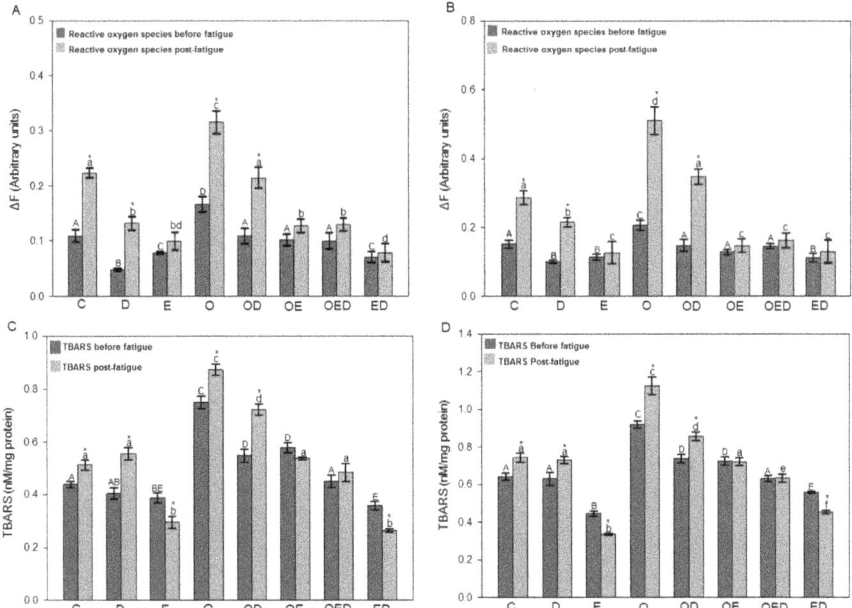

Figure 2. Effect of diazoxide and exercise on ROS levels and lipid peroxidation in slow and fast skeletal muscle of obese rats. (**A**) ROS levels in the soleus muscle, (**B**) ROS levels in the EDL muscle, (**C**) TBARS levels in the soleus muscle, and (**D**) TBARS levels in the EDL muscle, before and after fatigue. C: control; D: diazoxide; E: exercise; O: obese; OD: obese diazoxide; OE: obese exercise; OED, obese exercise with diazoxide; ED: exercise with diazoxide. Data are represented as the mean ± standard error. Different letters indicate statistically significant differences between the groups, capital letters compare the different groups before fatigue, and lowercase letters compare the different groups after fatigue. * indicates significant differences in the comparison of the same group before and after fatigue ($p < 0.05$) two-way ANOVA, Tukey's post-hoc test, $n = 8$.

3.4. Effect of Diazoxide and Exercise on Glutathione Redox Status in Slow and Fast Skeletal Muscle of Obese Rats

To evaluate the effect of diazoxide, exercise, and both on glutathione's redox status in obesity, Figures 3 and 4 show the results obtained for total glutathione levels (GSH + GSSG), reduced glutathione (GSH), and oxidized glutathione (GSSG) before and after inducing fatigue for the soleus muscle and EDL muscle, respectively. These results indicate that during obesity no differences were observed in the levels of GSH + GSSG before and after fatigue in relation to the control group in the soleus muscle (Figure 3A), while in the EDL muscle (Figure 4A) a decrease in GSH + GSSG levels of 16.74% before fatigue and 20.12% after fatigue was observed in relation to the control group. However, in the group of obese rats treated with diazoxide, a 45.31% increase was observed for soleus muscle and 49.38% for EDL muscle before fatigue and 88.90% for soleus muscle and 37.48% for EDL muscle after fatigue. The group of obese rats exercised showed an increase of 18.80% before fatigue for soleus muscle and 24.37% for EDL muscle and 71.42% for soleus muscle and 51.38% for EDL after fatigue. Finally, in the group of obese rats exercised with diazoxide, an increase of 28.16% was observed for soleus muscle and 27.27% for EDL muscle before fatigue and 92.23% for soleus muscle and 83.33% for EDL muscle after fatigue in relation to the group of obese rats without treatment. Regarding the GSH levels of the soleus muscle (Figure 3B) and EDL (Figure 4B), the group of obese rats showed a decrease in levels of 39.76% for soleus muscle and 57.84% for EDL muscle before fatigue and a decrease of 58.25% for soleus muscle and 52.20% for EDL muscle after fatigue to control. However, an increase in GSH levels could be seen in obese rats with each of the treatments, concerning obese rats without treatment.

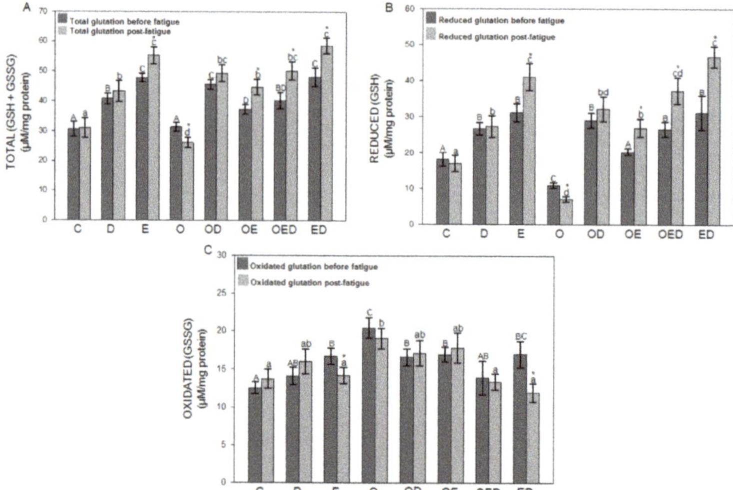

Figure 3. Effect of diazoxide and exercise on glutathione redox status in slow skeletal muscle of obese rats. (**A**) Total glutathione soleus muscle, (**B**) reduced soleus glutathione muscle, and (**C**) oxidized glutathione soleus muscle, before and after fatigue. C: control; D: diazoxide; E: exercise; O: obese; OD: obese diazoxide; OE: obese exercise; OED, obese exercise with diazoxide; ED: exercise with diazoxide. Data are represented as the mean ± standard error. Different letters indicate statistically significant differences between the groups, capital letters compare the different groups before fatigue, and lowercase letters compare the different groups after fatigue. * indicates significant differences in the comparison of the same group before and after fatigue ($p < 0.05$) two-way ANOVA, Tukey's post-hoc test, $n = 8$.

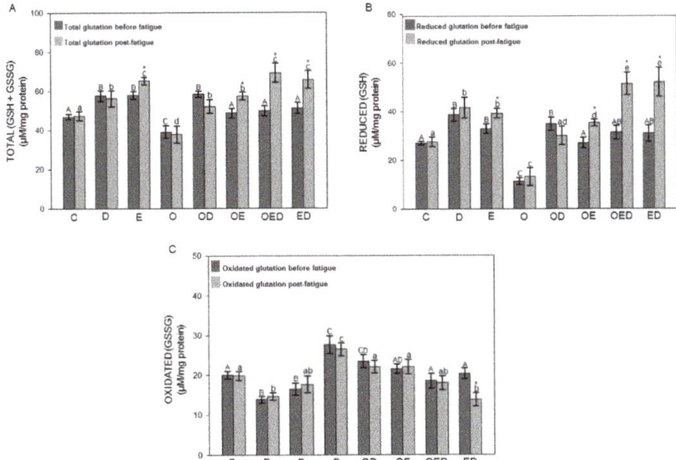

Figure 4. Effect of diazoxide and exercise on glutathione redox status in fast skeletal muscle of obese rats. (**A**) EDL muscle with total glutathione, (**B**) EDL muscle with reduced glutathione, and (**C**) EDL muscle with oxidized glutathione, before and after fatigue. C: control; D: diazoxide; E: exercise; O: obese; OD: obese diazoxide; OE: obese exercise; OED, obese exercise with diazoxide; ED: exercise with diazoxide. Data are represented as the mean ± standard error. Different letters indicate statistically significant differences between the groups, capital letters compare the different groups before fatigue, and lowercase letters compare the different groups after fatigue. * indicates significant differences in the comparison of the same group before and after fatigue ($p < 0.05$) two-way ANOVA, Tukey's post-hoc test, $n = 8$.

To conclude, it should be mentioned that the levels of oxidized glutathione did not show significant differences between the muscles analyzed before and after fatigue in both muscles of each group evaluated.

4. Discussion

More than a third of the world's adult population suffers from obesity [30], with more than 500 million people worldwide. Obesity is related to many pathologies [7] and is exacerbated by a sedentary lifestyle and lack of physical activity (WHO, 2016) [31]. Skeletal muscle dysfunction is a complication of obesity, as it causes significant atrophy, leading to a decrease in the muscle contraction force, reduced ability to support fatigue, and numerous metabolic alterations and increased oxidative stress [1,3,6].

Studies conducted by Alemzadeh et al. [17], Pompeani et al. [1], Bae et al. [5], and Lu et al. [32] showed that obesity-induced with a high-fat diet increases bodyweight and plasma glycemia in Wistar rats. The results in Table 3 show consistency with these studies for both parameters. However, in this study, it was observed that both the treatment with diazoxide, the exercise protocol, and the combination of both decreased bodyweight and plasma glycemia during obesity, effects that may be related to the reduction in the consumption of food, improvement of basal metabolic rate, reduction of lipogenesis, improvement of insulin sensitivity and glucose transport, and suppression of hyperinsulinemia [5,17,22,33].

A higher concentration of lipids in muscle tissue produces skeletal muscle dysfunction [6]. This experimental series observed how muscle contraction and fatigue resistance were decreased in the soleus muscle and EDL of obese rats (Figure 1). The results showed that obesity affects the normal contractile function of muscles, which is associated with an increase in intramuscular fat, causing a deficiency in the muscle ability to contract [6], an increase in proteolysis, change in fiber type, metabolic alterations, and increased oxidative stress [1,7,34].

Previous studies have shown that treatment with diazoxide prevents and reverses metabolic disorders, such as loss of insulin sensitivity, and has been associated with an improvement in glucose transport and the promotion of lipid metabolism [6], which directly or indirectly improves the functioning of muscle tissue.

In this study, it was observed that the treatment with diazoxide increased muscle contraction and promoted resistance to fatigue in the obese group treated with this drug (Figure 1); data are consistent with what was observed by García et al. [18] where diazoxide increased the post-fatigue tension of the mouse EDL muscle. This suggests that the protective effect of diazoxide in obesity may be due to the opening of KATP channels, leading to an increase in cellular respiration through the electron transport chain (CTE) and an increase in the synthesis of ATP [35], favoring muscle contraction and resistance to fatigue [36,37]. A very similar effect was observed with implementing the exercise protocol during obesity (Figure 1). Studies by Zigman et al. [38] and Kraljievic et al. [13] showed an exercise-induced increase in the expression of cardiac sarcolemmal KATP channels, which improved conditioning and contraction, a process that is not ruled out could be occurring in our results, as improvement in muscle contraction and fatigue resistance. Similarly, the positive effects of exercise are attributed to the decrease in intramuscular fat, as a consequence of the increase in skeletal muscle metabolism [3], or the regulation of numerous pathways of signaling in which exercise participates by improving its functioning, such as insulin sensitivity, glucose transport, lipid profile, and reduced stress markers [3,5].

The increase in fat stored in skeletal muscle has been associated with increased ROS levels, the appearance of oxidative stress, and eventual cell damage [7].

Abrigo et al. [2] evaluated ROS levels in obese mouse muscles induced by a high-fat diet. They observed in this group an increase in ROS levels 7.22 times that of the control group. This is consistent with the obtained results in this work since it was possible to observe increased ROS levels in the groups of obese rats, compared to the control group in both muscles before and after the fatigue was induced (Figure 2A,B), observing a higher level of ROS in EDL muscle because fast-twitch fibers are more susceptible to oxidative stress than slow-twitch fibers [39]. The increase in ROS levels during obesity is due to the increase in intramuscular fat causing a reduction in the muscle's ability to contract [6], which causes reduced ATP levels and decreases mitochondrial volume producing mitochondrial dysfunction and promoting muscle fatigue [36,37,39]. It has been shown that there is a dose-dependent relationship between ROS concentration and reduced muscle contractility and the occurrence of fatigue [40]. This work showed that the use of diazoxide, regular exercise, and the combination of both reduced ROS levels in the group of obese rats in both soleus muscle and EDL muscle was analyzed before and after inducing fatigue (Figure 2). A similar effect could be observed in the work of García et al. [18] in mouse muscle fibers where in the presence of diazoxide the ROS production rate was lower compared to the untreated group. This effect was attributed to the opening of mitoKATP channels, which improves contraction and decreases fatigue, thus reducing ROS levels [18,41].

On the other hand, diazoxide can also decrease ROS levels in obesity by preventing mitochondrial dysfunction produced by hyperinsulinemia and hyperglycemia triggered by increased intramuscular fat [5–7]. This protective effect results from inhibiting insulin secretion and improving insulin sensitivity and glucose transport [18], which is also attributed to the characteristic of increasing antioxidant defenses, which favors the decrease of ROS levels [42,43]. In the same way, it was observed how exercise decreased ROS levels in groups of obese rats (Figure 2A,B). This effect of exercise was also observed in the study by Ji et al. [23], where a group of rats was subjected to an exercise regimen where decreased ROS levels were observed compared to the untrained group, in response to an adaptation in their antioxidant systems. This protective role of exercise in obesity could be attributed to the production and activity of antioxidant enzymes, in addition to the activation of numerous signaling cascades involved in exercise [7,23,24].

Similarly, exercise could decrease ROS levels during obesity by increasing the expression and participation of sarcolemmal KATP channels, improving muscle contraction and resistance to fatigue [13,34,38]. In the same way, these positive effects of exercise are attributed to the decrease in intramuscular fat due to the increase in skeletal muscle metabolism [5], or the optimization of signaling cascades related to insulin sensitivity, glucose transport, and inflammatory processes, which in turn decreases ROS production [7,44]. In this experiment series, the muscles of the groups subjected to the exercise protocol analyzed after fatigue showed the same ROS levels as the muscles analyzed before fatigue, an effect that was not observed in the rest of the groups, which can be explained with the concept of hormesis, which describes the response related to a beneficial stress factor at moderate levels and harmful at high levels [4].

It has been observed that one of the main effects of the increase in ROS is the damage to biomolecules [45], so in this study, the levels of TBARS were quantified as an indicator of lipid peroxidation and oxidative stress. This work observed that lipid peroxidation levels were increased in obese rats, both in the soleus and in EDL muscles analyzed before and after inducing fatigue (Figure 2C,D). However, the group of obese rats treated with diazoxide showed decreased lipid peroxidation levels in both muscles (Figure 2C,D). Farahini et al. [46] observed that pretreatment with diazoxide contributed to muscle tissue resistance against ischemic damage by decreasing lipid peroxidation. The same was appreciated by Moghtadaei et al. [42], where they observed this protective effect of diazoxide in skeletal muscle ischemia, as the levels of lipid peroxidation were decreased in the group preconditioned with diazoxide.

Similarly, the experimental series results showed decreased lipid peroxidation levels in the group of obese-exercise rats (Figure 2C,D). This effect with the implementation of an exercise protocol was previously observed by Lambertucci et al. [24], where the levels of lipid peroxidation of muscle tissue from trained rats were decreased. The decrease in both ROS levels and lipid peroxidation with diazoxide treatment and exercise, and the combination of both is directly proportional to muscle contraction improvement and increased resistance time to fatigue.

Cells have antioxidant systems that protect them against oxidative damage. This includes the antioxidant enzyme glutathione peroxidase, the synthesis of which can be modified by exercise, diet, and age [40]. Our results showed that both the levels of GSH + GSSG of the soleus muscle (Figure 3A) and EDL muscle (Figure 4A), as well as of GSH of the soleus muscle (Figure 3B) and EDL muscle (Figure 4B) were decreased during obesity, while the GSSG levels of the soleus muscle (Figure 3C) and EDL muscle (Figure 4C) were increased. The decrease in the content of GSH + GSSG suggests that obesity affects the synthesis of this antioxidant, possibly because ROS that are overproduced during this pathology are involved in activating various signaling pathways that negatively affect the expression of genes involved. The synthesis and content of GSH decreased [47], suggesting that obesity affects the antioxidant capacity. Diazoxide treatment, exercise, and a combination of both increased GSH + GSSG and GSH levels and, to a lesser extent, reduced GSSG levels in obesity. Moghtadaei et al. [42] reported that glutathione activity was increased in skeletal muscle treated with this drug during ischemia-reperfusion, concerning the muscle without treatment.

On the other hand, exercise presented this same protective effect during obesity since it enhances antioxidant activity [24]. Both the protective effect of diazoxide and exercise may be because these treatments regulate signaling pathways for the activation of genes that regulate the synthesis of antioxidant enzymes [42], in addition to the fact that glutathione synthesis is widely dependent on ATP concentrations. Both diazoxide treatment and exercise favor its production [36].

5. Conclusions

This is the first study, which shows how diazoxide and exercise and the combination of both improve the contraction and functioning of muscle fibers in obesity by reducing oxidative stress. In conclusion, diazoxide and exercise improves muscle contraction in obesity by decreasing ROS levels, lipid peroxidation, and improving the redox status of glutathione.

Author Contributions: Conceptualization, M.G.-B., E.S.-D., and R.M.-P.; methodology, M.G.-B. and K.M.M.-C.; formal analysis, M.G.-B., C.C.-R., and R.M.-P.; investigation, M.G.-B., A.S.-M., and A.R.R.-O; resources, R.M.-P; writing—original draft preparation, M.G.-B., C.C.-R., and R.M.-P.; writing—review and editing, M.G.-B., C.C.-R., E.S.-D., and R.M.-P.; supervision, R.M.-P.; project administration, R.M.-P.; funding acquisition, R.M.-P. All authors have read and agreed to the published version of the manuscript.

Funding: This research was partially funded by Coordinación de la Investigación Científica—Universidad Michoacana de San Nicolás de Hidalgo, R.M.P. CIC-UMSNH2019".

Conflicts of Interest: The authors declare no conflict of interest. The funders had no role in the design of the study; in the collection, analyses, or interpretation of data; in the writing of the manuscript, or in the decision to publish the results.

References

1. Pompeani, N.; Rybalka, E.; Latchman, H.; Murphy, R.M.; Croft, K.D.; Hayes, A. Skeletal muscle atrophy in sedentary Zucker obese rats is not caused by calpain-mediated muscle damage or lipid peroxidation induced by oxidative stress. *J. Negat. Results Biomed.* **2014**, *13*, 1–11. [CrossRef]

2. Abrigo, J.; Rivera, J.C.; Aravena, J.; Cabrera, D.; Simon, F.; Ezquer, F.; Ezquer, M.; Cabello-Verrugio, C. High Fat Diet-Induced Skeletal Muscle Wasting Is Decreased by Mesenchymal Stem Cells Administration: Implications on Oxidative Stress, Ubiquitin Proteasome Pathway Activation, and Myonuclear Apoptosis. *Oxidative Med. Cell. Longev.* **2016**, *2016*, 1–13. [CrossRef] [PubMed]

3. Martinez-Huenchullan, S.F.; Maharjan, B.R.; Williams, P.F.; Tam, C.S.; McLennan, S.V.; Twigg, S.M. Differential metabolic effects of constant moderate versus high intensity interval training in high-fat fed mice: Possible role of muscle adiponectin. *Physiol. Rep.* **2018**, *6*, e13599. [CrossRef] [PubMed]

4. Espinosa, A.; Henríquez-Olguín, C.; Jaimovich, E. Reactive oxygen species and calcium signals in skeletal muscle: A crosstalk involved in both normal signaling and disease. *Cell Calcium* **2016**, *60*, 172–179. [CrossRef] [PubMed]

5. Bae, J.Y.; Shin, K.O.; Woo, J.; Woo, S.H.; Jang, K.S.; Lee, Y.H.; Kang, S. Exercise and dietary change ameliorate high fat diet induced obesity and insulin resistance via mTOR signaling pathway. *J. Exerc. Nutr. Biochem.* **2016**, *20*, 28–33. [CrossRef] [PubMed]

6. Choi, S.J.; Files, D.C.; Zhang, T.; Wang, Z.-M.; Messi, M.L.; Gregory, H.; Stone, J.; Lyles, M.F.; Dhar, S.; Marsh, A.P.; et al. Intramyocellular Lipid and Impaired Myofiber Contraction in Normal Weight and Obese Older Adults. *J. Gerontol. A Boil. Sci. Med. Sci.* **2015**, *71*, 557–564. [CrossRef] [PubMed]

7. Heo, J.-W.; No, M.-H.; Park, D.-H.; Kang, J.-H.; Seo, D.Y.; Han, J.; Neufer, P.D.; Kwak, H.-B. Effects of exercise on obesity-induced mitochondrial dysfunction in skeletal muscle. *Korean J. Physiol. Pharmacol.* **2017**, *21*, 567–577. [CrossRef]

8. Scott, K.; Benkhalti, M.; Calvert, N.D.; Paquette, M.; Zhen, L.; Harper, M.-E.; Al-Dirbashi, O.Y.; Renaud, J. KATP channel deficiency in mouse FDB causes an impairment of energy metabolism during fatigue. *Am. J. Physiol. Physiol.* **2016**, *311*, C559–C571. [CrossRef]

9. Hibino, H.; Inanobe, A.; Furutani, K.; Murakami, S.; Findlay, I.; Kurachi, Y. Inwardly Rectifying Potassium Channels: Their Structure, Function, and Physiological Roles. *Physiol. Rev.* **2010**, *90*, 291–366. [CrossRef]

10. Szabo, I.; Zoratti, M. Mitochondrial Channels: Ion Fluxes and More. *Physiol. Rev.* **2014**, *94*, 519–608. [CrossRef]

11. Constant-Urban, C.C.; Charif, M.M.; Goffin, E.; Van Heugen, J.-C.; Elmoualij, B.B.; Chiap, P.; Mouithys-Mickalad, A.; Serteyn, D.D.; Lebrun, P.; Pirotte, B.; et al. Triphenylphosphonium salts of 1,2,4-benzothiadiazine 1,1-dioxides related to diazoxide targeting mitochondrial ATP-sensitive potassium channels. *Bioorganic Med. Chem. Lett.* **2013**, *23*, 5878–5881. [CrossRef] [PubMed]

12. Garlid, A.O.; Jabůrek, M.; Jacobs, J.P.; Garlid, K.D. Mitochondrial reactive oxygen species: Which ROS signals cardioprotection? *Am. J. Physiol. Circ. Physiol.* **2013**, *305*, H960–H968. [CrossRef] [PubMed]

13. Kraljević, J.; Hoydal, M.A.; Ljubkovic, M.; Moreira, J.B.N.; Jørgensen, K.; Ness, H.O.; Bækkerud, F.H.; Dujić, Ž; Wisløff, U.; Marinovic, J. Role of KATP Channels in Beneficial Effects of Exercise in Ischemic Heart Failure. *Med. Sci. Sports Exerc.* **2015**, *47*, 2504–2512. [CrossRef]

14. Andrade, F.; Trujillo, X.; Sánchez-Pastor, E.; Montoya-Pérez, R.; Saavedra-Molina, A.; Ortiz-Mesina, M.; Huerta, M. Glibenclamide increases post-fatigue tension in slow skeletal muscle fibers of the chicken. *J. Comp. Physiol. B* **2011**, *181*, 403–412. [CrossRef] [PubMed]

15. Cahoon, N.J.; Naparus, A.; Ashrafpour, H.; Hofer, S.; Huang, N.; Lipa, J.E.; Forrest, C.R.; Pang, C.Y. Pharmacologic Prophylactic Treatment for Perioperative Protection of Skeletal Muscle from Ischemia-Reperfusion Injury in Reconstructive Surgery. *Plast. Reconstr. Surg.* **2013**, *131*, 473–485. [CrossRef]

16. Alemzadeh, R.; Tushaus, K.M. Modulation of Adipoinsular Axis in Prediabetic Zucker Diabetic Fatty Rats by Diazoxide. *Endocrinology* **2004**, *145*, 5476–5484. [CrossRef]

17. Alemzadeh, R.; Karlstad, M.D.; Tushaus, K.; Buchholz, M. Diazoxide enhances basal metabolic rate and fat oxidation in obese Zucker rats. *Metabolism* **2008**, *57*, 1597–1607. [CrossRef]

18. García, M.C.; Hernández, A.; Sánchez, J.A. Role of mitochondrial ATP-sensitive potassium channels on fatigue in mouse muscle fibers. *Biochem. Biophys. Res. Commun.* **2009**, *385*, 28–32. [CrossRef]

19. Montoya-Perez, R.; Sánchez-Duarte, E.; Trujillo, X.; Huerta, M.; Ortiz-Mesina, M.; Cortés-Rojo, C.; Manzo-Ávalos, S.; Saavedra-Molina, A. Mitochondrial KATP channels in skeletal muscle: Are protein kinases C and G, and nitric oxide synthase involved in the fatigue process? *Open Access Anim. Physiol.* **2012**, *4*, 21–28. [CrossRef]

20. Sánchez-Duarte, E.; Trujillo, X.; Cortés-Rojo, C.; Saavedra-Molina, A.; Camargo, G.; Hernández, L.; Huerta, M.; Montoya-Pérez, R. Nicorandil improves post-fatigue tension in slow skeletal muscle fibers by modulating glutathione redox state. *J. Bioenerg. Biomembr.* **2017**, *49*, 159–170. [CrossRef]

21. Pattanakuhar, S.; Pongchaidecha, A.; Chattipakorn, N.; Chattipakorn, S.C. The effect of exercise on skeletal muscle fibre type distribution in obesity: From cellular levels to clinical application. *Obes. Res. Clin. Pr.* **2017**, *11*, 112–132. [CrossRef] [PubMed]

22. Cartee, G.D.; Arias, E.B.; Yu, C.S.; Pataky, M.W. Novel single skeletal muscle fiber analysis reveals a fiber type-selective effect of acute exercise on glucose uptake. *Am. J. Physiol. Metab.* **2016**, *311*, E818–E824. [CrossRef] [PubMed]

23. Ji, L.L.; Gomez-Cabrera, M.; Vina, J. Exercise and Hormesis: Activation of Cellular Antioxidant Signaling Pathway. *Ann. N. Y. Acad. Sci.* **2006**, *1067*, 425–435. [CrossRef] [PubMed]

24. Lambertucci, R.H.; Levada-Pires, A.C.; Rossoni, L.V.; Curi, R.; Pithon-Curi, T.C. Effects of aerobic exercise training on antioxidant enzyme activities and mRNA levels in soleus muscle from young and aged rats. *Mech. Ageing Dev.* **2007**, *128*, 267–275. [CrossRef]

25. Gornall, A.G.; Baradawill, C.J.; David, M.M. Determination of serum proteins by means of the biuret reaction. *J. Biol. Chem.* **1949**, *177*, 751.

26. Ortiz-Avila, O.; Gallegos-Corona, M.A.; Sánchez-Briones, L.A.; Calderón-Cortés, E.; Montoya-Pérez, R.; Rodríguez-Orozco, A.R.; Campos-Garciía, J.; Saavedra-Molina, A.; Mejía-Zepeda, R.; Cortés-Rojo, C. Protective effects of dietary avocado oil on impaired electron transport chain function and exacerbated oxidative stress in liver mitochondria from diabetic rats. *J. Bioenerg. Biomembr.* **2015**, *47*, 337–353. [CrossRef]

27. Cortés-Rojo, C.; Calderón-Cortés, E.; Clemente-Guerrero, M.; Estrada-Villagómez, M.; Manzo-Ávalos, S.; Mejía-Zepeda, R.; Boldogh, I.; Saavedra-Molina, A. Elucidation of the effects of lipoperoxidation on the mitochondrial electron transport chain using yeast mitochondria with manipulated fatty acid content. *J. Bioenerg. Biomembr.* **2009**, *41*, 15–28. [CrossRef]

28. Akerboom, T.P.; Sies, H. Assay of glutathione, glutathione disulfide, and glutathione mixed disulfides in biological samples. *Immobil. Enzym.* **1981**, *77*, 373–382. [CrossRef]

29. Ortiz-Avila, O.; Sámano-García, C.A.; Calderón-Cortés, E.; Pérez-Hernández, I.H.; Mejía-Zepeda, R.; Rodríguez-Orozco, A.R.; Saavedra-Molina, A.; Cortés-Rojo, C. Dietary avocado oil supplementation attenuates the alterations induced by type I diabetes and oxidative stress in electron transfer at the complex II-complex III segment of the electron transport chain in rat kidney mitochondria. *J. Bioenerg. Biomembr.* **2013**, *45*, 271–287. [CrossRef]

30. Trewin, A.J.; Levinger, I.; Parker, L.; Shaw, C.S.; Serpiello, F.R.; Anderson, M.J.; McConell, G.K.; Hare, D.L.; Stepto, N.K. Acute exercise alters skeletal muscle mitochondrial respiration and H2O2 emission in response to hyperinsulinemic-euglycemic clamp in middle-aged obese men. *PLoS ONE* **2017**, *12*, e0188421. [CrossRef]

31. World Health Organization (WHO). Available online: https://www.who.int/es/news-room/fact-sheets/detail/obesity-and-overweight (accessed on 5 May 2020).

32. Lu, Y.; Li, H.; Shen, S.-W.; Shen, Z.-H.; Xu, M.; Yang, C.-J.; Li, F.; Feng, Y.-B.; Yun, J.-T.; Wang, L.; et al. Swimming exercise increases serum irisin level and reduces body fat mass in high-fat-diet fed Wistar rats. *Lipids Health Dis.* **2016**, *15*, 1–8. [CrossRef] [PubMed]

33. Botezelli, J.D.; Coope, A.; Ghezzi, A.C.; Cambri, L.T.; Moura, L.P.; Scariot, P.P.M.; Gaspar, R.S.; Mekary, R.A.; Ropelle, E.R.; Pauli, J.R. Strength Training Prevents Hyperinsulinemia, Insulin Resistance, and Inflammation Independent of Weight Loss in Fructose-Fed Animals. *Sci. Rep.* **2016**, *6*, 31106. [CrossRef]

34. Brown, D.A.; Chicco, A.J.; Jew, K.N.; Johnson, M.S.; Lynch, J.M.; Watson, P.A.; Moore, R.L. Cardioprotection afforded by chronic exercise is mediated by the sarcolemmal, and not the mitochondrial, isoform of the KATPchannel in the rat. *J. Physiol.* **2005**, *569*, 913–924. [CrossRef] [PubMed]

35. Debska, G.; Kicinska, A.; Skalska, J.; Szewczyk, A.; May, R.; E Elger, C.; Kunz, W.S. Opening of potassium channels modulates mitochondrial function in rat skeletal muscle. *Biochim. Biophys. Acta Bioenerg.* **2002**, *1556*, 97–105. [CrossRef]

36. Andrukhiv, A.; Costa, A.D.; West, I.C.; Garlid, K.D. Opening mitoKATP increases superoxide generation from complex I of the electron transport chain. *Am. J. Physiol. Heart Circ. Physiol.* **2006**, *291*, H2067–H2074. [CrossRef] [PubMed]

37. Costa, A.D.; Quinlan, C.L.; Andrukhiv, A.; West, I.C.; Jabůrek, M.; Garlid, K.D. The direct physiological effects of mitoKATP opening on heart mitocondria. *Am. J. Physiol. Heart Circ. Physiol.* **2006**, *290*, H406–H415. [CrossRef] [PubMed]

38. Zingman, L.V.; Zhu, Z.; Sierra, A.; Stepniak, E.; Burnett, C.M.-L.; Maksymov, G.; Anderson, M.E.; Coetzee, W.A.; Hodgson-Zingman, D.M. Exercise-induced expression of cardiac ATP-sensitive potassium channels promotes action potential shortening and energy conservation. *J. Mol. Cell. Cardiol.* **2011**, *51*, 72–81. [CrossRef] [PubMed]

39. Montoya-Pérez, R.; Saavedra-Molina, A.; Trujillo, X.; Huerta, M.; Andrade, F.; Sánchez-Pastor, E.; Ortiz, M. Inhibition of oxygen consumption in skeletal muscle-derived mitochondria by pinacidil, diazoxide, and glibenclamide, but not by 5-hydroxydecanoate. *J. Bioenerg. Biomembr.* **2010**, *42*, 21–27. [CrossRef]

40. Fernández, J.M.; Da Silva-Grigoletto, M.E.; Tunez-Fiñaca, I. Estrés oxidativo inducido por el ejercicio. *Rev. Andal. Med. Deporte.* **2009**, *2*, 19–34.

41. Ferranti, R.; Da Silva, M.M.; Kowaltowski, A.J. Mitochondrial ATP-sensitive K+ channel opening decreases reactive oxygen species generation. *FEBS Lett.* **2003**, *536*, 51–55. [CrossRef]

42. Moghtadaei, M.; Habibey, R.; Ajami, M.; Soleimani, M.; Ebrahimi, S.A.; Pazoki-Toroudi, H. Skeletal muscle post-conditioning by diazoxide, anti-oxidative and anti-apoptotic mechanisms. *Mol. Biol. Rep.* **2012**, *39*, 11093–11103. [CrossRef] [PubMed]

43. Sánchez-Duarte, E.; Cortés-Rojo, C.; Sánchez-Briones, L.A.; Campos-García, J.; Saavedra-Molina, A.; Delgado-Enciso, I.; A López-Lemus, U.; Montoya-Pérez, R. Nicorandil Affects Mitochondrial Respiratory Chain Function by Increasing Complex III Activity and ROS Production in Skeletal Muscle Mitochondria. *J. Membr. Biol.* **2020**, *253*, 309–318. [CrossRef] [PubMed]

44. McGarrah, R.W.; Slentz, C.A.; Kraus, W.E. The Effect of Vigorous- Versus Moderate-Intensity Aerobic Exercise on Insulin Action. *Curr. Cardiol. Rep.* **2016**, *18*, 117. [CrossRef] [PubMed]

45. Beckendorf, L.; Linke, W.A. Emerging importance of oxidative stress in regulating striated muscle elasticity. *J. Muscle Res. Cell Motil.* **2015**, *36*, 25–36. [CrossRef] [PubMed]

46. Farahini, H.; Habibey, R.; Ajami, M.; Davoodi, S.H.; Azad, N.; Soleimani, M.; Tavakkoli-Hosseini, M.; Pazoki-Toroudi, H. Late anti-apoptotic effect of K(ATP) channel opening in skeletal muscle. *Clin. Exp. Pharmacol. Physiol.* **2012**, *39*, 909–916. [CrossRef]

47. Marotte, C.; Zeni, S.N. Reactive oxygen species on bone cells activity. *Acta Bioquím. Clín. Lat.* **2013**, *47*, 661–674.

Publisher's Note: MDPI stays neutral with regard to jurisdictional claims in published maps and institutional affiliations.

 *antioxidants* 

Article

ApoE Genotype-Dependent Response to Antioxidant and Exercise Interventions on Brain Function

Kiran Chaudhari [1], Jessica M. Wong [1], Philip H. Vann [1], Tori Como [2], Sid E. O'Bryant [2] and Nathalie Sumien [1,*]

[1] Department of Pharmacology and Neuroscience, University of North Texas Health Science Center, Fort Worth, TX 76107, USA; kchaudha@live.unthsc.edu (K.C.); Jessica.Wong@unthsc.edu (J.M.W.); Philip.Vann@unthsc.edu (P.H.V.)

[2] Institute for Translational Research, Department of Pharmacology and Neuroscience, University of North Texas Health Science Center, Fort Worth, TX 76107, USA; tori.como@unthsc.edu (T.C.); sid.Obryant@unthsc.edu (S.E.O.)

* Correspondence: Nathalie.Sumien@unthsc.edu; Tel.: +1-817-735-2389

Received: 29 May 2020; Accepted: 23 June 2020; Published: 25 June 2020

Abstract: This study determined whether antioxidant supplementation is a viable complement to exercise regimens in improving cognitive and motor performance in a mouse model of Alzheimer's disease risk. Starting at 12 months of age, separate groups of male and female mice expressing human Apolipoprotein E3 (GFAP-ApoE3) or E4 (GFAP-ApoE4) were fed either a control diet or a diet supplemented with vitamins E and C. The mice were further separated into a sedentary group or a group that followed a daily exercise regimen. After 8 weeks on the treatments, the mice were administered a battery of functional tests including tests to measure reflex and motor, cognitive, and affective function while remaining on their treatment. Subsequently, plasma inflammatory markers and catalase activity in brain regions were measured. Overall, the GFAP-ApoE4 mice exhibited poorer motor function and spatial learning and memory. The treatments improved balance, learning, and cognitive flexibility in the GFAP-ApoE3 mice and overall the GFAP-ApoE4 mice were not responsive. The addition of antioxidants to supplement a training regimen only provided further benefits to the active avoidance task, and there was no antagonistic interaction between the two interventions. These outcomes are indicative that there is a window of opportunity for treatment and that genotype plays an important role in response to interventions.

Keywords: ApoE; exercise; antioxidants; oxidative stress; cognition; motor; vitamin E; vitamin C; aging; Alzheimer's disease

---

1. Introduction

Apolipoprotein E4 (ApoE4) is the most prevalent genetic risk factor for late-onset Alzheimer's disease (AD) [1]. In mice expressing human ApoE4, the cognitive deficits can be measured in terms of impairments in spatial learning and memory [2] and working memory [3]. Apart from cognitive declines, other behavioral effects are associated with AD and preclinical AD, such as increased anxiety [4], motor function disability, and the inability to learn new motor skills [5,6]. Interestingly, the presence of the ε4 allele was also associated with a two-fold increase in the rate of global motor function decline when compared with non-carriers with comparable age, sex, and education [7].

Oxidative stress has been associated with cognitive and motor declines and has been suggested as a major contributor to AD pathology. Oxidative stress has also been involved in vascular cognitive impairment, a risk for dementia [8]. The brains of AD patients are more vulnerable to oxidative stress, as evidenced in animal models and humans [9]. Furthermore, ApoE4 is associated with the aggravation of AD pathophysiology via increased oxidative stress [10]. Therefore, AD symptomology, especially in the presence of the ApoE4 allele, should be responsive to therapies reducing oxidative stress. Vitamin E is an example of such a therapeutic known to reduce oxidative stress and is able to improve cognitive function in AD patients either alone or in combination [11].

Another well-marketed healthy lifestyle modification is exercise. Physical activity has been associated with a reduced risk of AD [12], delayed AD onset [13], and improved AD symptoms [14], as well as improved vascular cognitive impairments [8]. Furthermore, exercise was even more beneficial in ApoE4 carriers than non-carriers [15,16]. Exercise lowered oxidative stress [17] and improved cognition [18], but also reduced anxiety in the elderly [19] and in rats [20]. Furthermore, it improved motor function in a cognitively impaired geriatric population [21,22], and motor training dramatically reduced injurious falls among AD patients [23,24]. Oxidative stress has also been implicated in the development of neuromuscular disorders (NMDs), and exercise in an intensity- and duration-dependent manner can alter NMDs, as reviewed by Siciliano et al. [25].

Both lines of therapy improved behavioral outcomes associated with AD, and seemed to at least partially involve oxidative stress as part of their mechanism of action. Therefore, it can be hypothesized that combining antioxidants with exercise training will lead to an additive beneficial effect, reducing impairments. While some reports have determined that there can be such a positive interaction [26–28], other studies have found a negative relationship, in which the presence of antioxidants negated the beneficial effects of exercise [29]. How such combinations affect behavioral measures have not been fully explored, and the influence of genotype and sex remain to be evaluated.

The goals of the current study were to determine (1) cognitive, motor, and anxiety phenotypes of middle-aged GFAP-ApoE3 and E4 male and female mice; (2) whether antioxidant intake or exercise training leads to functional improvements; (3) whether the combination of antioxidant and exercise yields an additive beneficial effect; (4) the involvement of oxidative stress and inflammation in behavioral outcomes. Our hypothesis was that moderate exercise and antioxidants would lead to additional beneficial effects when compared to each intervention alone, which would be exacerbated in the ApoE4 genotype. The outcomes are important in deciding the need for antioxidant supplementation in exercising individuals, and as a guiding parameter in genotype-based experiments.

2. Materials and Methods

2.1. Animals

All animal use protocols were approved by the Institutional Animal Care and Use Committee at UNT HSC (Protocol #: 2009/10-50-A04, approved 09/1/2010). Groups of male and female GFAP-ApoE3 (B6.Cg-Tg(GFAP-*APOE3*)37Hol *APOE*tm1Unc/J) and GFAP-ApoE4 (B6.Cg-Tg(GFAP-*APOE4*)1Hol *APOE*tm1Unc/J) mice were obtained from The Jackson Laboratory (Bar Harbor, ME, USA) (catalog numbers 004633 and 004631; we started with a total of 187 mice, and recorded 14 and 10 deaths for the ApoE3 and ApoE4 respectively) at 2 months. These mice express the human apolipoprotein E3 or E4 under the control of the glial fibrillary acidic protein (GFAP) promoter and do not express endogenous ApoE. The mice were group housed in standard housing conditions and received ad libitum access to food and water at an ambient temperature, under a 12-h light/dark cycle starting at 06:00.

2.2. Treatment

When the mice were 12 months old, they were randomly allotted to: (1) sedentary + control diet (Sed-Con), (2) sedentary + control diet supplemented with vitamins E and C (Sed-EC), (3) exercise + control diet (Ex-Con), (4) exercise + control diet supplemented with vitamins E and C (Ex-EC). The diets were obtained from TestDiet (St Louis, MO, USA). The control diet (LabDiet® R&M 5LG6 4F, #5S84) was modified by adding ascorbic acid (1.65 mg/g diet) and α-tocopheryl acetate (1.12 IU/g diet) (#5SH0). Using treadmills, the mice were progressively introduced to a moderate exercise regimen (AccuPacer Treadmill, Omnitech Electronics, Inc., Columbus, OH, USA). The training was increased in time and speed to reach a maximal exercise of 1 h over 12 days (6, 8, 10, and 12 m/min for 5 min each, and then at 14 m/min for 40 min). Compliance was achieved by a transient 0.29 mA electric foot shock to the feet. The number of shocks were tallied, and the paired control mice received the same number of shocks/training day.

2.3. Food Intake and Body Weights

The mice were weighed weekly and food intake (average of five consecutive days) was recorded the week before the start of behavioral testing.

2.4. Functional Testing

For 16 weeks, the mice were on their respective treatments, including 8 weeks prior to the behavioral assessments. The behavioral tasks were described in details previously [30,31].

2.4.1. Elevated Plus Maze

The position of a mouse in the maze (two open arms and two closed arms, 3ft above the floor) was determined by a tracking system (Any-maze, Stoelting Co., Wood Dale, IL, USA). Under dim light (60 W) and for 5 min, each mouse was left to explore. Time in open arms was used to measure anxiety levels.

2.4.2. Spontaneous Activity

Each mouse was left to explore an acrylic test chamber (40.5 cm × 40.5 cm × 30.5 cm) under dim light and with background noise (80 dB) for 16 min. Their activity was recorded using a Digiscan apparatus (Omnitech Electronics Inc., Columbus, OH, USA, model RXYZCM-16). Their horizontal, vertical, and spatial movements were detected by the photocells and processed by a software program.

2.4.3. Coordinated Running

A motor-driven cylinder rotating at increasing speed was used to measure motor learning and running performance (Rotorod, Omnitech Electronics Inc., Columbus, OH, USA, Model # AIO411RRT525M). The mice were given two training sessions per day (four trials/session with a 10 min inter-trial intervals (ITI)), which continued until improvements failed over three consecutive sessions. The average of latency to fall for the four trials in each session was used for motor learning and for the final session when stable performance was achieved.

2.4.4. Reflexive Musculoskeletal Responses

Walk initiation: record the latency to move one body length immediately after being placed on a smooth surface. Alley turn: record the latency to make a full turn in a dead-end alley. Negative geotaxis: record the latency to turn 180 ° in either direction when placed facing downward on a 45 ° tilted grid. Wire suspension: record the latency to grasp wire with hindpaws after being suspended from wire by front paws and the latency to fall (two trials/session). The mice received four sessions (one/day) and latencies were averaged across the four sessions. A maximum of 60 s for each test was used.

2.4.5. Bridge Walking

Each mouse was placed on one of four bridges (large square, small square, large round, small round) and latency to fall (maximum of 60 s) was recorded. Each bridge was presented three times and latency to fall was averaged for each bridge and across bridges.

2.4.6. Morris Water Maze (MWM)

Spatial learning and memory were measured using a modified Morris water maze test. Prior to any testing, the mice received pre-training using a covered straight alley to teach the mice to swim and climb onto a platform. Following this, the mice were given a maximum of 90 s to find a hidden platform from different starting positions (one session/day, five trials/session with a 2 min ITI). Their performance was tracked via Any-maze software (Stoelting Co., Wood Dale, IL, USA). Path length and swim speed were used as measures of performance.

2.4.7. Discriminated Avoidance

Using a T-maze apparatus set on a grid floor set to deliver 0.69 mA scrambled shock, we tested the mice for learning and cognitive flexibility using a discriminated avoidance task. During the information trial (first one), a shock is administered when the mouse enters their preference arm (first one) and allowed to escape the shock by running to the opposite arm (correct arm). Thereafter, the initiation of shock was 5 s upon start door opening or upon entry into an incorrect arm until the mouse entered the correct arm or a maximum of 60 s. After 10 s in the correct arm, the mouse was placed in a holding cage for 1 min (ITI). Training continued until a criterion of correct avoidance was reached (choosing the correct arm in under 5 s in four out of five trials, with the last two being correct avoidances). The reversal sessions followed the same training and switching of the correct arm. Learning ability was considered inversely proportional to the number of trials required to reach the avoidance criterion.

2.5. Biochemical Measurements

After behavioral testing, blood was collected via tail bleed prior to euthanization (by cervical dislocation). Brains were harvested and dissected into brain regions that were saved at $-80\ °C$ (cerebral cortex, striatum, cerebellum, and hippocampus).

2.5.1. Catalase Activity

The activity of the antioxidative enzyme, catalase, was evaluated using the Catalase Assay Kit from Cayman Chemical (Ann Arbor, MI, USA, catalog number 707002). Methanol was reacted with catalase in the presence of for the optimal concentration of H_2O_2. The production of the resulting formaldehyde was measured colorimetrically with Purpald (4-amino-3-hydrazino- 5-mercapto-1,2,4-triazole) at 540 nm.

2.5.2. Inflammatory Markers

Plasma levels for IL-10, IL-6, and TNFα were measured using an MSD Inflammatory Panel Kit from Meso Scale Diagnostics, LLC, Rockville, MD, USA) (10-plex Proinflammatory Panel 1 (mouse) kit (cat # C4048-1) and analyzed by MSD DISCOVERY WORKBENCH® analysis software (Meso Scale Diagnostics, Rockville, MD, USA). While the panel measures more than these four markers, the others were out of range based on the manual and literature searches. Therefore, we did not include them in our analyses.

2.6. Statistical Analysis

The functional performance of the mice on the behavioral tests, as well as the biochemical measurements, were assessed using three-way analyses of variance (ANOVA) with sex, genotype, and treatment as between-group factors (or genotype and treatment for catalase measurements due to low n). Planned individual comparisons between different sexes, genotypes, and treatments were performed using single degree-of-freedom F-tests involving the error term from the overall ANOVA. Some behavioral performances and catalase activity were also considered in four-way analyses with session or brain region as the repeated measure. The software used for the analyses was Systat 13 (San Jose, CA, USA) and p was set< 0.05.

3. Results

3.1. Body Weight and Food Intake

Percent changes in body weight from the week before behavioral testing started and at the end of the study are presented in Figure 1. Overall, male mice weighed more than female mice, and within the female groups, GFAP-ApoE4 mice weighed less than the GFAP-ApoE3 mice (not shown). These observations were supported by main effects of strain, sex, and an interaction between sex and strain (all $p < 0.02$). By week 7, most mice lost 3–7% of their body weight, and the only significant differences were between E3 and E4 Sed-Con females, and between Sed-Con and Sed-Aox in the GFAP-ApoE4 females (all $p < 0.05$). By week 12, the loss was more pronounced (up to 13%), and the only significant differences were Ex-Aox GFAP-ApoE4 females losing less weight than Sed-Con ($p < 0.05$), while in males, the Ex-Aox groups lost the most weight ($p < 0.05$ only for GFAP-ApoE3). Food intake was not significantly affected by sex, strain or treatment (all $p > 0.2$) (not shown).

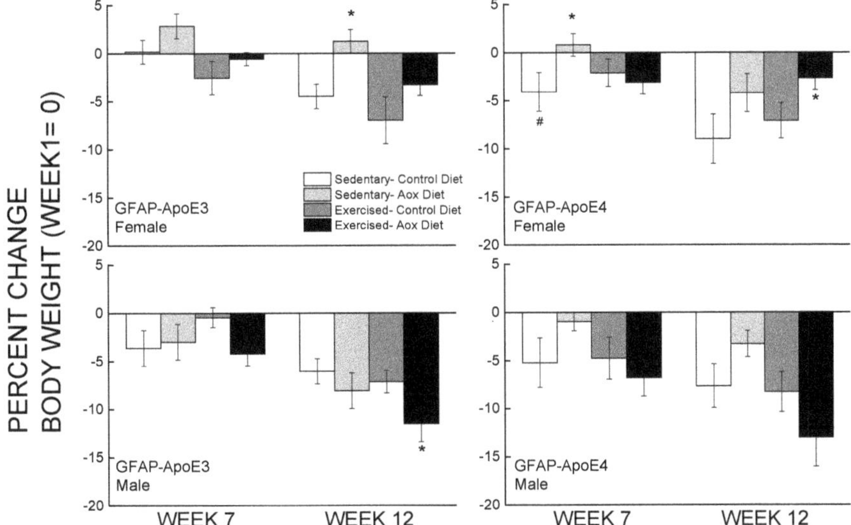

Figure 1. Minimal effects of exercise and/or antioxidant regimen over time on body weights of middle-aged GFAP-APOE3 and GFAP-APOE4 male and female mice (mice expressing the human apolipoprotein (Apo) E3 or E4 under glial fibrillary acidic protein (GFAP) promoter control). Each value represents mean ± SEM, $n = 5$–16 for body weights, and $n = 3$–7 for food intake. * $p < 0.05$ vs. sex- and strain-matched Sedentary-Control (Sed-Con) groups; # $p < 0.05$ comparing sex-matched Sed-Con GFAP-ApoE3 and GFAP-ApoE4.

3.2. Behavioral Measurements

3.2.1. Elevated Plus Maze

The performance of the mice in this test measuring affective function was analyzed and is presented in Figure S1. The GFAP-ApoE3 and GFAP-ApoE4 mice spent the same amount of time in the open arms, however, males spent more time in the open arm than females, supported by a main effect of sex ($p < 0.001$) and no effect of strain ($p = 0.366$). Treatments did not affect performance (all $p > 0.185$), with the exception of Sed-Aox group spending more time than the Sed-Con group in the open arms ($p = 0.046$). Overall, the GFAP-ApoE3 mice were more active than the GFAP-ApoE4 mice, and some treatments affected the activity of females (all $p < 0.003$). These observations were supported by the significant main effects of strain and treatment and a sex × strain × treatment interaction. (all $p < 0.01$).

3.2.2. Spontaneous Activity

Horizontal distance and rearing activity were analyzed (Figure S2). GFAP-ApoE4 mice travelled 40% shorter distances than the GFAP-ApoE3 mice, and GFAP-ApoE3 females traveled 18% more than their male counterparts. Treatment had no effect on distance traveled. An ANOVA yielded a main effect of strain ($p < 0.001$) and no effect for treatment, sex or any interactions between the factors (all $p > 0.300$). GFAP-ApoE4 mice also spent 20% less time rearing than the GFAP-ApoE3, supported by a main effect of strain ($p = 0.001$). Most treatments had no effect on the rearing activity of the mice with the exception of Sed-Aox, which increased it ($p = 0.046$).

3.2.3. Coordinated Running Performance

The data were averaged as learning (sessions 1–4) and plateau (sessions 5–7) performance (Figure 2). During the learning phase, GFAP-ApoE4 mice fell faster from the rod than the GFAP-ApoE3 mice, especially males, supported by the main effects of sex, strain and a sex x strain interaction (all $p < 0.003$). A main effect of treatment ($p = 0.014$) was due to Ex-Con and Ex-Aox GFAP-ApoE3 mice exhibiting better performance in males (all $p < 0.04$). The Ex-Con GFAP-ApoE4 group also seemed to have a better performance ($p = 0.051$) which was not seen in the Ex-Aox group. During the plateau phase, there was only an effect of strain, in which GFAP-ApoE4 mice had a poorer performance than the GFAP-ApoE3 mice ($p < 0.001$).

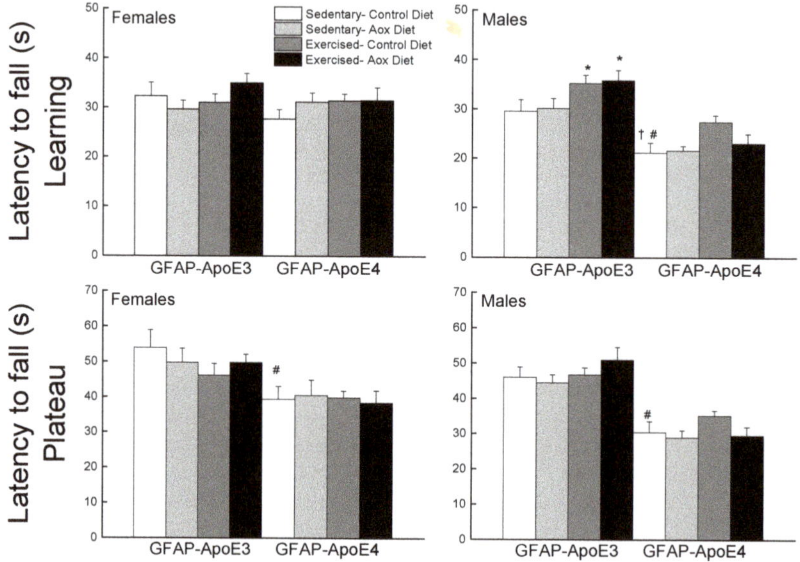

Figure 2. Exercise improved learning performance in male GFAP-ApoE3 and GFAP-ApoE4 mice, but not plateau performance. Each value represents mean ± SEM, $n = 8$–16. * $p < 0.05$ vs. sex- and strain-matched Sed-Con groups; # $p < 0.05$ comparing sex-matched Sed-Con GFAP-ApoE3 and GFAP-ApoE4 mice; † $p < 0.05$ comparing strain-matched Sed-Con males and females.

3.2.4. Reflexive Musculoskeletal Responses

Performance was averaged over four sessions and is presented in Figure S3. GFAP-ApoE4 mice took longer latencies than the GFAP-ApoE3 mice to initiate walking, supported by a main effect of strain

($p = 0.019$). There was no effect of treatment, sex, or interactions between any of the factors (all $p > 0.086$). Alley-turning was affected differentially by the treatments depending on sex and genotype (sex × strain × treatment interaction, $p = 0.004$). GFAP-ApoE4 females took longer latencies than the GFAP-ApoE3 females, while there was no difference in males. In GFAP-ApoE4 females, Sed-Aox took longer to turn, while in GFAP-ApoE3, the Ex-Aox males had the shortest latency. For negative geotaxis, males took longer latencies to turn, as well as GFAP-ApoE4 mice. The Sed-Aox mice had shorter latencies in the GFAP-ApoE3 groups (strain × treatment, $p = 0.058$). Latency to tread was affected by strain ($p = 0.01$), due mainly to treatments affecting each genotype differently (higher in GFAP-APoE4 males vs. lower latencies in GFAP-ApoE3 mice). There was no effect of treatment or interactions between any of the factors (all $p > 0.10$).

3.2.5. Bridge Walking

The latency to fall was analyzed for each bridge and is presented in Figure 3. On the easiest bridge (session 1), there was no effect of strain, sex, or treatment on the latency to fall (all $p > 0.13$). In session 2, females performed better than males leading to a main effect of strain. While there was no main effect of treatment, exercised groups had higher latencies in males regardless of strain but it only reached significance for the Ex-Aox GFAP-ApoE3 mice ($p = 0.029$). In session 3, a main effect of treatment was obtained ($p = 0.03$) due to all treatments in male GFAP-ApoE3 mice. In session 4, treatments, especially exercise, were associated with increased latencies in the GFAP-ApoE3 but not GFAP-ApoE4 groups. GFAP-ApoE4 females performed better than their male counterparts, driving the main effects of sex and strain (all $p < 0.03$).

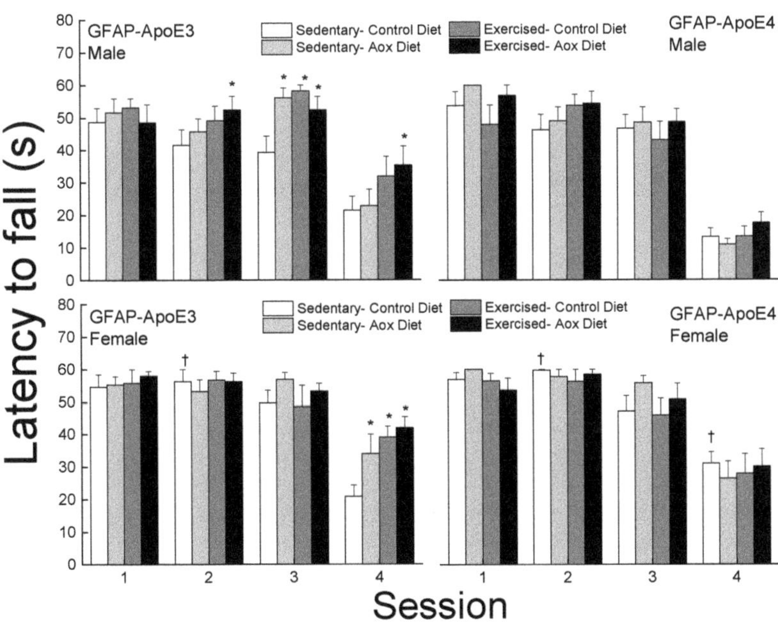

Figure 3. Exercise and antioxidants improved balance in male and female GFAP-ApoE3 mice but not in GFAP-ApoE4 mice. Each value represents mean ± SEM, $n = 8$–16. * $p < 0.05$ vs. sex- and strain-matched Sed-Con groups; † $p < 0.05$ comparing strain-matched Sed-Con males and females.

3.2.6. Morris Water Maze

The performance of the mice was separated into initial, learning phase, and maximum performances (Figure 4). During the initial session, there was no effect of strain or treatment for females, however, in the males, all treatment improved performance but it only reached significance for Ex-Con ($p = 0.044$). During the learning phase, female mice had higher path lengths than the males (main effect of sex; $p < 0.001$). Treated GFAP-ApoE4 females had lower latencies than the sex- and strain-matched Sed-Con group ($p = 0.064$), however, it only reached significance with the Ex-Aox mice. Maximum performance was impaired in the GFAP-ApoE4 males, and was reversed by all treatments. However, an ANOVA did not reach significance for strain ($p = 0.098$) or strain x treatment ($p = 0.096$). Females swam slower than males, GFAP-ApoE4 mice seemed slightly faster than GFAP-ApoE3 mice, and none of the treatments affected the swimming speed (not shown). ANOVA yielded a main effect of sex (all $p < 0.007$), but the effect of strain did not reach significance ($p = 0.057$).

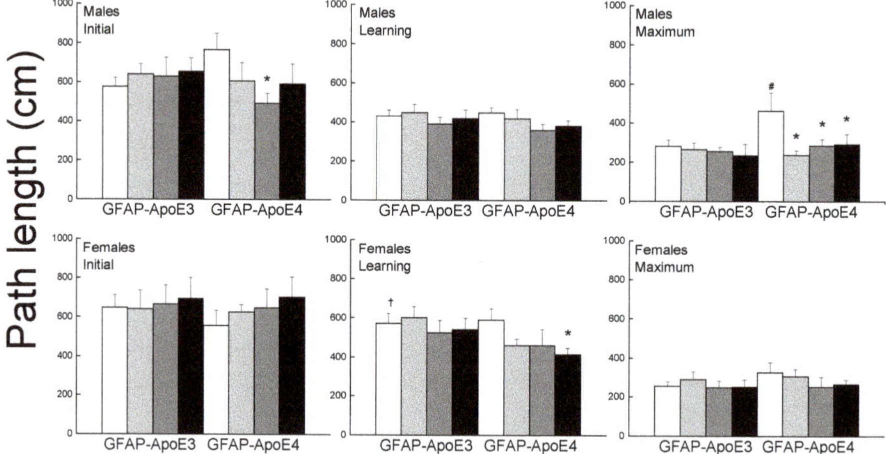

Figure 4. Exercise and antioxidants were associated with mild improvement in swim maze performance in male GFAP-ApoE4 mice but not in GFAP-ApoE3 mice. Each value represents mean ± SEM, $n = 5$–16. * $p < 0.05$ vs. sex- and strain-matched Sed-Con groups; # $p < 0.05$ comparing sex-matched Sed-Con GFAP-ApoE3 and GFAP-ApoE4 mice; † $p < 0.05$ comparing strain-matched Sed-Con males and females.

3.2.7. Discriminated Avoidance Test

Performance during acquisition and reversal is presented in Figure 5. During acquisition, most treatments improved performance (main effect of treatment; $p = 0.005$). Sed-Aox ($p = 0.035$) and Ex-Con ($p = 0.052$) groups took fewer trials to reach criterion in the GFAP-ApoE3 males and there was no significant effect in the GFAP-ApoE4 mice. In females, Ex-Con ($p = 0.058$) and Ex-Aox ($p = 0.019$) performed better than the Sed-Con group in the GFAP-ApoE3 group, while only Ex-Aox ($p = 0.018$) took fewer trials than the controls in the GFAP-ApoE4 group.

During reversal, treatments improved performance (main effect of treatment; $p = 0.007$), mainly in the GFAP-ApoE3 mice, but this observation was not supported by an interaction between strain and treatment ($p = 0.23$). There was no significant effect of any of the treatments in the GFAP-ApoE4 mice. In the GFAP-ApoE3 group, all treated mice took fewer trials compared to Sed-Con in females and Ex-Aox was the only significantly different group in males ($p = 0.029$).

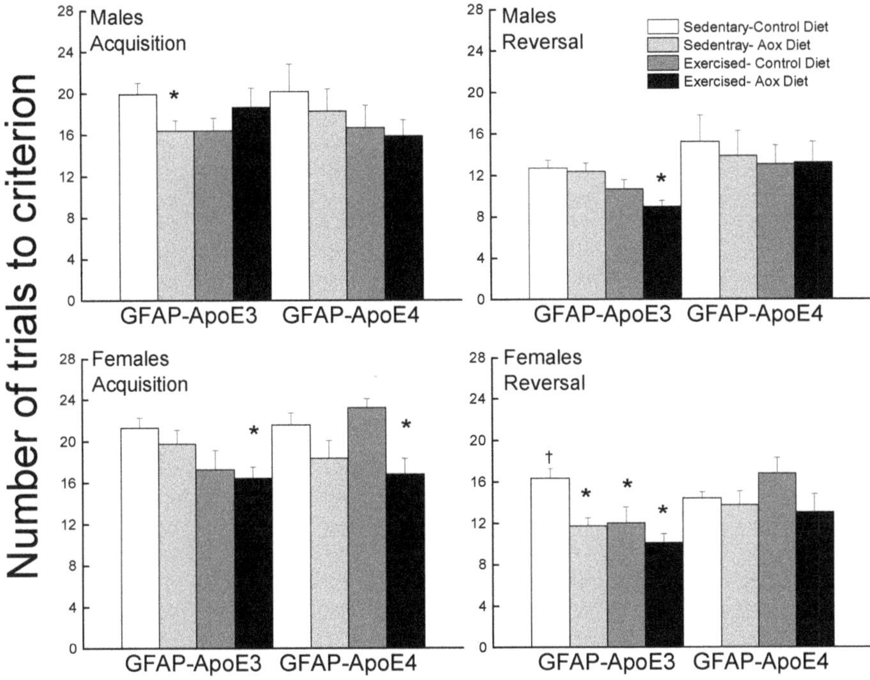

Figure 5. Exercise and antioxidants improved learning and cognitive flexibility in male and female GFAP-ApoE3 mice but not in GFAP-ApoE4 mice. Each value represents mean ± SEM, n = 8–16. * $p < 0.05$ vs. sex- and strain-matched Sed-Con groups; † $p < 0.05$ comparing strain-matched Sed-Con males and females.

3.3. Biochemical Measurements

3.3.1. Catalase

Catalase activity was measured in the cerebral cortex, hippocampus, cerebellum, and midbrain and is presented in Figure S4 (due to low n, sexes were combined). Overall, the activity was ranked as follows: hippocampus > cerebellum, midbrain > cortex, supported by a main effect of region ($p < 0.001$). There was an overall effect of treatment ($p = 0.003$) which was not region-dependent (all $p > 0.96$), but no effect of genotype (all $p > 0.35$). The effect of treatment was due mostly to effects of Ex-Aox in all regions (not significant for cerebellum).

3.3.2. Inflammatory Markers

Levels of IL6, TNFα, and IL10 were measured in the plasma and are reported in Figure 6. For IL6, the effects of treatments were seen in the GFAP-ApoE3 females, with a sex by treatment interaction approaching significance ($p = 0.055$). The female groups that exercised had lower levels of IL6 than the control groups. There was no effect of sex, strain, or treatment on TNFα levels (all $p > 0.09$), though the Ex-Aox GFAP-ApoE4 females had higher levels than the controls ($p < 0.05$).

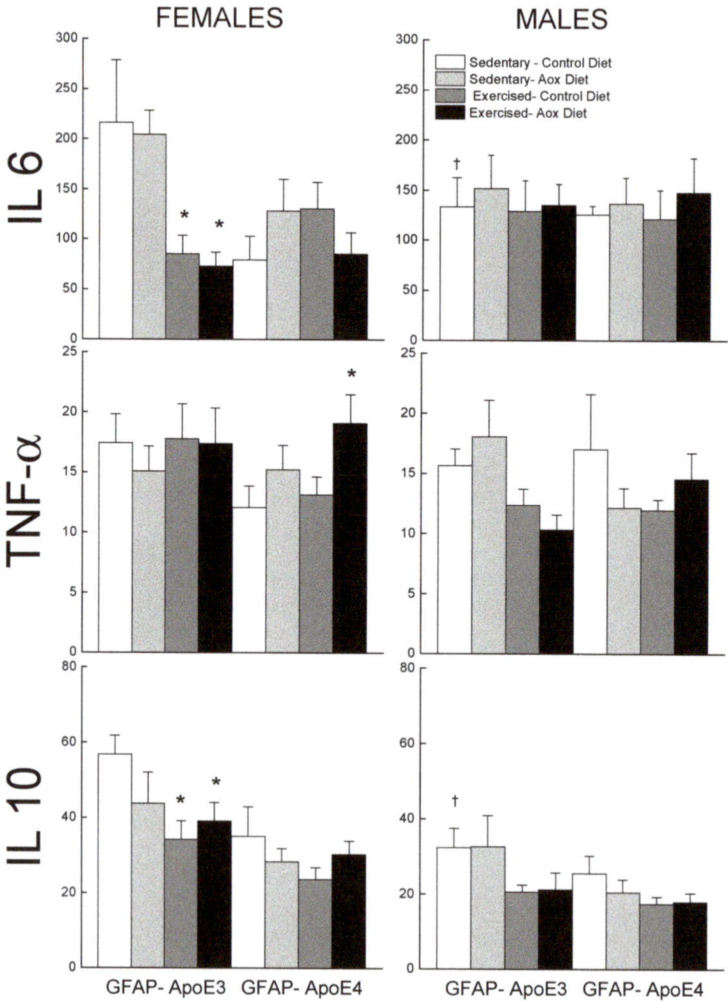

Figure 6. Effect of exercise and antioxidants on markers of inflammation in male and female GFAP-ApoE3 and GFAP-ApoE4 mice. Each value represents mean ± SEM, $n = 5$–9. * $p < 0.05$ vs. sex- and strain-matched Sed-Con groups; † $p < 0.05$ comparing strain-matched Sed-Con males and females.

4. Discussion

The main findings of this study were: (1) there were strain differences for most motor functions, but no major differences for strength, balance, or cognition; (2) there were strain and test-dependent differences in response to the treatments: GFAP-ApoE3 mice were responsive on bridge and active avoidance, while GFAP-ApoE3 mice were on spatial learning and memory; (4) the most effective treatment was exercise, and no major additive or antagonistic effects were observed with antioxidant intake.

The current study provided a comprehensive phenotype of this mouse model, and its response to non-conventional therapeutic interventions. Anxiety, the most common non-cognitive symptom of AD [32] and associated with impaired daily activities [33], is often managed with benzodiazepines, which can lead to further cognitive and motor function declines [34,35].

Therefore, identifying non-pharmacotherapeutic agents to reduce anxiety could be an overall positive approach towards managing anxiety symptoms in AD without the added risk of furthering the functional declines. Previously, we determined that at 4 months of age, GFAP-ApoE4 mice were less anxious than GFAP-ApoE3 mice [30]; however, in the current study of 14-month-old mice, that difference subsided. This may indicate a pleiotropic effect of the ApoE genotype on anxiety levels, as has been described with cognitive function [30,36]. These data contrast with a previous study reporting higher anxiety among adult GFAP-ApoE4 mice [37]. Neither antioxidant nor exercise treatment affected the anxiety levels of the mice. In young mice, antioxidants increased the anxiety of the GFAP-ApoE4 mice, suggesting an age-dependent response to interventions [30]. Overall, the GFAP-ApoE3 mice were more active than the GFAP-ApoE4 mice, which is consistent with a previous report [38], and the treatments had no to minimal effects. Other complementary and alternative medicines have also been studied as non-traditional therapies alone or in combination with exercise. More specifically, in a pilot study, depression in AD patients was reduced when they combined Shiatsu with physical activity for 10 months [39].

During its early phase, AD is often associated with motor function impairments. Pathological changes in the motor cortex, striatum, cerebellum, or substantia nigra might be responsible for motor decline in AD [40,41]. The presence of ApoE4 doubles the rate of motor decline associated with aging [7]. Motor declines were observed in our mouse model with GFAP-ApoE4 mice exhibiting decreased activity, reflexes, and coordination. Similar effects on coordinated running were reported, in which the *APOE*4 mice performed poorly [42]. Repetitive transcranial magnetic stimulation suggested that ApoE4 is a critical determinant of the response to conditioning insults, especially for motor and cognitive brain networks [43]. Exercise and/or antioxidants improved the motor learning of the GFAP-ApoE3 males but not of the females or GFAP-ApoE4 mice. At younger ages, exercise training was associated with a reversal of the ApoE4-associated deficits in coordinated running [31]. The lack of response to treatment at an older age again suggests an age-dependent response to treatment. Interestingly, treatments improved the balance of the GFAP-ApoE3 mice, but not of the GFAP-ApoE4 mice. This strain-dependent effect was previously observed in younger mice [31], though the effects were not as large, most likely due to better performance of the young mice. Furthermore, exercise was more successful at improving motor function than antioxidants alone, indicative of a more promising therapeutic. Combining exercise and antioxidants did not lead to additive or antagonistic effects, as seen at younger ages on motor function [31].

Different mouse models expressing human ApoE4 exhibited poor spatial learning and memory [36,38,42]. Our data also suggest a significant decline in spatial learning and memory in GFAP-ApoE4 mice compared to GFAP-ApoE3 mice, a difference that was not observed in younger mice [30]. Interestingly, the majority of that effect comes from the males, as females did not show any differences. This may reflect sex differences in performance or that perhaps a water maze is not useful to detect cognitive differences in females. The treatments did not improve the performance of the GFAP-ApoE3 mice but did improve that of the GFAP-ApoE4 mice. Exercise improved more aspects of water maze performance than antioxidants alone, again supporting the fact that exercise is a better therapy. Improvements in cognition with exercise associated with the ApoE4 genotype have been shown previously [44]. Recent studies have implicated a potential cross-talk between the brain and muscles, with exercise-induced mediators being released, leading to the neuroprotection and brain function improvements associated with exercise. FNDC5/irisin is a myokine released upon exercising that has been associated with improvements in synaptic activity and memory in the APP/PS1 AD mouse model [45]. While our exercise paradigm differed from that study, it would be of interest to determine in future studies using our training paradigm if the FNDC5/irisin pathway is also activated in our regimen.

AD symptoms start with the loss of non-spatial short-term/working memory [46]. While we found strain differences in the spatial task, both GFAP-ApoE3 and ApoE4 mice performed similarly in the non-spatial task used. This is in contrast to previous studies that have identified genotype differences in working memory, which deteriorated in an ε4 allele dose-dependent manner in humans [47].

The GFAP-ApoE3 mice were more responsive to the treatments than the GFAP-ApoE4 mice, and additive effects were observed in the reversal phase for males. Other studies have also determined that exercise can improve short-term/working memory task performance [48,49]. These treatment effects were also observed previously with young GFAP-ApoE3 mice [30], however, while the treatments improved the performance of the young GFAP-ApoE4 mice [30], they did not have any effect on the older mice in the current study. These outcomes on cognitive function suggest that the type of memory, along with the age and genotype, can affect the outcome of interventions.

Contrary to expectation, catalase activity was reduced in brain regions in mice that were exercised, especially the ones combined with antioxidants. Reports on the effects of exercise on catalase activity in the brain are conflicting: from no effect [50] to up-regulation as a result of the oxidative bursts associated with acute exercise [51]. Catalase may be activated upon acute but not chronic exercise. Furthermore, supplementation with antioxidants may have led to a sparing of catalase or a feedback down-regulation of catalase. Plasma-based inflammatory biomarkers are often suggested to be future biomarkers for AD progression and treatment response. The effects of exercise on inflammation are ambiguous, and depend on the focus of inflammation (peripheral (plasma) vs. central (brain)). Studies focused on long-term exercise observed higher IL6 in the brain [52] compared to plasma, while no such change was noticed in TNFα. Some studies demonstrated lower IL6 in the brain after prolonged exercise without affecting plasma IL6 levels [53]. IL6 was positively related to IL10 in the exercise training context [54]. We also observed similar a relation of lowered IL6 in the presence of chronic exercise and associated lowered IL10 levels in GFAP-ApoE3 female mice. On the contrary, treadmill training previously reported higher IL10 and lower TNFα in plasma [55]. Differences between the sexes may be due to hormone influences, as females see a decrease in circulating hormones, which can have repercussion on the inflammatory response [56], and/or may be due to differences in oxidative stress levels and homeostasis [57].

The presence of ApoE4 is associated with an increased risk of developing late-onset AD, however, clinically, it does not seem to always be true with AD patients not carrying the ε4 and vice versa, suggesting a more complicated picture and gene involvement in AD etiology [58]. Furthermore, the presence of one ε4 allele may be insufficient to increase the risk in individuals 65 and younger [59]. Lastly, ApoE4 and other mutations of the known AD genes, such as APP, PSEN1, and PSEN2, only account for 5% of early-onset AD cases [58]. Therefore, the outcomes of this study would need to be further studied in other models of AD, especially relating to early-onset dementia in order to generalize them to all AD patients.

While it remains under debate that these non-pharmacological therapies are effective in reducing neurodegeneration, it is possible that they affect the vascular contribution to these diseases. By managing and controlling the vascular risk factor via a healthy lifestyle, such as exercising and improving one's diet, it is likely to reduce vascular cognitive impairments and reduce the advance to neurodegenerative diseases and dementias [60]. These non-pharmacological therapies may be working via the control and management of vascular risk, leading to improved cognition and diminished neurodegenerative disorders, such as AD.

Our study has several limitations: (1) the model used may not be generalizable to all AD patients, as it focused on the ApoE genotype and on a glial model of ApoE expression; (2) the limited oxidative stress measurements, as such glutathione levels or protein damage measurements would have strengthened the role of oxidative stress in the study [61]; (3) the involvement of other pathways, such as the irisin-dependent pathway involving muscle–brain cross-talks, were not studied and may be of importance to determine the mechanism of action of our exercise paradigm; (4) a dose–response component of our exercise paradigm would be of interest to determine the translational impact of our training regimen [62].

5. Conclusions

Exercise was the most consistent treatment effective at improving motor and cognitive function, and the addition of antioxidants did not lead to major additive or antagonistic effects. ApoE4 mice were less responsive to the treatments than the ApoE3 mice, suggesting a genotype-dependent response to interventions. Therefore, factors such as sex, age, genotype, and chosen tests need to be carefully incorporated into preclinical studies of interventions to improve brain function during aging or neurodegenerative diseases.

Supplementary Materials: The following are available online at http://www.mdpi.com/2076-3921/9/6/553/s1, Figure S1: Effect of exercise and antioxidant on performance on the elevated plus maze in male and female GFAP-ApoE3 and GFAP-ApoE4 mice, Figure S2: Effect of exercise and antioxidant on spontaneous activity in male and female GFAP-ApoE3 and GFAP-ApoE4 mice, Figure S3: Effect of exercise and antioxidant on reflexes in male and female GFAP-ApoE3 and GFAP-ApoE4 mice, Figure S4: Effect of exercise and antioxidant on catalase activity from different brain regions from GFAP-ApoE3 and GFAP-ApoE4 mice.

Author Contributions: Conceptualization, K.C. and N.S.; methodology, K.C., J.M.W., T.C., S.E.O., P.H.V.; formal analysis, K.C.; writing—original draft preparation, K.C.; writing—review and editing, N.S.; supervision, N.S.; project administration, N.S.; funding acquisition, S.E.O. and N.S. All authors have read and agreed to the published version of the manuscript.

Funding: This research was funded by the Alzheimer's Association, grant number NIRG-10-173988, by the Pine Family Foundation (gift) and by UNT HSC Bridge Funding, grant number RI6096.

Conflicts of Interest: The authors declare no conflict of interest.

References

1. Riedel, B.C.; Thompson, P.M.; Brinton, R.D. Age, APOE and sex: Triad of risk of Alzheimer's disease. *J. Steroid. Biochem. Mol. Biol.* **2016**, *160*, 134–147. [CrossRef]
2. Yin, J.X.; Turner, G.H.; Lin, H.J.; Coons, S.W.; Shi, J. Deficits in spatial learning and memory is associated with hippocampal volume loss in aged apolipoprotein E4 mice. *J. Alzheimers Dis. JAD* **2011**, *27*, 89–98. [CrossRef] [PubMed]
3. Hartman, R.; Wozniak, D.; Nardi, A.; Olney, J.; Sartorius, L.; Holtzman, D. Behavioral Phenotyping of GFAP-ApoE3 and -ApoE4 Transgenic Mice: ApoE4 Mice Show Profound Working Memory Impairments in the Absence of Alzheimer's-like Neuropathology. *Exp. Neurol.* **2001**, *170*, 326–344. [CrossRef] [PubMed]
4. Johnson, L.A.; Olsen, R.H.; Merkens, L.S.; DeBarber, A.; Steiner, R.D.; Sullivan, P.M.; Maeda, N.; Raber, J. Apolipoprotein E-low density lipoprotein receptor interaction affects spatial memory retention and brain ApoE levels in an isoform-dependent manner. *Neurobiol. Dis.* **2014**, *64*, 150–162. [CrossRef] [PubMed]
5. Buchman, A.S.; Bennett, D.A. Loss of motor function in preclinical Alzheimer's disease. *Expert Rev. Neurother.* **2011**, *11*, 665–676. [CrossRef]
6. Hebert, L.E.; Bienias, J.L.; McCann, J.J.; Scherr, P.A.; Wilson, R.S.; Evans, D.A. Upper and lower extremity motor performance and functional impairment in Alzheimer's disease. *Am. J. Alzheimers Dis. Other Demen.* **2010**, *25*, 425–431. [CrossRef] [PubMed]
7. Buchman, A.S.; Boyle, P.A.; Wilson, R.S.; Beck, T.L.; Kelly, J.F.; Bennett, D.A. Apolipoprotein E e4 allele is associated with more rapid motor decline in older persons. *Alzheimer Dis. Assoc. Disord.* **2009**, *23*, 63–69. [CrossRef]
8. Luca, M.; Luca, A. Oxidative Stress-Related Endothelial Damage in Vascular Depression and Vascular Cognitive Impairment: Beneficial Effects of Aerobic Physical Exercise. *Oxidative Med. Cell. Longev.* **2019**, *2019*, 8067045. [CrossRef]
9. Pratico, D.; Sung, S. Lipid peroxidation and oxidative imbalance: Early functional events in Alzheimer's disease. *J. Alzheimers Dis. JAD* **2004**, *6*, 171–175. [CrossRef]
10. Zito, G.; Polimanti, R.; Panetta, V.; Ventriglia, M.; Salustri, C.; Siotto, M.C.; Moffa, F.; Altamura, C.; Vernieri, F.; Lupoi, D.; et al. Antioxidant status and APOE genotype as susceptibility factors for neurodegeneration in Alzheimer's disease and vascular dementia. *Rejuvenation Res.* **2013**, *16*, 51–56. [CrossRef]

11. Yasuno, F.; Tanimukai, S.; Sasaki, M.; Ikejima, C.; Yamashita, F.; Kodama, C.; Mizukami, K.; Asada, T. Combination of antioxidant supplements improved cognitive function in the elderly. *J. Alzheimers Dis. JAD* **2012**, *32*, 895–903. [CrossRef] [PubMed]

12. Roitto, H.M.; Kautiainen, H.; Ohman, H.; Savikko, N.; Strandberg, T.E.; Raivio, M.; Laakkonen, M.L.; Pitkala, K.H. Relationship of Neuropsychiatric Symptoms with Falls in Alzheimer's Disease—Does Exercise Modify the Risk? *J. Am. Geriatr. Soc.* **2018**, *66*, 2377–2381. [CrossRef]

13. Lin, T.W.; Tsai, S.F.; Kuo, Y.M. Physical Exercise Enhances Neuroplasticity and Delays Alzheimer's Disease. *Brain Plast.* **2018**, *4*, 95–110. [CrossRef]

14. Cui, M.Y.; Lin, Y.; Sheng, J.Y.; Zhang, X.; Cui, R.J. Exercise Intervention Associated with Cognitive Improvement in Alzheimer's Disease. *Neural. Plast.* **2018**, *2018*, 9234105. [CrossRef] [PubMed]

15. Head, D.; Bugg, J.M.; Goate, A.M.; Fagan, A.M.; Mintun, M.A.; Benzinger, T.; Holtzman, D.M.; Morris, J.C. Exercise Engagement as a Moderator of the Effects of APOE Genotype on Amyloid Deposition. *Arch. Neurol.* **2012**, *69*, 636–643. [CrossRef] [PubMed]

16. Brown, B.M.; Peiffer, J.J.; Martins, R.N. Multiple effects of physical activity on molecular and cognitive signs of brain aging: Can exercise slow neurodegeneration and delay Alzheimer's disease? *Mol. Psychiatry* **2013**, *18*, 864–874. [CrossRef]

17. Bouzid, M.A.; Hammouda, O.; Matran, R.; Robin, S.; Fabre, C. Low intensity aerobic exercise and oxidative stress markers in older adults. *J. Aging Phys. Act.* **2014**, *22*, 536–542. [CrossRef]

18. Marosi, K.; Bori, Z.; Hart, N.; Sarga, L.; Koltai, E.; Radak, Z.; Nyakas, C. Long-term exercise treatment reduces oxidative stress in the hippocampus of aging rats. *Neuroscience* **2012**, *226*, 21–28. [CrossRef] [PubMed]

19. Verrusio, W.; Andreozzi, P.; Marigliano, B.; Renzi, A.; Gianturco, V.; Pecci, M.T.; Ettorre, E.; Cacciafesta, M.; Gueli, N. Exercise training and music therapy in elderly with depressive syndrome: A pilot study. *Complement. Med.* **2014**, *22*, 614–620. [CrossRef]

20. Marais, L.; Stein, D.J.; Daniels, W.M. Exercise increases BDNF levels in the striatum and decreases depressive-like behavior in chronically stressed rats. *Metab. Brain Dis.* **2009**, *24*, 587–597. [CrossRef]

21. Schwenk, M.; Dutzi, I.; Englert, S.; Micol, W.; Najafi, B.; Mohler, J.; Hauer, K. An intensive exercise program improves motor performances in patients with dementia: Translational model of geriatric rehabilitation. *J. Alzheimers Dis. JAD* **2014**, *39*, 487–498. [CrossRef] [PubMed]

22. Zieschang, T.; Schwenk, M.; Oster, P.; Hauer, K. Sustainability of motor training effects in older people with dementia. *J. Alzheimers Dis. JAD* **2013**, *34*, 191–202. [CrossRef] [PubMed]

23. Hauer, K.; Hildebrandt, W.; Sehl, Y.; Edler, L.; Oster, P.; Droge, W. Improvement in muscular performance and decrease in tumor necrosis factor level in old age after antioxidant treatment. *J. Mol. Med.* **2003**, *81*, 118–125. [CrossRef] [PubMed]

24. Faber, M.J.; Bosscher, R.J.; Chin, A.P.M.J.; van Wieringen, P.C. Effects of exercise programs on falls and mobility in frail and pre-frail older adults: A multicenter randomized controlled trial. *Arch. Phys. Med. Rehabil.* **2006**, *87*, 885–896. [CrossRef] [PubMed]

25. Siciliano, G.; Chico, L.; Lo Gerfo, A.; Simoncini, C.; Schirinzi, E.; Ricci, G. Exercise-Related Oxidative Stress as Mechanism to Fight Physical Dysfunction in Neuromuscular Disorders. *Front. Physiol.* **2020**, *11*, 451. [CrossRef]

26. Cetin, E.; Top, E.C.; Sahin, G.; Ozkaya, Y.G.; Aydin, H.; Toraman, F. Effect of vitamin E supplementation with exercise on cognitive functions and total antioxidant capacity in older people. *J. Nutr. Health Aging* **2010**, *14*, 763–769. [CrossRef]

27. Jolitha, A.B.; Subramanyam, M.V.; Asha Devi, S. Modification by vitamin E and exercise of oxidative stress in regions of aging rat brain: Studies on superoxide dismutase isoenzymes and protein oxidation status. *Exp. Gerontol.* **2006**, *41*, 753–763. [CrossRef]

28. Wu, A.; Ying, Z.; Gomez-Pinilla, F. Docosahexaenoic acid dietary supplementation enhances the effects of exercise on synaptic plasticity and cognition. *Neuroscience* **2008**, *155*, 751–759. [CrossRef]

29. Ristow, M.; Zarse, K.; Oberbach, A.; Kloting, N.; Birringer, M.; Kiehntopf, M.; Stumvoll, M.; Kahn, C.R.; Bluher, M. Antioxidants prevent health-promoting effects of physical exercise in humans. *Proc. Natl. Acad. Sci. USA* **2009**, *106*, 8665–8670. [CrossRef]

30. Chaudhari, K.; Wong, J.M.; Vann, P.H.; Sumien, N. Exercise training and antioxidant supplementation independently improve cognitive function in adult male and female GFAP-*APOE* mice. *J. Sport Health Sci.* **2014**, *3*, 196–205. [CrossRef]

31. Chaudhari, K.; Wong, J.M.; Vann, P.H.; Sumien, N. Exercise, but not antioxidants, reversed ApoE4-associated motor impairments in adult GFAP-ApoE mice. *Behav. Brain Res.* **2016**, *305*, 37–45. [CrossRef] [PubMed]

32. Ferretti, L.; McCurry, S.M.; Logsdon, R.; Gibbons, L.; Teri, L. Anxiety and Alzheimer's disease. *J. Geriatr. Psychiatry Neurol.* **2001**, *14*, 52–58. [CrossRef] [PubMed]

33. Teri, L.; Ferretti, L.E.; Gibbons, L.E.; Logsdon, R.G.; McCurry, S.M.; Kukull, W.A.; McCormick, W.C.; Bowen, J.D.; Larson, E.B. Anxiety of Alzheimer's disease: Prevalence, and comorbidity. *J. Gerontol. A Biol. Sci. Med. Sci.* **1999**, *54*, M348–M352. [CrossRef]

34. Hetland, A.; Carr, D.B. Medications and impaired driving. *Ann. Pharmacother.* **2014**, *48*, 494–506. [CrossRef] [PubMed]

35. Stewart, S.A. The effects of benzodiazepines on cognition. *J. Clin. Psychiatry* **2005**, *66* (Suppl. 2), 9–13.

36. Raber, J.; Wong, D.; Buttini, M.; Orth, M.; Bellosta, S.; Pitas, R.E.; Mahley, R.W.; Mucke, L. Isoform-specific effects of human apolipoprotein E on brain function revealed in ApoE knockout mice: Increased susceptibility of females. *Proc. Natl. Acad. Sci. USA* **1998**, *95*, 10914–10919. [CrossRef]

37. Raber, J. Role of apolipoprotein E in anxiety. *Neural. Plast.* **2007**, *2007*, 91236. [CrossRef]

38. Bour, A.; Grootendorst, J.; Vogel, E.; Kelche, C.; Dodart, J.C.; Bales, K.; Moreau, P.H.; Sullivan, P.M.; Mathis, C. Middle-aged human apoE4 targeted-replacement mice show retention deficits on a wide range of spatial memory tasks. *Behav. Brain Res.* **2008**, *193*, 174–182. [CrossRef]

39. Lanza, G.; Centonze, S.S.; Destro, G.; Vella, V.; Bellomo, M.; Pennisi, M.; Bella, R.; Ciavardelli, D. Shiatsu as an adjuvant therapy for depression in patients with Alzheimer's disease: A pilot study. *Complement. Med.* **2018**, *38*, 74–78. [CrossRef]

40. Schneider, J.A.; Li, J.L.; Li, Y.; Wilson, R.S.; Kordower, J.H.; Bennett, D.A. Substantia nigra tangles are related to gait impairment in older persons. *Ann. Neurol.* **2006**, *59*, 166–173. [CrossRef]

41. Mavroudis, I.A.; Fotiou, D.F.; Adipepe, L.F.; Manani, M.G.; Njau, S.D.; Psaroulis, D.; Costa, V.G.; Baloyannis, S.J. Morphological changes of the human purkinje cells and deposition of neuritic plaques and neurofibrillary tangles on the cerebellar cortex of Alzheimer's disease. *Am. J. Alzheimers Dis. Other Demen.* **2010**, *25*, 585–591. [CrossRef] [PubMed]

42. van Meer, P.; Acevedo, S.; Raber, J. Impairments in spatial memory retention of GFAP-apoE4 female mice. *Behav. Brain Res.* **2007**, *176*, 372–375. [CrossRef] [PubMed]

43. Mkopi, A.; Range, N.; Lwilla, F.; Egwaga, S.; Schulze, A.; Geubbels, E.; van Leth, F. Adherence to tuberculosis therapy among patients receiving home-based directly observed treatment: Evidence from the United Republic of Tanzania. *PLoS ONE* **2012**, *7*, e51828. [CrossRef] [PubMed]

44. Nichol, K.; Deeny, S.P.; Seif, J.; Camaclang, K.; Cotman, C.W. Exercise improves cognition and hippocampal plasticity in APOE epsilon4 mice. *Alzheimers Dement. J. Alzheimers Assoc.* **2009**, *5*, 287–294. [CrossRef]

45. Lourenco, M.V.; Frozza, R.L.; de Freitas, G.B.; Zhang, H.; Kincheski, G.C.; Ribeiro, F.C.; Goncalves, R.A.; Clarke, J.R.; Beckman, D.; Staniszewski, A.; et al. Exercise-linked FNDC5/irisin rescues synaptic plasticity and memory defects in Alzheimer's models. *Nat. Med.* **2019**, *25*, 165–175. [CrossRef]

46. Stopford, C.L.; Thompson, J.C.; Neary, D.; Richardson, A.M.; Snowden, J.S. Working memory, attention, and executive function in Alzheimer's disease and frontotemporal dementia. *Cortex* **2012**, *48*, 429–446. [CrossRef]

47. Greenwood, P.M.; Lambert, C.; Sunderland, T.; Parasuraman, R. Effects of apolipoprotein E genotype on spatial attention, working memory, and their interaction in healthy, middle-aged adults: Results From the National Institute of Mental Health's BIOCARD study. *Neuropsychology* **2005**, *19*, 199–211. [CrossRef]

48. Kim, S.E.; Ko, I.G.; Shin, M.S.; Kim, C.J.; Jin, B.K.; Hong, H.P.; Jee, Y.S. Treadmill exercise and wheel exercise enhance expressions of neutrophic factors in the hippocampus of lipopolysaccharide-injected rats. *Neurosci. Lett.* **2013**, *538*, 54–59. [CrossRef]

49. Deeny, S.P.; Poeppel, D.; Zimmerman, J.B.; Roth, S.M.; Brandauer, J.; Witkowski, S.; Hearn, J.W.; Ludlow, A.T.; Contreras-Vidal, J.L.; Brandt, J.; et al. Exercise, APOE, and working memory: MEG and behavioral evidence for benefit of exercise in epsilon4 carriers. *Biol. Psychol.* **2008**, *78*, 179–187. [CrossRef]

50. Radak, Z.; Kumagai, S.; Taylor, A.W.; Naito, H.; Goto, S. Effects of exercise on brain function: Role of free radicals. *Appl. Physiol. Nutr. Metab. Physiol. Appl. Nutr. Metab.* **2007**, *32*, 942–946. [CrossRef]

51. Bogdanis, G.C.; Stavrinou, P.; Fatouros, I.G.; Philippou, A.; Chatzinikolaou, A.; Draganidis, D.; Ermidis, G.; Maridaki, M. Short-term high-intensity interval exercise training attenuates oxidative stress responses and improves antioxidant status in healthy humans. *Food Chem. Toxicol. Int. J. Publ. Br. Ind. Biol. Res. Assoc.* **2013**, *61*, 171–177. [CrossRef] [PubMed]

52. Nybo, L.; Nielsen, B.; Pedersen, B.K.; Moller, K.; Secher, N.H. Interleukin-6 release from the human brain during prolonged exercise. *J. Physiol.* **2002**, *542*, 991–995. [CrossRef] [PubMed]

53. Chennaoui, M.; Drogou, C.; Gomez-Merino, D. Effects of physical training on IL-1beta, IL-6 and IL-1ra concentrations in various brain areas of the rat. *Eur. Cytokine Netw.* **2008**, *19*, 8–14. [CrossRef] [PubMed]

54. Pedersen, B.K. The anti-inflammatory effect of exercise: Its role in diabetes and cardiovascular disease control. *Essays Biochem.* **2006**, *42*, 105–117. [CrossRef] [PubMed]

55. Goldhammer, E.; Tanchilevitch, A.; Maor, I.; Beniamini, Y.; Rosenschein, U.; Sagiv, M. Exercise training modulates cytokines activity in coronary heart disease patients. *Int. J. Cardiol.* **2005**, *100*, 93–99. [CrossRef] [PubMed]

56. Villa, A.; Vegeto, E.; Poletti, A.; Maggi, A. Estrogens, Neuroinflammation, and Neurodegeneration. *Endocr. Rev.* **2016**, *37*, 372–402. [CrossRef]

57. Torrens-Mas, M.; Pons, D.G.; Sastre-Serra, J.; Oliver, J.; Roca, P. Sexual hormones regulate the redox status and mitochondrial function in the brain. Pathological implications. *Redox Biol.* **2020**, *31*, 101505. [CrossRef]

58. D'Argenio, V.; Sarnataro, D. New Insights into the Molecular Bases of Familial Alzheimer's Disease. *J. Pers. Med.* **2020**, *10*, 26. [CrossRef]

59. Cacace, R.; Sleegers, K.; Van Broeckhoven, C. Molecular genetics of early-onset Alzheimer's disease revisited. *Alzheimers Dement. J. Alzheimers Assoc.* **2016**, *12*, 733–748. [CrossRef]

60. Ngandu, T.; Lehtisalo, J.; Solomon, A.; Levalahti, E.; Ahtiluoto, S.; Antikainen, R.; Backman, L.; Hanninen, T.; Jula, A.; Laatikainen, T.; et al. A 2 year multidomain intervention of diet, exercise, cognitive training, and vascular risk monitoring versus control to prevent cognitive decline in at-risk elderly people (FINGER): A randomised controlled trial. *Lancet* **2015**, *385*, 2255–2263. [CrossRef]

61. Fisher-Wellman, K.; Bloomer, R. Acute exercise and oxidative stress: A 30 year history. *Dyn. Med.* **2009**, *8*, 1–25. [CrossRef] [PubMed]

62. Gronwald, T.; de Bem Alves, A.C.; Murillo-Rodriguez, E.; Latini, A.; Schuette, J.; Budde, H. Standardization of exercise intensity and consideration of a dose-response is essential. Commentary on "Exercise-linked FNDC5/irisin rescues synaptic plasticity and memory defects in Alzheimer's models", by Lourenco et al., published 2019 in Nature Medicine. *J. Sport Health Sci.* **2019**, *8*, 353–354. [CrossRef] [PubMed]

antioxidants MDPI

Review

Effects of Dietary Strategies on Exercise-Induced Oxidative Stress: A Narrative Review of Human Studies

Zhen Zeng [1,2,*], Christoph Centner [1,3], Albert Gollhofer [1] and Daniel König [4,5]

[1] Department of Sport and Sport Science, University of Freiburg, 79117 Freiburg, Germany; christoph.centner@sport.uni-freiburg.de (C.C.); ag@sport.uni-freiburg.de (A.G.)
[2] School of Sports Medicine and Health, Chengdu Sport University, Chengdu 610041, China
[3] Praxisklinik Rennbahn, 4132 Muttenz, Switzerland
[4] Centre of Sports Science, Department for Nutrition, Exercise and Health, University of Vienna, 1150 Vienna, Austria; daniel.koenig@univie.ac.at
[5] Faculty of Life Sciences, Department for Nutrition, Exercise and Health, University of Vienna, 1090 Vienna, Austria
* Correspondence: zhen.zeng@sport.uni-freiburg.de

Abstract: Exhaustive exercise can induce excessive generation of reactive oxygen species (ROS), which may enhance oxidative stress levels. Although physiological levels are crucial for optimal cell signaling and exercise adaptations, higher concentrations have been demonstrated to damage macromolecules and thus facilitate detrimental effects. Besides single dosages of antioxidants, whole diets rich in antioxidants are gaining more attention due to their practicality and multicomponent ingredients. The purpose of this narrative review is to summarize the current state of research on this topic and present recent advances regarding the antioxidant effects of whole dietary strategies on exercise-induced oxidative stress in humans. The following electronic databases were searched from inception to February 2021: PubMed, Scope and Web of Science. Twenty-eight studies were included in this narrative review and demonstrated the scavenging effects of exercise-induced ROS generation, oxidative stress markers, inflammatory markers and antioxidant capacity, with only one study not confirming such positive effects. Although the literature is still scarce about the effects of whole dietary strategies on exercise-induced oxidative stress, the majority of the studies demonstrated favorable effects. Nevertheless, the protocols are still very heterogeneous and further systematically designed studies are needed to strengthen the evidence.

Keywords: diet; antioxidants; exercise; oxidative stress; reactive oxygen species

Citation: Zeng, Z.; Centner, C.; Gollhofer, A.; König, D. Effects of Dietary Strategies on Exercise-Induced Oxidative Stress: A Narrative Review of Human Studies. *Antioxidants* **2021**, *10*, 542. https://doi.org/10.3390/antiox10040542

Academic Editor: Gareth Davison

Received: 16 March 2021
Accepted: 29 March 2021
Published: 31 March 2021

Publisher's Note: MDPI stays neutral with regard to jurisdictional claims in published maps and institutional affiliations.

1. Background

The term oxidative stress is defined as a disturbance in the homeostatic balance between pro-oxidants and antioxidants with a subsequent excessive generation of free radicals [1–3]. Free radicals are highly reactive compounds that contain one or more unpaired electrons in their outer atomic or molecular orbital [1,4], and thus readily react with various organic substrates in order to make themselves more stable [3]. Species derived from oxygen are generally referred to as reactive oxygen species (ROS) and are naturally occurring byproducts of the human metabolism. Thereby, redox reactions represent fundamental components of organic and biological chemistry [5]. While low to moderate ROS concentrations seem to be involved in cell signaling and muscle remodulation [5–7], prolonged exposure to high doses of ROS induces oxidative damage [3]. In case of an insufficient ROS scavenging by antioxidants, high ROS concentrations can lead to modification and damage of cellular molecules including deoxyribonucleic acid (DNA), proteins or lipids [2]. Previous studies have also shown that oxidative stress is involved in the pathophysiology of a wide range of chronic diseases including cancer [5,8], cardiovascular [3,9] and neurological diseases [10–13].

During exercise, the amount of generated ROS seems to be intensity-dependent, with higher exercise intensities leading to supraphysiological ROS formations [14,15]. Mitochondrial hormesis (mitohormesis) was proposed to describe that sublethal mitochondrial stress can trigger a favorable cellular response, resulting in an improved mitochondrial and nonmitochondrial adaptation, and thus maintain redox homeostasis [16] (Figure 1). As depicted in Figure 1, high-intensity exercise might induce mitochondrial stress, leading the mitochondria to emit ROS in order to facilitate adaptations and thus protect against subsequent cellular stress [17]. In case of excessive ROS production, this might lead to oxidative damage.

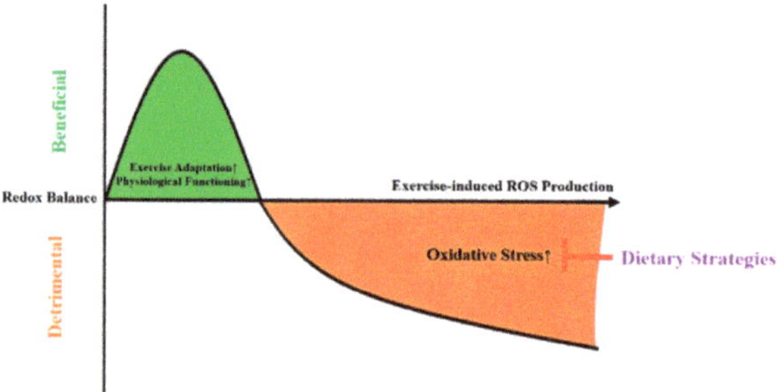

Figure 1. The mitohormesis-based model to explain the effects of dietary strategies on exercise-induced oxidative stress. ROS = reactive oxygen species.

From another perspective, aerobic exercise has been suggested to be beneficial in ROS-associated diseases, including cardiovascular pathologies [18]. Previous studies have demonstrated that regular and moderate exercise can enhance endothelia function by increasing the bioavailability of NO and improving redox states in subjects with pre-existing cardiovascular risk factors or diseases [19,20]. Nevertheless, a few studies also revealed conflicting results by showing that intense aerobic exercise could injure endothelia cells [21,22]. These results might be explained by the different exercise intensities and the resulting varying levels of oxidative damage.

As a potential countermeasure against excessive oxidative stress during exercise, antioxidative supplementations, which aim to protect against muscle damage and thus improve exercise performance, have been frequently discussed [4,22]. Nonetheless, many studies have indicated that large-dose antioxidant supplementation can interfere with intrinsic adaptive responses and may abolish the benefit of exercise [23,24]. These highly purified antioxidants can negatively affect ROS-mediated physiological processes through prooxidant mechanisms [22]. In a previously published meta-analysis, Stepanyan et al. [25] demonstrated that supplementation with vitamin E did not protect against exercise-induced lipid peroxidation or muscle damage. Instead of single antioxidative sources, it might be speculated that the intake of natural foods rich in antioxidants of phytochemicals (e.g., fruits and vegetables) might represent a more beneficial approach for enhancing the antioxidant status during exercise [26]. Along with their high antioxidant content, specific diets, including products such as oatmeal, dark chocolate, and mixed fruit beverages may also contain additional bioactive compounds which are not found in single-dose pharmacological antioxidant supplements but can act synergistically to reveal more beneficial effects than a single dose of antioxidant supplements [27,28]. Additionally, these compounds are more accessible than specific isolated antioxidants. Until now, few studies have investigated the clinical effects on exercise-induced oxidative stress by using a whole dietary strategy and consistent evidence from human study remains scarce.

2. Methods

The article search was conducted at the following electronic databases: Pubmed, Scopus and Web of Science. Searches were performed up to February 2021. The search term was developed with three segments: the first segment encompassed synonyms of diet; the second and third segments included synonyms of ROS and exercise, respectively. All segments were connected with the Boolean operator "AND". The respective MeSH terms were used for each keyword. In order to avoid the potential bias caused by different baseline values of redox status, only the untrained, nonathlete, healthy population were included in this review. Animal models were not included.

3. Dietary Strategies

The majority of currently available studies addressed the effects of phenol-rich foods on exercise-induced oxidative stress, including dark chocolate [29–31], high-flavanol cocoa drink [32], green tea [33], mate tea [34], New Zealand blueberry smoothie [35], blueberries [36,37], grape juice [38,39], Montmorency cherry juice [40], tart cherry juice [41], oatmeal [42], avenanthramides (AVA)-rich cookie [43,44], juçara juice [45], Sanguinello cultivar red orange juice [46], and purple sweet potato leaves [47]. Frequently, the effects of dietary strategies on exercise-induced stress are evaluated within short-term [29,32,35,36,38,42,45], as well as long-term interventions [30,31,33,34,37,39–41,43,44,46,47]. Across all studies, there is a compelling amount of evidence suggesting that different dietary regimens are viable tools for decreasing exercise-induced oxidative stress. However, the different biomarkers of oxidative stress do not allow a direct comparison between studies. Therefore, the individual effects of these dietary strategies on different redox systems will be discussed in the following section.

4. Effects on Biomarkers of Exercise-Induced Oxidative Stress

High intensity exercise has repeatedly been demonstrated to induce excessive amounts of ROS, which may react with macromolecules such as proteins, lipids, and DNA [2]. To date, the in vivo detection of free radicals remains a challenge due to their short lifetime and the low rates of formation. Numerous techniques and assays have been used to measure oxidative stress production directly or indirectly. Accordingly, the included studies will be categorized according to whether the main effects observed were in ROS generation, oxidative stress markers, inflammatory markers or antioxidant activity (Table 1).

Table 1. Effects of dietary strategies on exercise-induced oxidative stress.

Main Result Category	Study	Type of Diet	Nutritional Protocol	Type and Intensity of Exercise	Detection Method			
					ROS Generation	Oxidative Stress Marker	Inflammation Marker	Antioxidant Activity
ROS Generation	Zeng et al. [42]	Oatmeal	Oat flake + skim milk versus Fasting; 2 h before exercise	Body weight HIIT, 30 min	↓	N/A	N/A	N/A
ROS-induced Macromolecule Damage	Davison et al. [29]	Dark chocolate	Dark chocolate versus cocoa-liquor-free control bar versus neither, 2 h before exercise	Cycling, 2.5 h	N/A	F2-isoprostane↓	Circulating leucocyte↔, IL-6↔	N/A
	Wiswedel et al. [32]	High-flavanol cocoa drink (HFCD)	HFCD versus low-flavanol cocoa drink (LFCD), 2 h before exercise	Cycling, 29 min	N/A	F2-isoprostane↓	N/A	N/A
	Allgrove et al. [30]	Dark chocolate	Dark chocolate versus isocarbohydrate-fat control cocoa-liquor-free chocolate, twice/d, 2 weeks	Cycling for 90 min followed by 25 min exhaustion time trial	N/A	F2-isoprostane↓	Circulating leucocyte↔, IL-6↔, IL-10↔, IL-1Ra↔	N/A
	Taub et al. [31]	High-flavanol dark chocolate (HFCHO)	HFCHO versus Low-flavanol dark chocolate (LFCHO), 3 months	Ramped exercise on stationary bicycle (Cardiopulmonary exercise testing), ~10 min	N/A	PC↓	N/A	GSH/GSSH↑
	McAnulty et al. [37]	Blueberry	Blueberries versus blueberry-flavored shake, 7 days	Running, until a core temperature of 39.5 °C was reached	N/A	LH↓, F2-isoprostanes↔	IL-6↔, IL-8↔, IL-10↔	FRAP↔
	Bowtell et al. [40]	Montmorency cherry juice	Montmorency cherry juice versus isoenergetic fruit concentrate, 7 d before and 48 h after exercise	Two trials of 10 sets of 10 single-leg knee extensions	N/A	PC↓	N/A	N/A
	Pittaluga et al. [46]	Fresh red orange juice (ROJ)	ROJ versus nothing extra, thrice/day, 4 weeks	A single bout of exhaustive exercise by cycle ergometer (3 min warm-up, an initial load of 25 W, and further increments of 15 W/3 min)	N/A	MDA↓, ascorbic acid↓, hypoxanthine/xanthine↓	N/A	N/A
	Chang et al. [47]	Purple sweet potato leaves (PSPL)	Standard cooked PSPL versus low-polyphenols diet, 7 days	Treadmill running at 70% VO_{2max}, 1 h	N/A	PC↓	IL-6↓, HSP72↔	TAC (FRAP assay)↑, polyphenols↑
	Mazani et al. [48]	Probiotic yoghurt	Probiotic yoghurt versus ordinary yoghurt, 2 weeks	Exhaustive exercise (Bruce test)	N/A	MDA↓	TNF-α↓, MMP2↓, MMP9↓	SOD↑, GPX↑, TAC↑,
	Harms-Ringdahl et al. [49]	Tomato juice	Tomato juice versus nothing extra, 5 weeks	Cycle ergometer at 80% of HRmax, 20 min	N/A	8-oxodG↓	N/A	N/A
	Kawamura et al. [50]	Mixed diet	Salmon flakes + green and yellow vegetable juice + lingonberry jam versus normal diet, 10 weeks	Resistance training twice/week, 10 weeks	N/A	PC↓	N/A	N/A

Table 1. *Cont.*

Main Result Category	Study	Type of Diet	Nutritional Protocol	Type and Intensity of Exercise	Detection Method			
					ROS Generation	Oxidative Stress Marker	Inflammation Marker	Antioxidant Activity
	Sureda et al. [51]	Mixed beverage	Almond-based isotonic and energetic beverage with vitamin C and E versus Nonenriched beverage, 1 month	A half-marathon race (21 km-run)	N/A	MDA↓	N/A	N/A
	Carrera-Quintanar et al. [52]	Mixed beverage	Mixed beverage with extra *Lippia citriodora* extract versus mixed beverage enriched with vitamins C and E, 22 days	2000-m running exercise trial	N/A	PC↑	N/A	SOD↓, GRD↓
	M Daly et al. [53]	Multinutrient-fortified milk (MFMD)	MFMD versus placebo milk, twice/d, 4 months	Resistance exercise 3 d/week, 4 months	N/A	PC↔, 8-isoprostane↔	N/A	N/A
Inflammatory Markers	Koenig et al. [43], Zhang et al. [44]	AVA-enrich cookies	4.6 mg AVA/cookie versus 0.2 mg AVA/cookie, 2 cookies/day, 8 weeks	Downhill running, 1 h	N/A	N/A	NRB [43]; NF-κB↓ and IL-6 [43]; G-CSF, IL-1Ra and sVCAM-1 [44]	N/A
	Prasertsri et al. [54]	Cashew apple juice (CAJ)	CAJ versus placebo (isocaloric), 4 weeks	Cycling at 85% of VO$_{2max}$, 20 min	N/A	MDA↓, 8-isoprostane↓,	Leukocyte count↑	N/A
	Panza et al. [34]	Mate tea	Mate tea versus water, 11days, exercise and blood test were performed at 8th day	Three sets of twenty maximal eccentric elbow flexion exercises	N/A	N/A	N/A	GSH↑, GSSG↔, GSH/GSSG↔, LOOH↔
	Panze et al. [33]	Green tea	Green tea versus Water, three times/day, 7 days	A bench press exercise, four sets, 10 to 4 repetitions	N/A	LH↓	N/A	TAC (FRAP assay)↑, total polyphenol↑, GSH↑
Antioxidant Activity	McLeay et al. [35]	New Zealand blueberry	Blueberry + banana + commercial apple juice versus Shake dextrose + banana + commercial apple juice (isocaloric); 5 and 10 h pre, immediately, 12 and 36 h after exercise	3 sets × 100 eccentric repetitions of quadriceps muscle	N/A	PC↓	IL-6↓	TAC (FRAP assay)↑, ROS-GP↓
	Park et al. [36]	Blueberry	Blueberry + aronia + sugar + refined water versus nothing extra, before exercise	Treadmill exercise (Bruce test)	N/A	N/A	IL-6↓, CRP↓	TAS↑
	Toscano et al. [38,39]	Grape	Integral grape juice versus isocaloric, isoglycemic and isovolumetric control beverage, 10 mL/kg/day, 2 h before exercise [38], or for 28 days [39]	Time-to exhaustion exercise test, anaerobic threshold test and aerobic capacity test	N/A	N/A	N/A	TAC↑ [38,39], UA↑ and vitamin A↑ [39]
	Howatson et al. [41]	Tart cherry	Tart cherry juice versus control, before, on the day of, and 48 h following exercise	A marathon run	N/A	MDA↓	IL-6↓, CRP↓, UA↓	TAS↑
	Copetti et al. [45]	Juçara (*Euterpe edulis Martius*)	Juçara juice versus Water, 1 h before exercise	HIIT, 17 min	N/A	N/A	N/A	OSI↓

Table 1. *Cont.*

Main Result Category	Study	Type of Diet	Nutritional Protocol	Type and Intensity of Exercise	Detection Method			
					ROS Generation	Oxidative Stress Marker	Inflammation Marker	Antioxidant Activity
	Iwasa et al. [55]	Fermented milk	Fermented milk (*Lactobacillus helveticus*) versus equivalent dose of unfermented milk, 1 h before and 2 h after exercise	Resistance exercise consisting of five sets of leg and bench presses	N/A	N/A	hsCRP$\downarrow$, TNF-a$\leftrightarrow$	TAC (ORAC assay)$\uparrow$
	Beavers et al. [56]	Soy	Soy versus dairy milk, 3 serving/d, 4 weeks	Downhill-running at 60% VO$_{2max}$ and -10% grade, 45 min	N/A	N/A	TNF-$\alpha\leftrightarrow$, IL-1$\beta\leftrightarrow$, IL-6$\leftrightarrow$	GPx$\leftrightarrow$, COX-2$\leftrightarrow$

Legend: the arrows represent increase ($\uparrow$), decrease ($\downarrow$), no change ($\leftrightarrow$). Abbreviation list: high intensity interval training (HIIT), lipid hydroperoxide (LH), malondialdehyde (MDA), protein carbonyls (PC), superoxide dismutase (SOD), tumor necrosis factor-α (TNF-α), interleukin-6 (IL-6), interleukin-8 (IL-8), interleukin-10 (IL-10), interleukin-1 receptor antagonist (IL-1Ra), nterleukin-1β (IL-1β), matrix metalloproteinase 2 (MMP2), matrix metalloproteinase 9 (MMP9), neutrophil respiratory burst (NRB), nuclear factor-kappa B (NF-κB), total antioxidant status (TAS), Avenanthramides (AVA), soluble vascular cell adhesion molecule-1 (sVCAM-1), granulocyte-colony stimulating factor (G-CSF), lipid hydroperoxides (LOOH), 8-hydroxy-2′-deoxyguanosine (8-oxodG), ferric reducing ability of plasma (FRAP), glutathione (GSH), glutathione/oxidized glutathione (GSH/GSSH), uric acid (UA), radical oxygen species-generating potential (ROS-GP), heat shock proteins (HSP72), oxidative stress index (OSI), glutathione peroxidase (GPx), cyclooxygenase-2 (COX-2), total antioxidant capacity (TAC), glutathione reductase (GRD), C-reactive protein (CRP).

4.1. Effects of Dietary Interventions on Direct ROS Generation

To date, electron paramagnetic resonance (EPR) technology is the only method that can directly detect ROS generation in in vivo conditions [57]. Short-lived ROS can be added to the spin trap to form a spin-adduct that has a comparatively longer half-life to be detected using EPR spectroscopy [58]. Zeng et al. [42] revealed that consumption of AVA-rich oatmeal before high-intensity interval training (HIIT) significantly mitigates exercise-induced ROS generation compared to the control group, by using the EPR method. AVA, as one of the major components of polyphenolic amides (nonflavonoids), is considered the most important antioxidant found in oats [59,60]. Therefore, it can be speculated that the hydroxyl groups of AVA contribute to antioxidant defense through their ability to trap ROS in vitro [61]. Another assumption is that AVA can activate the nuclear factor erythroid 2-related factor 2 (NRF2) defense system against oxidative stress by attacking the α, β-unsaturated carbonyl moiety [62]. However, the underlying mechanisms for these effects are still unclear.

Indeed, only the study of Zeng et al., [42] applied direct ROS measurements using EPR technology, whereas the other experiments in this review used oxidative stress markers, inflammatory markers and antioxidant activity levels to interpret the changes in ROS production, as will be discussed below.

4.2. Effects of Dietary Interventions on ROS-Induced Macromolecule Damage

In the majority of studies, F2-isoprostanes, 8-isoprostanes, lipid hydroperoxides (LH), thiobarbituric acid-reactive substances (TBARS) and malondialdehydes (MDA) were used as the oxidative markers, which result from lipoperoxidation by oxidative damage. Similarly, protein carbonylation (PC) was used as a marker of protein damage, and 8-Hydroxydeoxyguanosine (8-oxodG) as a specific marker of $2'$-deoxyguanosine damage after ROS attack to DNA. In this narrative review, $n = 14$ articles ([29–32,37,40,46–53]) showed the antioxidant effects diets on oxidative stress markers.

Davison et al. [29], Wiswedel et al., [32] and Allgrove et al., [30] observed the beneficial antioxidant effects of dark chocolate by detecting the plasma levels of F2-isoprostane, while Davison et al., [29] and Wiswedel et al., [32] confirmed the acute antioxidant effects of dark chocolate due to its polyphenolic properties, Allgrove et al., [30] and Taub et al., [31] showed that these beneficial effects can also be seen following long-term dietary interventions. The derivatives of catechin and epicatechin, which can both be defined as monomeric flavanols, are the major antioxidant components in cacao beans (chocolate) [63]. The acute antioxidant effects of flavanols in cocoa were evaluated by Davison et al., [29] who investigated the association between the increased plasma epicatechin levels and F2-isoprostanes and found decreased oxidative stress markers in a group given dark chocolate compared to a control group. In this study, after consuming 100 g dark chocolate or an isomacronutrient control bar, each healthy male subject cycled for 2.5 h at ~60% maximal oxygen uptake. Blood samples were analyzed at pre-exercise and immediately postexercise. Plasma F2-isoprostane, also showed a decline after ingestion of a high-flavanol cocoa drink combined with strenuous cycling exercise in the study of Wiswedel et al., [32].

Despite the acute antioxidant effects, Allgrove et al., [30] found that consuming 40 g of dark chocolate twice daily for two weeks significantly decreased plasma F2-isoprostane levels at exhaustion and after one hour of recovery in a prolonged exercise trial.

Differently, the study of Taub et al., [31] explored the mechanisms underlying the long-term antioxidant effects of dark chocolate by examining human muscle samples (quadriceps femoris). After consuming 20 g of dark chocolate or placebo for three months, the $VO_{2\,max}$ and total work of each sedentary subject was assessed on a stationary bicycle. After exercise, the skeletal muscle evidenced significant decreases in PC and increased glutathione (GSH) levels only in the dark chocolate group [31]. Furthermore, the protein levels of liver kinase B1(LKB1), adenosine monophosphate (AMP)-activated protein kinase (AMPK), peroxisome proliferator-activated receptor gamma coactivator 1-alpha (PGC1α), and their active forms (phosphorylated AMPK and LKB1), along with citrated synthase

(CS) activity, were found significantly elevated [31]. Accordingly, the dark chocolate might activate upstream control systems and improve mitochondria performance in skeletal muscle, contributing to the improvements in maximum work achieved and $VO_{2\,max}$.

It has to be mentioned that the studies investigating the effects of cocoa had heterogeneous designs, which prevents us from making definite conclusions. Davison et al., [29] used 100 g dark chocolate (39.1 mg catechin, 96.8 mg epicatechin, 58.4 mg Dimer B2, 7.3 mg Dimer B5, 34.7 mg Trimer C and 10.5 mg tetramer D); Wiswedel et al., [32] chose a high-flavanol cocoa drink (187 mg flavan-3-ols/100 mL); Allgrove et al. [30] applied the 40 g 70% chocolate; Taub et al., [31] provided the dark chocolate at total of 20 g and ~100 calories per day.

In addition to cocoa, other phenol-rich fruits also exhibited antioxidant effects during exercise by detecting oxidative stress markers, including blueberry [37], cherry [40] and red orange [46]. In the study by McAnulty et al., [37], participants consumed 150g of blueberries in a milkshake every day for one week prior to one session of high-intensity training in hyperthermic conditions. The results showed that the blueberry diet attenuated an increase in LH concentration caused by exercise stress but not F2-isoprostane levels, compared with a blueberry-flavored shake as a placebo [37]. Montmorency cherry juice (30 mL twice per day) was provided for one week before and 48 h after a bout of strength exercise in the study of Bowtell et al., [40]. The recovery of isometric muscle strength after high-intensity exercise was improved. PC were lower in the Montmorency cherry juice group compared with the isoenergetic fruit concentrate (placebo) group. Red oranges are a cultivar of the Citrus sinensis family which are generally rich in vitamin C, anthocyanins, and flavanones. The Sanguinello cultivar red orange juice (ROJ) was provided as the intervention diet in a study by Pittaluga et al., [46] due to the remarkable antioxidant ability of anthocyanin family, such as cyanidine-3-*O*-β-glucoside (C3G). In this elderly human trial, the intervention group (250 mL ROJ thrice a day for 4 weeks) had lower exhausted exercise-induced MDA, lower hypoxanthine/xanthine system activation, and less ascorbic acid consumed.

Purple sweet potato leaves (PSPL), as another phenol-rich diet, showed decreases in oxidative stress markers in an exercise trial [47]. Chang et al., [47] investigated the effects of a 7-day PSPL-diet on running exercise-induced oxidative stress in a nontrained, young male population. PSPL consumption significantly increased total polyphenols concentrations, and significantly decreased plasma PC and TBARS in the PSPL group [47]. PSPLs, botanically identified as Ipomoea batatas (L.) Lam, have the highest levels of polyphenols and flavonoids (33.4 ± 0.5 mg gallic acid/g and 426.8 ± 8.9 µg/g dry wt) [64]. This research group also proved that the PSPL diet could modulate antioxidative status [65] and immune responses [66] in basketball players during a training period.

Besides phenol-rich foods, probiotic-rich dairy also showed promising antioxidant effects on the levels of oxidative stress markers. Mazani et al., [48] described the antioxidant effect of 450 g of probiotic yogurt taken daily for two weeks by young females. Compared with regular yogurt, after intense physical activity, probiotic yogurt consumption significantly decreased serum levels of MDA, and some inflammatory factors (tumor necrosis factor-α (TNF-α), matrix metalloproteinase 2 (MMP2), matrix metallopeptidase 9 (MMP9)), and increased the levels of superoxide dismutase (SOD), glutathione peroxidase (GP_X), and total antioxidant capacity (TAC). This result might be explained by the previous assumptions that some strains of probiotics positively prevent and correct oxidative stress in humans due to their direct antioxidative activity and positive effect on the immune system [67,68].

Lycopene is a carotenoid that is mainly found in tomatoes [69]. Among dietary carotenoids, lycopene is one of the most active antioxidants with a singlet-oxygen-quenching capacity twice as high as that of β-carotene and ten times greater than that of α-tocopherol [70]. However, the underlying mechanism of how it resists oxidative stress in vivo is still unclear. Two major potential hypotheses to explain the antioxidant abilities of lycopene are oxidative and nonoxidative mechanisms [69]. In a human trial by Harms-Ringdahl et al., [49], a daily intake of tomato juice, equal to 15 mg lycopene per day for five weeks significantly reduced the serum levels of 8-oxodG after extensive physical exercise.

A few dietary strategies that have been described as mixed foods have shown antioxidant effects on markers of oxidative stress. To investigate the effects of mixed antioxidant foods on resistance training-induced oxidative stress, a diet containing salmon flakes, green and yellow vegetable juice, and lingonberry jam, which contain astaxanthin, β-carotene, and resveratrol was provided by Kawamura et al., [50]. This mixed diet was consumed by the intervention group twice a week for 10 weeks. The results showed that serum PC levels tended to be lower immediately after exercise than before exercise in the intervention group only. The mixture of these nutrients might collectively enhance the antioxidant effects in this trial.

However, the combined antioxidant mechanisms of mixed foods are complex, and previous studies have shown inconsistent results for some mixed diets. Sureda et al., [51] demonstrated that a mixed beverage with vitamin C and E reduced the plasma oxidative damage induced by a half-marathon. After this, Carrera-Quintanar et al., [71] proved that ingestion of a mixed beverage that included an excess of polyphenolic antioxidants (*Lippia citriodora*) for 21 days could interfere with antioxidant activities and reduce the gene expression of specific enzymes (e.g., Cu-Zn-SOD, Mn-SOD and glutathione reductase (GRD)) in neutrophils, in a human study without exercise intervention. Recently, Carrera-Quintanar et al., [52] compared the antioxidant effects of two mixed beverages and one control beverage on exercise-induced oxidative damage: a mixed beverage enriched with vitamins C and E; the same beverage with extra *Lippia citriodora* extract; and the control beverage. This study was performed in a 2000-m running exercise trial. However, the results showed that all the oxidative stress markers increased in the control group, plasma PC significantly increased only in the mixed beverage with *Lippia citriodora*, and no significant changes in oxidative stress levels were detected for the mixed beverage which only added vitamins C and E. Accordingly, further studies are needed to explore the mechanisms by which certain antioxidants in the *Lippia citriodora* extract were less effective at combating oxidative stress than their components in isolation.

In contrast, one study did not find positive effects by testing oxidative stress markers [53]. In a four-month study of 216 women, a twice-daily multi-nutrient-fortified milk drink (MFMD), containing added protein, milk fat globule membrane (phospholipids and other bioactives), vitamin D, calcium, and other micronutrients, did not enhance the effects of an exercise program on markers of oxidative stress (marker: 8-isoprostane, PC) and the primary outcome measure of stair climbing ascent power [53]. However, the MFMD did elicit greater improvements in various secondary outcomes of physical functions compared to an energy-matched placebo [53].

4.3. Effects of Dietary Interventions on Inflammatory Markers

Exercise-induced oxidative stress can activate a range of transcription factors that contribute to the differential expression of certain genes involved in inflammatory pathways [72]. In this review, diets with antioxidant effects have demonstrated to reduce inflammatory markers including neutrophil respiratory burst (NRB), interleukin-6 (IL-6), nuclear factor-kappa B (NF-κB), granulocyte-colony stimulating factor (G-CSF), interleukin-1 receptor antagonist (IL-1Ra), soluble vascular cell adhesion molecule-1 (sVCAM-1). In this narrative review, n = 3 studies ([43,44,54]) showed decreases in inflammatory markers from the diet interventions.

Koenig et al., [43] and Zhang et al., [44] investigated the eight-week effects of AVA-rich cookies on exercise-induced oxidative stress by detecting the inflammatory markers. Both found that this AVA-rich diet decreased ROS production from the NRB after high intensity downhill training when compared to control group. Additionally, plasma IL-6 and NF-κB activity significantly decreased in AVA group versus control group in the study of young women by Koenig et al., [43]. Zhang et al., [44] further found that the neutrophil stimulating cytokine G-CSF, IL-1Ra, sVCAM-1 was significantly lower in AVA group compared to the control group after exercise stress. Similar to the study by Zeng et al., [42], the main antioxidant effects of these two investigations were described as being attributed to AVA components. However, Zeng et al., [42] examined the direct ROS generation by EPR method, whereas these two studies detected indirect inflammatory markers.

The role of vitamin C in mitigating the overproduction of ROS caused by high-intensity exercise is assumed to occur by helping to preserve the redox integrity of the immune cells and reduce the inflammation [73,74]. A four-week cashew apple juice (CAJ) supplementation was shown to enhance leukocyte count by reducing oxidative stress after high-intensity exercise in trained and untrained men [54]. The CAJ contained significant amounts of vitamin C (3.36 mg/100 g) and further antioxidants such as anacardia acids. The anacardia acids in CAJ may enhance the ability of vitamin C to prevent the generation of superoxide radicals by inhibiting xanthine oxidase and increasing heme oxygenase-1 [75]. However, the previously elaborated study by McAnulty et al., [37] compared the antioxidant effects of a blueberry diet and vitamin C supplements in hot training conditions. In contrast, the results supported the assumption of a prophylactic effect of polyphenol on exercise-induced oxidative stress, but not of vitamin C.

4.4. Effects of Dietary Interventions on Antioxidant Activity

In concert with alterations affecting levels of oxidative stress markers and inflammatory markers, exercise-induced oxidative stress could attenuate the endogenous antioxidant defense including enzymatic antioxidant activity (catalase (CAT), SOD, GPx, cyclooxygenase-2 (COX-2)) and nonenzymatic antioxidant activity (GSH, oxygen radical absorbance capacity (ORAC), total antioxidant capacity (TAC), total antioxidant status (TAS), ferric reducing antioxidant power (FRAP), vitamins C and E, and reduced glutathione content). In this current review, $n = 10$ articles ([33–36,38,39,41,45,55,56]) found that dietary strategies increased antioxidant activity.

Some included studies have demonstrated that phenol-rich foods could increase antioxidant capacity during high intensity exercise. As for cocoa, the major antioxidant properties in tea leaves are flavanol compounds, such as catechin and epicatechin [76]. Two studies by Panza et al., [33,34] investigated the antioxidant activity following the consumption of green tea or mate tea for one week in young men undergoing resistance exercise. Green tea increased the values of total polyphenols, GSH, FRAP and diminished the plasma levels of LH after a bench press exercise [33]. Similarly, mate tea increased the concentration of total polyphonic compounds at all time points and the levels of GSH after twenty maximal eccentric elbow flexion exercises [34].

McLeay et al. [35] and Park et al., [36] demonstrated the short-term effects of blueberries by detecting antioxidant activity. McLeay et al., [35] researched the antioxidant effects of New Zealand blueberries on exercise-induced muscle damage (EIMD) after strenuous eccentric exercise. This study showed that ingestion of a blueberry smoothie before and after EIMD accelerates recovery of muscle peak isometric strength, which might be due to the decreased ROS-generating potential and the gradual increase in plasma antioxidant capacity [35]. Similar results were reported by Park et al., [36]—increased TAS levels and significantly decreased IL-6 and C-reactive protein (CRP) levels were found in the blueberry supplementation period following exercise. Meanwhile, VO_2 max and exercise performance time were grown during the blueberry supplementation period.

Integral grape juice was used as the dietary strategy against exercise-induced oxidative stress in an acute study [38] and a 28-day study [39] by Toscano et al. A single-dose grape juice (10 mL/kg/day) taken 2 h before running to exhaustion showed an ergogenic effect by significantly increasing TAC at the postexercise time point compared to the baseline level [38]. After taking the same daily dose for 28 days, the grape juice group exhibited significant increases in plasma levels of TAC, vitamin A and uric acid compared to control group [39]. These improvements in antioxidant capacity found in both studies were accompanied by an increased time to exhaustion in recreational runners [38,39].

Tart cherry juice showed subchronic positive effects on antioxidant activity caused by high-intensity exercises in the study of Howatson et al. [41]. Tart cherry juice (two 8 oz bottles per day) was given for five days before, on the day of, and for 48 h following, a marathon run [41]. In the study of tart cherry juice, significantly increased TAS levels, and significantly reduced inflammation (IL-6, CRP, uric acid) and MDA levels were found in the intervention group compared with the placebo group [41]. One 8 oz bottle of tart cherry juice, which contains the equivalent of 50–60 cherries, provided at least 600 mg of phenolic compounds [41].

Juçara juice (*Euterpe edulis Martius*), with a similar chemical composition to açai fruit (*Euterpe oleracea Martius*), has strong antioxidant activity due to its high anthocyanins content [77]. Copetti et al. [45] evaluated the acute antioxidant effect of juçara juice during HIIT by observing antioxidant status. The HIIT was performed 1 h after drinking 250 mL of juçara juice or water (control). Compared to the control group, juçara juice intake promoted a decrease in oxidative stress index (OSI) immediately post exercise and an increase in reduced glutathione 1 h after exercise [45]. OSI was defined as the ratio of serum total oxidant status (TOS) to serum TAC in this study. These enhancements came with a significant increase in total plasma phenols content [45].

In addition to phenols-rich foods, Iwasa et al., [55] found that fermented milk (Lactobacillus helveticus) inhibited the reduction of antioxidant capacity (ORAC assay) induced by acute resistance exercise in a clinical trial. In the processing and manufacturing of fermented milk, *Lactobacillus* digests the proteins and transforms them into small peptides, which are more readily absorbed by the intestines than amino acids or large oligopeptides. Although the evidence for potential mechanisms is still lacking, the small peptides might contribute to the increasing level of antioxidants in contracting muscles [55].

Nevertheless, one of the dietary strategies included in this review showed no antioxidant effects. Beavers et al., [56] found that soy foods, as a source of high-quality protein and isoflavones, did not elevate antioxidant capacity (GPx, COX-2) after high intensity exercise stress.

In summary, the included studies elucidated the antioxidant effects of different dietary strategies by detecting ROS generation, oxidative stress markers, inflammatory markers and antioxidant activity. Among them, most studies included in this narrative review found that phenol-rich foods reduced exercise-induced oxidative stress, by short-term consumption [29,32,35,36,38,42,45] or long-term intake [30,31,33,34,37,39–41,43,44,46,47]. The potential antioxidant ability of dietary polyphenols has been widely demonstrated in both in vitro and in vivo studies [78]. As secondary plant metabolites, the majority of polyphenols have at least one aromatic ring and typically occur in the form of glycosides in their molecules. According to the chemical structures of the aglycones, polyphenols have been classified into flavonoid polyphenols (e.g., flavanols, anthocyanidins) and nonflavonoid polyphenols (e.g., phenolic acid, polyphenolic amides (e.g., AVA), resveratrol, curcumin, ellagic acid) [78]. Over 8000 polyphenolic compounds have been identified and more than 4000 flavonoids have been found among them. The functional hydroxyl group (OH) of polyphenolic compounds is assumed to play a key role in antioxidant defense [78]. It may inhibit the ROS synthesis, chelate with trace elements responsible for ROS generation, scavenge excessive ROS production, and improve the antioxidant defense [2,79]. The Phenol-Explorer Database (www.phenol-explorer.eu, accessed on 11 February 2021)) offers data on the presence of 502 polyphenols in 452 foods and provides an analysis of the volume of polyphenols included in a food serving [80]. From the database, the 100 richest dietary sources of polyphenols were identified [80]. The main rich sources of polyphenols are cocoa, fruits, vegetables, whole grains and tea in this review [78,80–82]. Besides flavanols, another category of flavonoid polyphenols, plant anthocyanidins, are found in the red, blue, and purple pigments of a plurality of flower petals, vegetables, fruits and some special types of grains (e.g., black rice). In this review, blueberries, grapes, cherries, and citrus fruits are all the main fruit sources. In addition to AVA, other nonflavonoid compounds have shown potential antioxidant effects in vitro [78].

Capsaicin, mainly found in chili peppers, is another polyphenolic amide compound that belongs to the nonflavonoids group [83]. Curcumin is a potent antioxidant in turmeric [84]. Resveratrol is a unique component of red wine and grapes [78]. Lignans are present in bound forms in sesame, flax and several grains [78]. Ellagic acid and its derivatives are contained in berry fruits (e.g., strawberries and raspberries) and the skins of some different tree nuts [78]. However, the mechanisms of their antioxidant effects in vitro and in vivo are still unclear.

A potential limitation of this narrative review needs to be mentioned. It is reported that endurance training could influence the oxidative stress response to acute exercise [85]. Variations in training type, intensity, and duration can activate different patterns of oxidant–antioxidant balance, resulting in different transcriptome responses for regulatory and metabolic processes [85,86]. In order to avoid the potential bias caused by different baseline values of redox status, only the untrained general population were included in this review.

When interpreting the results of diet interventions on exercise-induced oxidative stress, it is important to note that some studies included methods with questionable validity. One such method includes the assessment of TBARS, which was commonly regarded as a quantification assay for lipid peroxidation [87]. This assay, however, is no longer recommended in redox research, since TBA-reactive material in human body fluids is not related to lipid peroxidation [87]. TBARS lacks specificity since it reacts with numerous substrates in the assay medium to form MDA. Thereby, most MDA is produced artificially [88]. Another assay which is equally flawed is the measurement of total antioxidant capacity (TAC) [89]. TAC, for instance, is greatly dependent on plasma albumin or urate levels [87] and therefore, exercise-induced changes in urate concentrations can bias the assay by urate reacting with the peroxyl radical [88].

The only method that can directly measure free radicals is electron paramagnetic resonance (EPR), because it identifies the presence of unpaired electrons. However, EPR alone is limited to detecting only fairly unreactive radicals, since highly reactive compounds do not accumulate to measurable levels [87]. Therefore, specific agents (e.g., spin traps or spin probes) have been developed which allow forming a more stable radical which can then be detected by EPR [57,87]. Future studies are thus needed which combine valid direct and indirect measures of oxidative stress in order to further investigate the effects of different dietary strategies on exercise-induced oxidative stress.

5. Perspectives

The biological actions of antioxidant properties from an antioxidant-rich diet are complex. Previous studies have led to a few contrasting results, probably due to differences in antioxidant composition and actual bioavailability. Moreover, only Zeng et al., [42] used a direct measurement technique for assessing ROS generation, whereas the majority of studies used indirect markers of oxidative stress as a surrogate marker. Furthermore, the available studies show a high level of heterogeneity in their study designs. Consistent and standardized research procedures may be essential to obtain convincing evidence in future studies.

In this narrative review, most studies found positive effects of dietary strategies on exercise-induced ROS generation. Especially, phenol-rich diets showed effects in combating exercise-induced oxidative stress in the greater proportion of the articles. Accordingly, while dietary strategies might help to keep ROS generation in a physiological range during exercise, the use of the antioxidant-rich diets may upregulate the endogenous antioxidants' defense system, which may have important implications for preventing excessive damage and facilitating recovery. Nevertheless, consistent evidence is still lacking, and the underlying mechanisms in human trials are not well understood.

In future research, antioxidant dietary regimens for different individuals should developed with consideration of individual physiological characteristics and style. Moreover, a standardized assay as well as a study design protocol needs to be established. Further research is necessary to explore optimal antioxidant diets and to elucidate the potential mechanisms, by using standard detection assays and research protocols.

6. Conclusions

Although the literature about the effects of whole dietary strategies on exercise-induced oxidative stress is still scarce, the majority of the studies demonstrated favorable effects. Within this context, most of the included studies showed that phenol-rich foods had positive effects on exercise-induced oxidative stress in short-term and long-term experimental designs. Nevertheless, the protocols are still very heterogeneous and further systematically designed studies are needed to strengthen the evidence.

Author Contributions: Conceptualization, Z.Z., C.C., D.K. and A.G.; methodology, Z.Z. and C.C.; writing—original draft preparation, Z.Z.; writing—review and editing, Z.Z., C.C., D.K. and A.G.; supervision, D.K. and A.G. All authors have read and agreed to the published version of the manuscript.

Funding: The article processing charge was funded by the Baden-Wuerttemberg Ministry of Science, Research and Art and the University of Freiburg in the funding programme Open Access Publishing.

Conflicts of Interest: The authors declare no conflict of interest.

References

1. Halliwell, B. Free Radicals, Antioxidants, and Human Disease: Curiosity, Cause, or Consequence? *Lancet* **1994**, *344*, 721–724. [CrossRef]
2. Halliwell, B.; Gutteridge, J.M.C. *Free Radicals in Biology and Medicine*, 5th ed.; Oxford University Press: New York, NY, USA, 2015; ISBN 978-0198717485.
3. Dröge, W. Free Radicals in the Physiological Control of Cell Function. *Physiol. Rev.* **2002**, *82*, 47–95. [CrossRef] [PubMed]
4. Peternelj, T.T.; Coombes, J.S. Antioxidant Supplementation during Exercise Training: Beneficial or Detrimental? *Sports Med.* **2011**, *41*, 1043–1069. [CrossRef]
5. Valko, M.; Leibfritz, D.; Moncol, J.; Cronin, M.T.D.; Mazur, M.; Telser, J. Free Radicals and Antioxidants in Normal Physiological Functions and Human Disease. *Int. J. Biochem. Cell Biol.* **2007**, *39*, 44–84. [CrossRef] [PubMed]
6. Powers, S.K.; Duarte, J.; Kavazis, A.N.; Talbert, E.E. Reactive Oxygen Species Are Signalling Molecules for Skeletal Muscle Adaptation. *Exp. Physiol.* **2010**, *95*, 1–9. [CrossRef]
7. Powers, S.K.; Ji, L.L.; Kavazis, A.N.; Jackson, M.J. Reactive Oxygen Species: Impact on Skeletal Muscle. *Compr. Physiol.* **2011**, *1*, 941–969. [CrossRef]
8. Waris, G.; Ahsan, H. Reactive Oxygen Species: Role in the Development of Cancer and Various Chronic Conditions. *J. Carcinog.* **2006**, *5*, 14. [CrossRef]
9. Maritim, A.C.; Sanders, R.A.; Watkins, J.B. Diabetes, Oxidative Stress, and Antioxidants: A Review. *J. Biochem. Mol. Toxicol.* **2003**, *17*, 24–38. [CrossRef]
10. Ozkul, A.; Akyol, A.; Yenisey, C.; Arpaci, E.; Kiylioglu, N.; Tataroglu, C. Oxidative Stress in Acute Ischemic Stroke. *J. Clin. Neurosci.* **2007**, *14*, 1062–1066. [CrossRef]
11. Allen, C.L.; Bayraktutan, U. Oxidative Stress and Its Role in the Pathogenesis of Ischaemic Stroke. *Int. J. Stroke* **2009**, *4*, 461–470. [CrossRef]
12. Bloomer, R.J.; Schilling, B.K.; Karlage, R.E.; Ledoux, M.S.; Pfeiffer, R.F.; Callegari, J. Effect of Resistance Training on Blood Oxidative Stress in Parkinson Disease. *Med. Sci. Sports Exerc.* **2008**, *40*, 1385–1389. [CrossRef] [PubMed]
13. Brieger, K.; Schiavone, S.; Miller, F.J.; Krause, K.H. Reactive Oxygen Species: From Health to Disease. *Swiss Med. Wkly.* **2012**, *142*, w13659. [CrossRef] [PubMed]
14. Alessio, H.M.; Hagerman, A.E.; Fulkerson, B.K.; Ambrose, J.; Rice, R.E.; Wiley, R.L. Generation of Reactive Oxygen Species after Exhaustive Aerobic and Isometric Exercise. *Med. Sci. Sports Exerc.* **2000**, *32*, 1576–1581. [CrossRef] [PubMed]
15. Finaud, J.; Lac, G.; Filaire, E. Oxidative Stress: Relationship with Exercise and Training. *Sport. Med.* **2006**, *36*, 327–358. [CrossRef] [PubMed]
16. Tapia, P.C. Sublethal Mitochondrial Stress with an Attendant Stoichiometric Augmentation of Reactive Oxygen Species May Precipitate Many of the Beneficial Alterations in Cellular Physiology Produced by Caloric Restriction, Intermittent Fasting, Exercise and Dietary. *Med. Hypotheses* **2006**, *66*, 832–843. [CrossRef]
17. Merry, T.L.; Ristow, M. Mitohormesis in Exercise Training. *Free Radic. Biol. Med.* **2016**, *98*, 123–130. [CrossRef] [PubMed]
18. Pinckard, K.; Baskin, K.K.; Stanford, K.I. Effects of Exercise to Improve Cardiovascular Health. *Front. Cardiovasc. Med.* **2019**, *6*, 69. [CrossRef] [PubMed]
19. Russomanno, G.; Corbi, G.; Manzo, V.; Ferrara, N.; Rengo, G.; Puca, A.A.; Latte, S.; Carrizzo, A.; Calabrese, M.C.; Andriantsito-haina, R.; et al. The Anti-Ageing Molecule Sirt1 Mediates Beneficial Effects of Cardiac Rehabilitation. *Immun. Ageing* **2017**, *14*, 1–9. [CrossRef]

20. Korsager Larsen, M.; Matchkov, V.V. Hypertension and Physical Exercise: The Role of Oxidative Stress. *Med.* **2016**, *52*, 19–27. [CrossRef]

21. Sun, M.-W.; Zhong, M.-F.; Gu, J.; Qian, F.-L.; Gu, J.-Z.; Chen, H. Effects of Different Levels of Exercise Volume on Endothelium-Dependent Vasodilation: Roles of Nitric Oxide Synthase and Heme Oxygenase. *Hypertens. Res.* **2008**, *31*, 805–816. [CrossRef] [PubMed]

22. Pingitore, A.; Lima, G.P.P.; Mastorci, F.; Quinones, A.; Iervasi, G.; Vassalle, C. Exercise and Oxidative Stress: Potential Effects of Antioxidant Dietary Strategies in Sports. *Nutrition* **2015**, *31*, 916–922. [CrossRef] [PubMed]

23. Paulsen, G.; Hamarsland, H.; Cumming, K.T.; Johansen, R.E.; Hulmi, J.J.; Børsheim, E.; Wiig, H.; Garthe, I.; Raastad, T. Vitamin C and E Supplementation Alters Protein Signalling after a Strength Training Session, but Not Muscle Growth during 10 Weeks of Training. *J. Physiol.* **2014**, *592*, 5391–5408. [CrossRef]

24. Gomez-Cabrera, M.C.; Domenech, E.; Romagnoli, M.; Arduini, A.; Borras, C.; Pallardo, F.V.; Sastre, J.; Viña, J. Oral Administration of Vitamin C Decreases Muscle Mitochondrial Biogenesis and Hampers Training-Induced Adaptations in Endurance Performance. *Am. J. Clin. Nutr.* **2008**, *87*, 142–149. [CrossRef]

25. Stepanyan, V.; Crowe, M.; Haleagrahara, N.; Bowden, B. Effects of Vitamin E Supplementation on Exercise-Induced Oxidative Stress: A Meta-Analysis. *Appl. Physiol. Nutr. Metab.* **2014**, *39*, 1029–1037. [CrossRef]

26. Bruce, B.; Spiller, G.A.; Klevay, L.M.; Gallagher, S.K. A Diet High in Whole and Unrefined Foods Favorably Alters Lipids, Antioxidant Defenses, and Colon Function. *J. Am. Coll. Nutr.* **2000**, *19*, 61–67. [CrossRef]

27. Ristow, M.; Zarse, K.; Oberbach, A.; Klöting, N.; Birringer, M.; Kiehntopf, M.; Stumvoll, M.; Kahn, C.R.; Blüher, M. Antioxidants Prevent Health-Promoting Effects of Physical Exercise in Humans. *Proc. Natl. Acad. Sci. USA* **2009**, *106*, 8665–8670. [CrossRef]

28. Higgins, M.R.; Izadi, A.; Kaviani, M. Antioxidants and Exercise Performance: With a Focus on Vitamin e and c Supplementation. *Int. J. Environ. Res. Public Health* **2020**, *17*, 8452. [CrossRef] [PubMed]

29. Davison, G.; Callister, R.; Williamson, G.; Cooper, K.A.; Gleeson, M. The Effect of Acute Pre-Exercise Dark Chocolate Consumption on Plasma Antioxidant Status, Oxidative Stress and Immunoendocrine Responses to Prolonged Exercise. *Eur. J. Nutr.* **2012**, *51*, 69–79. [CrossRef] [PubMed]

30. Allgrove, J.; Farrell, E.; Gleeson, M.; Williamson, G.; Cooper, K. Regular Dark Chocolate Consumption's Reduction of Oxidative Stress and Increase of Free-Fatty-Acid Mobilization in Response to Prolonged Cycling. *Int. J. Sport Nutr. Exerc. Metab.* **2011**, *21*, 113–123. [CrossRef]

31. Taub, P.R.; Ramirez-Sanchez, I.; Patel, M.; Higginbotham, E.; Moreno-Ulloa, A.; Román-Pintos, L.M.; Phillips, P.; Perkins, G.; Ceballos, G.; Villarreal, F. Beneficial Effects of Dark Chocolate on Exercise Capacity in Sedentary Subjects: Underlying Mechanisms. A Double Blind, Randomized, Placebo Controlled Trial. *Food Funct.* **2016**, *7*, 3686–3693. [CrossRef]

32. Wiswedel, I.; Hirsch, D.; Kropf, S.; Gruening, M.; Pfister, E.; Schewe, T.; Sies, H. Flavanol-Rich Cocoa Drink Lowers Plasma F2-Isoprostane Concentrations in Humans. *Free Radic. Biol. Med.* **2004**, *37*, 411–421. [CrossRef] [PubMed]

33. Panza, V.S.P.; Wazlawik, E.; Ricardo Schütz, G.; Comin, L.; Hecht, K.C.; da Silva, E.L. Consumption of Green Tea Favorably Affects Oxidative Stress Markers in Weight-Trained Men. *Nutrition* **2008**, *24*, 433–442. [CrossRef] [PubMed]

34. Panza, V.P.; Diefenthaeler, F.; Tamborindeguy, A.C.; Camargo, C.D.Q.; De Moura, B.M.; Brunetta, H.S.; Sakugawa, R.L.; De Oliveira, M.V.; Puel, E.D.O.; Nunes, E.A.; et al. Effects of Mate Tea Consumption on Muscle Strength and Oxidative Stress Markers after Eccentric Exercise. *Br. J. Nutr.* **2016**, *115*, 1370–1378. [CrossRef]

35. McLeay, Y.; Barnes, M.J.; Mundel, T.; Hurst, S.M.; Hurst, R.D.; Stannard, S.R. Effect of New Zealand Blueberry Consumption on Recovery from Eccentric Exercise-Induced Muscle Damage. *J. Int. Soc. Sports Nutr.* **2012**, *9*, 19. [CrossRef] [PubMed]

36. Park, C.H.; Kwak, Y.S.; Seo, H.K.; Kim, H.Y. Assessing the Values of Blueberries Intake on Exercise Performance, TAS, and Inflammatory Factors. *Iran J. Public Health* **2018**, *47*, 27–32. [PubMed]

37. McAnulty, S.R.; McAnulty, L.S.; Nieman, D.C.; Dumke, C.L.; Morrow, J.D.; Utter, A.C.; Henson, D.A.; Proulx, W.R.; George, G.L. Consumption of Blueberry Polyphenols Reduces Exercise-Induced Oxidative Stress Compared to Vitamin C. *Nutr. Res.* **2004**, *24*, 209–221. [CrossRef]

38. de Lima Tavares Toscano, L.; Silva, A.S.; de França, A.C.L.; de Sousa, B.R.V.; de Almeida Filho, E.J.B.; da Silveira Costa, M.; Marques, A.T.B.; da Silva, D.F.; de Farias Sena, K.; Cerqueira, G.S.; et al. A Single Dose of Purple Grape Juice Improves Physical Performance and Antioxidant Activity in Runners: A Randomized, Crossover, Double-Blind, Placebo Study. *Eur. J. Nutr.* **2020**, *59*, 2997–3007. [CrossRef]

39. Toscano, L.T.; Tavares, R.L.; Toscano, L.T.; da Silva, C.S.O.; de Almeida, A.E.M.; Biasoto, A.C.T.; Gonçalves, M.d.C.R.; Silva, A.S. Potential Ergogenic Activity of Grape Juice in Runners. *Appl. Physiol. Nutr. Metab.* **2015**, *40*, 899–906. [CrossRef]

40. Bowtell, J.L.; Sumners, D.P.; Dyer, A.; Fox, P.; Mileva, K.N. Montmorency Cherry Juice Reduces Muscle Damage Caused by Intensive Strength Exercise. *Med. Sci. Sports Exerc.* **2011**, *43*, 1544–1551. [CrossRef]

41. Howatson, G.; McHugh, M.P.; Hill, J.A.; Brouner, J.; Jewell, A.P.; Van Someren, K.A.; Shave, R.E.; Howatson, S.A. Influence of Tart Cherry Juice on Indices of Recovery Following Marathon Running. *Scand. J. Med. Sci. Sport.* **2010**, *20*, 843–852. [CrossRef] [PubMed]

42. Zeng, Z.; Jendricke, P.; Centner, C.; Storck, H.; Gollhofer, A.; König, D. Acute Effects of Oatmeal on Exercise-Induced Reactive Oxygen Species Production Following High-Intensity Interval Training in Women: A Randomized Controlled Trial. *Antioxidants* **2021**, *10*, 3. [CrossRef]

43. Koenig, R.T.; Dickman, J.R.; Kang, C.H.; Zhang, T.; Chu, Y.F.; Ji, L.L. Avenanthramide Supplementation Attenuates Eccentric Exercise-Inflicted Blood Inflammatory Markers in Women. *Eur. J. Appl. Physiol.* **2016**, *116*, 67–76. [CrossRef]

44. Zhang, T.; Zhao, T.; Zhang, Y.; Liu, T.; Gagnon, G.; Ebrahim, J.; Johnson, J.; Chu, Y.F.; Ji, L.L. Avenanthramide Supplementation Reduces Eccentric Exercise-Induced Inflammation in Young Men and Women. *J. Int. Soc. Sports Nutr.* **2020**, *17*. [CrossRef]

45. Copetti, C.L.K.; Orssatto, L.B.R.; Diefenthaeler, F.; Silveira, T.T.; da Silva, E.L.; de Liz, S.; Mendes, B.C.; Rieger, D.K.; Vieira, F.G.K.; Hinnig, P.F.; et al. Acute Effect of Juçara Juice (Euterpe Edulis Martius) on Oxidative Stress Biomarkers and Fatigue in a High-Intensity Interval Training Session: A Single-Blind Cross-over Randomized Study. *J. Funct. Foods* **2020**, *67*, 103835. [CrossRef]

46. Pittaluga, M.; Sgadari, A.; Tavazzi, B.; Fantini, C.; Sabatini, S.; Ceci, R.; Amorini, A.M.; Parisi, P.; Caporossi, D. Exercise-Induced Oxidative Stress in Elderly Subjects: The Effect of Red Orange Supplementation on the Biochemical and Cellular Response to a Single Bout of Intense Physical Activity. *Free Radic. Res.* **2013**, *47*, 202–211. [CrossRef]

47. Chang, W.H.; Hu, S.P.; Huang, Y.F.; Yeh, T.S.; Liu, J.F. Effect of Purple Sweet Potato Leaves Consumption on Exercise-Induced Oxidative Stress and IL-6 and HSP72 Levels. *J. Appl. Physiol.* **2010**, *109*, 1710–1715. [CrossRef]

48. Mazani, M.; Nemati, A.; Baghi, A.N.; Amani, M.; Haedari, K.; Alipanah-Mogadam, R. The Effect of Probiotic Yoghurt Consumption on Oxidative Stress and Inflammatory Factors in Young Females after Exhaustive Exercise. *J. Pak. Med. Assoc.* **2018**, *68*, 1748–1754. [PubMed]

49. Harms-Ringdahl, M.; Jenssen, D.; Haghdoost, S. Tomato Juice Intake Suppressed Serum Concentration of 8-OxodG after Extensive Physical Activity. *Nutr. J.* **2012**, *11*, 29. [CrossRef]

50. Kawamura, A.; Aoi, W.; Abe, R.; Kobayashi, Y.; Kuwahata, M.; Higashi, A. Astaxanthin-, β-Carotene-, and Resveratrol-Rich Foods Support Resistance Training-Induced Adaptation. *Antioxidants* **2021**, *10*, 113. [CrossRef]

51. Sureda, A.; Tauler, P.; Aguiló, A.; Cases, N.; Llompart, I.; Tur, J.A.; Pons, A. Antioxidant Supplementation Influences the Neutrophil Tocopherol Associated Protein Expression, but Not the Inflammatory Response to Exercise. *Cent. Eur. J. Biol.* **2007**, *2*, 56–70. [CrossRef]

52. Carrera-Quintanar, L.; Funes, L.; Herranz-López, M.; Martínez-Peinado, P.; Pascual-García, S.; Sempere, J.M.; Boix-Castejón, M.; Córdova, A.; Pons, A.; Micol, V.; et al. Antioxidant Supplementation Modulates Neutrophil Inflammatory Response to Exercise-Induced Stress. *Antioxidants* **2020**, *9*, 1242. [CrossRef]

53. Daly, R.M.; Gianoudis, J.; de Ross, B.; O'Connell, S.L.; Kruger, M.; Schollum, L.; Gunn, C. Effects of a Multinutrient-Fortified Milk Drink Combined with Exercise on Functional Performance, Muscle Strength, Body Composition, Inflammation, and Oxidative Stress in Middle-Aged Women: A 4-Month, Double-Blind, Placebo-Controlled, Randomized Trial. *Am. J. Clin. Nutr.* **2020**, *112*, 427–446. [CrossRef]

54. Prasertsri, P.; Roengrit, T.; Kanpetta, Y.; Tong-Un, T.; Muchimapura, S.; Wattanathorn, J.; Leelayuwat, N. Cashew Apple Juice Supplementation Enhances Leukocyte Count by Reducing Oxidative Stress after High-Intensity Exercise in Trained and Untrained Men. *J. Int. Soc. Sports Nutr.* **2019**, *16*, 31. [CrossRef]

55. Iwasa, M.; Aoi, W.; Mune, K.; Yamauchi, H.; Furuta, K.; Sasaki, S.; Takeda, K.; Harada, K.; Wada, S.; Nakamura, Y.; et al. Fermented Milk Improves Glucose Metabolism in Exercise-Induced Muscle Damage in Young Healthy Men. *Nutr. J.* **2013**, *12*, 83. [CrossRef]

56. Beavers, K.M.; Serra, M.C.; Beavers, D.P.; Cooke, M.B.; Willoughby, D.S. Soy and the Exercise-Induced Inflammatory Response in Postmenopausal Women. *Appl. Physiol. Nutr. Metab.* **2010**, *35*, 261–269. [CrossRef]

57. Dikalov, S.I.; Polienko, Y.F.; Kirilyuk, I. Electron Paramagnetic Resonance Measurements of Reactive Oxygen Species by Cyclic Hydroxylamine Spin Probes. *Antioxid. Redox Signal.* **2018**, *28*, 1433–1443. [CrossRef]

58. Suzen, S.; Gurer-Orhan, H.; Saso, L. Detection of Reactive Oxygen and Nitrogen Species by Electron Paramagnetic Resonance (EPR) Technique. *Molecules* **2017**, *22*, 181. [CrossRef]

59. Collins, F.W. Oat Phenolics: Avenanthramides, Novel Substituted N-Cinnamoylanthranilate Alkaloids from Oat Groats and Hulls. *J. Agric. Food Chem.* **1989**, *37*, 60–66. [CrossRef]

60. Bratt, K.; Sunnerheim, K.; Bryngelsson, S.; Fagerlund, A.; Engman, L.; Andersson, R.E.; Dimberg, L.H. Avenanthramides in Oats (Avena Sativa L.) and Structure-Antioxidant Activity Relationships. *J. Agric. Food Chem.* **2003**, *51*, 594–600. [CrossRef]

61. Lee-Manion, A.M.; Price, R.K.; Strain, J.J.; Dimberg, L.H.; Sunnerheim, K.; Welch, R.W. In Vitro Antioxidant Activity and Antigenotoxic Effects of Avenanthramides and Related Compounds. *J. Agric. Food Chem.* **2009**, *57*, 10619–10624. [CrossRef]

62. Fu, J.; Zhu, Y.; Yerke, A.; Wise, M.L.; Johnson, J.; Chu, Y.; Sang, S. Oat Avenanthramides Induce Heme Oxygenase-1 Expression via Nrf2-Mediated Signaling in HK-2 Cells. Mol. *Nutr. Food Res.* **2015**, *59*, 2471–2479. [CrossRef] [PubMed]

63. Andújar, I.; Recio, M.C.; Giner, R.M.; Ríos, J.L. Cocoa Polyphenols and Their Potential Benefits for Human Health. *Oxid. Med. Cell. Longev.* **2012**, *2012*, 906252. [CrossRef]

64. Yan-Hwa, C.; Chang, C.L.; Hsu, H.F. Flavonoid Content of Several Vegetables and Their Antioxidant Activity. *J. Sci. Food Agric.* **2000**, *80*, 561–566. [CrossRef]

65. Chang, W.H.; Chen, C.M.; Hu, S.P.; Kan, N.W.; Chiu, C.C.; Liu, J.F. Effect of Purple Sweet Potato Leaf Consumption on the Modulation of the Antioxidative Status in Basketball Players during Training. *Asia Pac. J. Clin. Nutr.* **2007**, *16*, 455–461. [CrossRef]

66. Chang, W.H.; Chen, C.M.; Hu, S.P.; Kan, N.W.; Chiu, C.C.; Liu, J.F. Effect of Purple Sweet Potato Leaves Consumption on the Modulation of the Immune Response in Basketball Players during the Training Period. *Asia Pac. J. Clin. Nutr.* **2007**, *16*, 609–615. [CrossRef] [PubMed]

67. Mikelsaar, M.; Zilmer, M. Lactobacillus Fermentum ME-3—An Antimicrobial and Antioxidative Probiotic. *Microb. Ecol. Health Dis.* **2009**, *21*, 1–27. [CrossRef]
68. Uskova, M.A.; Kravchenko, L.V. Antioxidant properties of lactic acid bacteria—Probiotic and yogurt strains. *Vopr. Pitan.* **2009**, *78*, 18–23. [PubMed]
69. Abegaz, E.G.; Tandon, K.S.; Scott, J.W.; Baldwin, E.A.; Shewfelt, R.L. Partitioning Taste from Aromatic Flavor Notes of Fresh Tomato (Lycopersicon Esculentum, Mill) to Develop Predictive Models as a Function of Volatile and Nonvolatile Components. *Postharvest Biol. Technol.* **2004**, *34*, 227–235. [CrossRef]
70. Di Mascio, P.; Kaiser, S.; Sies, H. Lycopene as the Most Efficient Biological Carotenoid Singlet Oxygen Quencher. *Arch. Biochem. Biophys.* **1989**, *274*, 532–538. [CrossRef]
71. Carrera-Quintanar, L.; Funes, L.; Vicente-Salar, N.; Blasco-Lafarga, C.; Pons, A.; Micol, V.; Roche, E. Effect of Polyphenol Supplements on Redox Status of Blood Cells: A Randomized Controlled Exercise Training Trial. *Eur. J. Nutr.* **2015**, *54*, 1081–1093. [CrossRef]
72. Hussain, T.; Tan, B.; Yin, Y.; Blachier, F.; Tossou, M.C.B.; Rahu, N. Oxidative Stress and Inflammation: What Polyphenols Can Do for Us? *Oxid. Med. Cell. Longev.* **2016**, *2016*, 7432797. [CrossRef] [PubMed]
73. Kawamura, T.; Muraoka, I. Exercise-Induced Oxidative Stress and the Effects of Antioxidant Intake from a Physiological Viewpoint. *Antioxidants* **2018**, *7*, 119. [CrossRef] [PubMed]
74. Wintergerst, E.S.; Maggini, S.; Hornig, D.H. Immune-Enhancing Role of Vitamin C and Zinc and Effect on Clinical Conditions. *Ann. Nutr. Metab.* **2006**, *50*, 85–94. [CrossRef]
75. Kubo, I.; Masuoka, N.; Ha, T.J.; Tsujimoto, K. Antioxidant Activity of Anacardic Acids. *Food Chem.* **2006**, *99*, 555–562. [CrossRef]
76. Si, W.; Gong, J.; Tsao, R.; Kalab, M.; Yang, R.; Yin, Y. Bioassay-Guided Purification and Identification of Antimicrobial Components in Chinese Green Tea Extract. *J. Chromatogr.* **2006**, *1125*, 204–210. [CrossRef]
77. Schulz, M.; da Silva Campelo Borges, G.; Gonzaga, L.V.; Oliveira Costa, A.C.; Fett, R. Juçara Fruit (Euterpe Edulis Mart.): Sustainable Exploitation of a Source of Bioactive Compounds. *Food Res. Int.* **2016**, *89*, 14–26. [CrossRef]
78. Tsao, R. Chemistry and Biochemistry of Dietary Polyphenols. *Nutrients* **2010**, *2*, 1231–1246. [CrossRef]
79. Elejalde, E.; Villarán, M.C.; Alonso, R.M. Grape Polyphenols Supplementation for Exercise-Induced Oxidative Stress. *J. Int. Soc. Sports Nutr.* **2021**, *18*, 1–12. [CrossRef]
80. Pérez-Jiménez, J.; Neveu, V.; Vos, F.; Scalbert, A. Identification of the 100 Richest Dietary Sources of Polyphenols: An Application of the Phenol-Explorer Database. *Eur. J. Clin. Nutr.* **2010**, *64*, 112–120. [CrossRef]
81. Kim, K.H.; Tsao, R.; Yang, R.; Cui, S.W. Phenolic Acid Profiles and Antioxidant Activities of Wheat Bran Extracts and the Effect of Hydrolysis Conditions. *Food Chem.* **2006**, *95*, 466–473. [CrossRef]
82. Adom, K.K.; Liu, R.H. Antioxidant Activity of Grains. *J. Agric. Food Chem.* **2002**, *50*, 6182–6187. [CrossRef] [PubMed]
83. Davis, C.B.; Markey, C.E.; Busch, M.A.; Busch, K.W. Determination of Capsaicinoids in Habanero Peppers by Chemometric Analysis of UV Spectral Data. *J. Agric. Food Chem.* **2007**, *55*, 5925–5933. [CrossRef]
84. Suhett, L.G.; de Miranda Monteiro Santos, R.; Silveira, B.K.S.; Leal, A.C.G.; de Brito, A.D.M.; de Novaes, J.F.; Lucia, C.M. Effects of Curcumin Supplementation on Sport and Physical Exercise: A Systematic Review. *Crit. Rev. Food Sci. Nutr.* **2021**, *61*, 946–958. [CrossRef]
85. Schmutz, S.; Däpp, C.; Wittwer, M.; Vogt, M.; Hoppeler, H.; Flück, M. Endurance Training Modulates the Muscular Transcriptome Response to Acute Exercise. *Pflugers Arch. Eur. J. Physiol.* **2006**, *451*, 678–687. [CrossRef] [PubMed]
86. Nocella, C.; Cammisotto, V.; Pigozzi, F.; Borrione, P.; Fossati, C.; D'Amico, A.; Cangemi, R.; Peruzzi, M.; Gobbi, G.; Ettorre, E.; et al. Impairment between Oxidant and Antioxidant Systems: Short- and Long-term Implications for Athletes' Health. *Nutrients* **2019**, *11*, 1353. [CrossRef] [PubMed]
87. Halliwell, B.; Whiteman, M. Measuring Reactive Species and Oxidative Damage in Vivo and in Cell Culture: How Should You Do It and What Do the Results Mean? *Br. J. Pharmacol.* **2004**, *142*, 231–255. [CrossRef]
88. Cobley, J.N.; Close, G.L.; Bailey, D.M.; Davison, G.W. Exercise Redox Biochemistry: Conceptual, Methodological and Technical Recommendations. *Redox Biol.* **2017**, *12*, 540–548. [CrossRef] [PubMed]
89. Arts, M.J.T.J.; Haenen, G.R.M.M.; Voss, H.-P.; Bast, A. Antioxidant Capacity of Reaction Products Limits the Applicability of the Trolox Equivalent Antioxidant Capacity (TEAC) Assay. *Food Chem. Toxicol.* **2004**, *42*, 45–49. [CrossRef]

Review

A Brief Overview of Oxidative Stress in Adipose Tissue with a Therapeutic Approach to Taking Antioxidant Supplements

Shima Taherkhani [1,*], Katsuhiko Suzuki [2,*] and Ruheea Taskin Ruhee [3,*]

[1] Department of Exercise Physiology, Faculty of Sport Sciences, University of Guilan, Rasht 4199843653, Iran
[2] Faculty of Sport Sciences, Waseda University, 2-579-15 Mikajima, Tokorozawa 359-1192, Japan
[3] Gradute School of Sport Sciences, Waseda University, 2-579-15 Mikajima, Tokorozawa 359-1192, Japan
* Correspondence: shimataherkhani@msc.guilan.ac.ir (S.T.); katsu.suzu@waseda.jp (K.S.); ruhee@fuji.waseda.jp (R.T.R.); Tel./Fax: +98-910-092-7732 (S.T.); +81-4-2947-6898 (K.S. & R.T.R.)

Abstract: One of the leading causes of obesity associated with oxidative stress (OS) is excessive consumption of nutrients, especially fast-foods, and a sedentary lifestyle, characterized by the ample accumulation of lipid in adipose tissue (AT). When the body needs energy, the lipid is broken down into glycerol (G) and free fatty acids (FFA) during the lipolysis process and transferred to various tissues in the body. Materials secreted from AT, especially adipocytokines (interleukin (IL)-1β, IL-6, and tumor necrosis factor-α (TNF-α)) and reactive oxygen species (ROS), are impressive in causing inflammation and OS of AT. There are several ways to improve obesity, but researchers have highly regarded the use of antioxidant supplements due to their neutralizing properties in removing ROS. In this review, we have examined the AT response to OS to antioxidant supplements focusing on animal studies. The results are inconsistent due to differences in the study duration and diversity in animals (strain, age, and sex). Therefore, there is a need for different studies, especially in humans.

Keywords: oxidative stress; adipose tissue; obesity; antioxidant supplement

Citation: Taherkhani, S.; Suzuki, K.; Ruhee, R.T. A Brief Overview of Oxidative Stress in Adipose Tissue with a Therapeutic Approach to Taking Antioxidant Supplements. *Antioxidants* **2021**, *10*, 594. https://doi.org/10.3390/antiox10040594

Academic Editors: Gareth Davison and Conor McClean

Received: 5 March 2021
Accepted: 8 April 2021
Published: 13 April 2021

Publisher's Note: MDPI stays neutral with regard to jurisdictional claims in published maps and institutional affiliations.

1. Introduction

Since 1998, the National Institutes of Health (NIH) has recognized obesity as a disease due to the impact of individuals' health on society and the high economic and social costs incurred [1]. There is an adjacent link between obesity and metabolic disorders, including Alzheimer's disease, respiratory problems, cardiovascular disease (CVD), type 2 diabetes (T2D), cancer, and non-alcoholic fatty liver disease (NAFLD) [2]. Body mass index (BMI) $\geq$ 30 has been accepted in many studies as one of the critical indicators of obesity. Still, this index is less valid than measuring the waist-to-hip ratio (WHR) due to the inability to count the lean body mass (LBM) [3].

Obesity is the result of overconsumption of nutrients and a sedentary lifestyle. As the consumption of nutrients increases, an imbalance is created between energy intake and expenditure, leading to fat accumulation in adipose tissue (AT) and obesity [4]. The World Health Organization (WHO) estimates the number of obese people globally at 650 million [2]. Several studies have shown that obesity depends on the regional distribution of excess body fat, not excess body weight. Thus, one of the most critical risk factors for obesity and related diseases is abdominal fat, which leads to the stimulation of pro-inflammatory and pro-oxidant states [5], the overproduction of free radicals, and pursuant oxidative stress (OS) in AT [6].

Scientists have made several efforts to control this disease. Various treatment methods, such as medication, surgery, exercise, and diet, have been considered in this regard. However, control of the disease is still far from expected. Increasing energy expenditure and subsequent weight loss is a smart way to control and prevent obesity [7]. In this regard, although authoritative articles have approved anti-obesity drugs, such as orlistat, and the use of weight-loss surgeries, the use of these methods is associated with many side effects [4]. Health researchers have identified diet, especially antioxidant supplements, as the most appropriate treatment for obesity [8]. Antioxidants affect the body's endocrine and metabolic functions, leading to increased exothermic process and energy expenditure to reduce OS and body weight and improve obesity [9].

In this study, we have reviewed the effect of antioxidant supplements on AT changes under OS. The authors have allocated the content of this review article to the introduction, sources, and tools for measuring reactive oxygen species (ROS) in AT, a brief description of AT and related disorders, and finally, the relationship between antioxidant supplementation and obesity, respectively.

2. Overview of ROS

Living organisms need oxygen (O_2) molecules to survive on earth. Therefore, these molecules' presence is necessary to produce energy by the electron transfer chain (ETC) [10,11]. Under stressful conditions, O_2 molecules in the body are converted into two separate atoms with unpaired electrons, named free radicals. These radicals are derived from O_2 and are known as ROS [12]. ROS include superoxide anion ($O_2^{\bullet-}$), hydrogen peroxide (H_2O_2), and hydroxyl radical ($OH\cdot$), which play a vital role in causing pathophysiological damage, especially cellular damage to lipids, proteins, and deoxyribonucleic acid (DNA) [13–15]. When an unpaired electron is added to free radicals, $O_2^{\bullet-}$ is formed. $O_2^{\bullet-}$ has shown various behaviors in different environments. For example, in aqueous perimeters, this radical is reduced first to H_2O_2 by superoxide dismutase (SOD) and then converted to H_2O and O_2 by catalase (CAT). However, H_2O_2 may be converted to $OH\bullet$ in the presence of molecules containing ferrous iron (Fe^{2+}) [16]. Various factors such as ROS concentration, time, and location of cells exposed to these species can determine the extent of these molecules' damage. ROS is not harmful in low to moderate concentrations and has beneficial effects on cellular responses and signaling, gene expression, regulation of muscle power fluctuations, mitogenic responses, apoptosis, and protection against infections [11,12]. On the other hand, $OH\cdot$ is highly reactive and harmful due to its very short half-life of only a few nanoseconds. Although H_2O_2 can be stable for a more extended period and does not damage cells, in higher concentrations, it has highly detrimental potency. H_2O_2 in aquatic environments, especially the human body, has a shorter half-life due to its neutralizing enzymes, which quickly cause irreparable damage to cells [17].

Since the discovery of ROS in 1970, most tissues in the body have been found to be affected by these reactive species, including cellular redox imbalance, OS, and cell dysfunction. ROS's breakdown and production imbalance cause OS to alter cell function by damaging various molecules in the body [18,19]. In addition to ROS, other reactive species such as reactive nitrogen (RNS) and sulfur species (RSS) are also known as free radicals, although they are not derived from O_2 [20]. When nitric oxide (NO) is added to $O_2^{\bullet-}$, it creates a highly damaging radical named RNS that can cause the formation of peroxinitrate ($ONOO^-$). This molecule causes nitrosative stress to various cells in the body [21]. When ROS is overproduced in the body, the antioxidant defense systems cannot eliminate or neutralize these species, and components such as proteins and lipids are damaged. Following this damage, pathological conditions such as vascular diseases (atherosclerosis, hypertension, and diabetes), respiratory disease, cell death, premature aging, neurological disorders, and degradation of skin enzymes (hyaluronidase and collagenase), platelet aggregation in vessels, and mutations and damage occur [22–25].

Many molecules play an essential role in maintaining the body's homeostasis. However, one of the most important natural products of metabolism is ROS, which participate in numerous cellular signaling pathways in the body. Of course, these products have few effects on the cellular system, but their excessive production may have irreversible effects on the body's various physiological systems [26]. The body uses endogenous antioxidant defenses against these molecules. However, in stressful situations, endogenous protection alone may not be enough to eliminate or neutralize ROS. In such cases, various exogenous factors such as diet, lifestyle, medication, and physical activity play an essential role in maintaining ROS balance [27]. One of the tissues that are severely affected by ROS imbalance is AT. Under similar conditions, adipokines secreted by AT such as leptin and adiponectin increase and decrease, respectively [16]. ROS production in AT occurs due to excessive consumption of nutrients [20,23,28]. It is noteworthy that the hormone adiponectin acts as an anti-inflammatory hormone in AT. Since obesity is an inflammatory disease, this hormone's concentration in obesity decreases due to increased inflammatory cytokines. By reducing this hormone's expression in obese people, its influential role in improving insulin sensitivity also diminishes. As a result, obese people face a complication named insulin resistance (IR), which predisposes them to T2D [29].

3. ROS Manufacturer Resources

Multiple factors are responsible for the production of ROS, both endogenous and exogenous. Endogenous sources are: mitochondria, cellular oxidases (xanthine oxidase (XO), nicotinamide adenine dinucleotide phosphate (NADPH) oxidase (NOX)), nitric oxide synthase (NOS), myeloperoxidase (MPO), processes related to peroxisomes, cellular respiration, cytochrome P450 oxidases, microsomal cyclooxygenase (COX), and catalyzed metal reactions. ROS is also produced exogenously through sources such as chemical drugs, pollutants, nutrient overdose, mutagens, xenobiotics, and ionizing radiation (Figure 1) [24]. Several studies have shown that ROS-derived mitochondria and NOX are critical sources of ROS production in adipocytes [30].

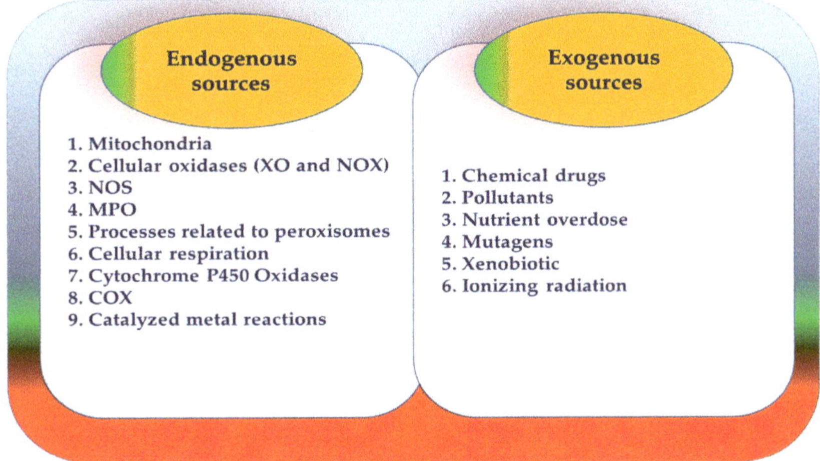

Figure 1. Reactive oxygen species (ROS) resources [24].

3.1. ROS-Derived from Mitochondria

The main energy production source in the body is the mitochondria, which do this by oxidative phosphorylation. Interestingly, the $O_2^{\bullet-}$ radical is mainly produced by oxidative phosphorylation. $O_2^{\bullet-}$ is made in the mitochondrial ETC complex due to not being metabolized by about 0.15% and 2% of oxygen consumption in complexes I and III. Hence, mitochondria are one of the main sources of ROS and oxidative stress. After producing $O_2^{\bullet-}$ mitochondrial manganese SOD (MnSOD) converts it to H_2O_2 [16,21]. On the other hand, ROS is mainly produced by the respiratory chain and during the formation of adenosine triphosphate (ATP). O_2 is created by activating the oxygen molecule's base state by transferring electrons or energy in the form of a single O_2 [11].

3.2. NOX

Various cellular oxidases such as NOX and XO can produce ROS by reducing electrons from O_2. Endothelial cells, chondrocytes, fibroblasts, myocytes, and phagocytes are the sites of NOX that produce ROS, particularly $O_2^{\bullet-}$ and H_2O_2 to regulate cellular responses [31]. NOX initially produces $O_2^{\bullet-}$, followed by produces H_2O_2 by the action of the antioxidant enzyme SOD. Scientists have confirmed that H_2O_2 at low concentrations can modulate the signaling pathway and metabolism and have a similar function to ATP and calcium (Ca^{2+}). Because this radical crosses the cell membrane by aquaporins (AQPS) or proxy purines it can cause effects such as proliferation and recruitment of immune cells [32].

When germs attack these cells, NOX enzymes are activated during a respiratory burst. The enhanced products then absorb NADPH and O_2. Thus, NADPH can act as an electron donor. This action starts the NOX enzyme complex in the plasma membrane by producing $O_2^{\bullet-}$ from O_2 molecules. In general, the production of $O_2^{\bullet-}$ by NOX is related to the time when an electron is taken from NADPH in the cytoplasm and transferred to an O_2 molecule [33].

NOX consists of a total of seven isoforms of catalytic subunits, including NOX 1-5 and dual oxidase 1 (Duox1) and dual oxidase 2 (Duox2). It should be noted that the main isoform of NOX in fat cells is NOX4. In response to the excessive consumption of glucose or palmitate, this isoform concentration in AT increases [21]. On the other hand, classical cytosolic subunits are not required for NOX4 activation, and only P22 ^{phox} is needed. Furthermore, the modulation of NOX4 activity is responsible for Polymerase delta-interacting protein 2 (Poldip2), which ultimately produces $O_2^{\bullet-}$ and H_2O_2. NOX5 and Duoxs 1 and 2 do not require cytosolic subunits for activation. These three members of the NOX family must bind to intracellular N-terminal EF hand motifs via Ca^{2+} for activation. The EF hand has a helix-loop-helix structure, which is mainly found in calcium-bound proteins. This eventually leads to the production of $O_2^{\bullet-}$ and H_2O_2, respectively [34–37]. In short, all NOX members except NOX5 need the P22 ^{phox} subunit to form. This subunit is usually regulated by the mineralocorticoid receptor (MR). It should be noted that all NOX components look at NADPH as an electron donor for the production of $O_2^{\bullet-}$ and H_2O_2 [16]. NOX enzyme complexes play an important role in the production of $O_2^{\bullet-}$ by transferring electrons from NADPH to O_2. H_2O_2 is known as a highly absorbent radical in cell membranes. Finally, H_2O_2 is reduced to H_2O and O_2 by the enzyme CAT [34].

Mitochondria can produce ROS in both direct and indirect forms. Mitochondria can indirectly serve as a target for ROS production by the NOX enzyme complex, indicating a cross-link between NOX and mitochondria. In addition to acting as a potential source of ROS, mitochondria can also be responsible for NOX stimulation under certain conditions. This is especially important when ROS is neutralized by target mitochondrial antioxidant enzymes. By inhibiting ROS production, these enzymes can also partially alleviate NOX activity [33].

NADH and 1,5-dihydroflavin adenine dinucleotide (FADH2) are the products of glucose metabolism as electron donors in the tricarboxylic acid (TCA) cycle. This process eventually accelerates ROS production. On the other hand, the oxidation of free fatty acids (FFA) by mitochondria increases FFA intake. In this case, NADH and FADH2 are also produced by the oxidation of FFA-derived acetyl-CoA and the beta-oxidation of fatty acids (FAs) as electron donors. On the other hand, NOX is present on plasma membranes and can convert molecular O_2 to $O_2^{\bullet-}$. NOX may be closely related to ROS production associated with nutrient overdose [38]. Excessive FFA accumulation in adipocytes increases ROS production. On the other hand, ROS overproduction is reversed by NOX inhibitors such as diphenyleneiodonium or apocynin. This indicates NOX's role in the production of ROS due to excessive consumption of fatty acids. Activation of NADPH oxidase by excessive consumption of fatty acids stimulates the synthesis of diacylglycerol and subsequent activation of protein kinase C (PKC) by FFA, especially palmitate [39]. FFA's molecular mechanism that activates the NOX enzyme complex is closely related to the stimulation of diacylglycerol synthesis and subsequently activated PKC [21].

4. ROS Measuring Tools

ROS levels' evaluation and measurement are important, practical steps to improve these reactive species' effects. By measuring these species, a more accurate view of them can be achieved, and the appropriate treatment method can be used for each of them. Of course, direct measurement of ROS has its problems and difficulties [40]. This is important because some ROS, such as $O_2^{\bullet-}$ and $^{\bullet}OH$, while having very short half-lives of 5^{-10} and 9^{-10} s, respectively, also have very high reactivity. Over the years, countless indicators have been discovered to measure these species, but many of them did not provide consistent and reliable results and were easily discarded. However, in the following years, valid indicators were calculated to measure the oxidation of various tissues and cells of lipids, proteins, and DNA [41].

In general, the grouping of OS indicators is very important. In one group, the biochemical nature of molecules such as proteins, lipids, carbohydrates, and DNA is considered. While in the second group, products of oxidation of cellular compounds are formed to balance cellular mechanisms (oxidation-reduction). This group itself is divided into several subgroups [42]. The first group to free radicals leads to the change of various biomolecules such as malondialdehyde (MDA) from lipid oxidation, 4-hydroxy-2-nonenal (4-HNE), and reactive carbonyls from protein oxidation and 8-hydroxy-2'-deoxyguanosine (8-OHdG) of nucleic acid oxidation have been noted. In the second group, the relationship between free radical metabolism and physiological antioxidant defense molecules such as reduced glutathione (GSH) and CAT is considered. Furthermore, in the third group, modulation of free radicals with transcription factors such as c-Myc and Nuclear factor-κB (NF-κB) is important [16]. Here are the most common and commonly used indicators of OS:

4.1. MDA

When lipid molecules are exposed to OS, various products are produced in low-density lipoproteins (LDL) or cell membranes. One of its end products is MDA. The reactive substance, thiobarbituric acid reactive substances (TBARS), is responsible for measuring MDA levels [13].

4.2. 8-OHdG

When DNA is exposed to ROS, compounds such as 8-hydroxy guanine, 8-OHdG, and MDA-DNA are formed to break the DNA strands. 8-OHdG is composed of guanine oxidation, which plays an important role in mutagenic DNA damage and is used as a suitable indicator of oxidative damage to DNA [42].

4.3. 8-Nitroguanine (8-NO$_2$-Gua)

In addition to producing 8-OHdG, guanine nitration also makes 8-NO$_2$-Gua. These DNA metabolites are used to measure OS to DNA. In fact, in inflamed epithelial cells, 8-NO$_2$-Gua levels increase. The amino acid polypeptides are separated by the reaction of amino acid side groups with ROS, and proteins are oxidized. With the oxidation of proteins, reactive carbonyl groups (aldehydes and ketones) are formed, and their tracing is known as an indicator of oxidative damage in protein molecules. It should also be noted that carbonyl groups are formed due to ROS reactions with proteins, carbohydrates, and lipids [43].

4.4. Oxidative Products of Sugars

Among the products produced by the oxidation of carbohydrates are advanced glycation end products (AGEs), which are formed due to non-enzymatic glycosylation of proteins. The highest presence of AGEs is in plasma and tissues, leading to diabetes, kidney failure, and aging [44].

4.5. Reduced Glutathione (GSH): Oxidized Glutathione (GSSG) Ratio

One of the most sensitive indicators of oxidative damage is the redox ratio GSH: GSSG. Influence on regulating gene expression, signaling conduction, NO metabolism, apoptosis, and impact on free radical scavenging are among the effective roles of GSH. On the other hand, the removal of $ONOO^-$ is highly dependent on the formation of oxidized glutathione (GS-SG) by GSH, which is eventually converted to GSH through NADPH-dependent glutathione reductase [45]. Various signaling pathways such as C-Jun N-terminal kinase, protein kinase B, mitogen-activated protein kinase, apoptosis signal-regulated kinase 1, NF-κB, and protein phosphatase 1 and 2A are affected due to changes in the GSH/GSSG ratio [46].

5. A Brief Look at AT

The definition given to AT today is very different from what it used to be. It was previously thought that this tissue was just a tissue with the property of storing energy in the form of lipids. Today, a new perspective has emerged on it as an endocrine tissue [21,47–49]. Of course, the feature of this tissue's storage source has helped many living things throughout history. When there is a lack or excessive consumption of nutrients, it has always been AT that has been able to help maintain the body's energy homeostasis with hyperplasia or hypertrophy in different conditions [50–53].

On the other hand, the new look at this tissue owes much to discovering the hormone leptin in 1994 as a food controller. Even earlier, in 1987, it was found that sex steroids were metabolized in this tissue, followed by the production of adipsin [54,55]. Adipsin was one of the first adipokines to be identified in cultured adipocytes based on differentiation-dependent expression of its mRNA. Adipsin is an endocrine factor secreted by 3T3 fat cells [56–58]. Other important roles of adipose tissue in the body include effects on lipid and glucose metabolism, maintaining energy balance, appetite control, glucose homeostasis, insulin sensitivity, energy expenditure, inflammation, and repair of AT [59,60]. Various factors such as FFA supply, FFA esterification to triglycerides (TG), and TG degradation through the lipolysis process determine fat stores in AT. In general, the two enzymes of hormone-sensitive lipase (HSL) and adipose triglyceride lipase (ATGL) play an important role in the lipolysis process, which indicates the formation of FFA and glycerol as a result of the separation of the TG ester bond [61]. In general, AT contains high levels of stromal vascular cells, immunity, stem, endothelial, lymphocytes, adipocytes, preadipocytes, connective tissue matrix, and nerve tissue [55,62,63]. This tissue is also divided into brown adipose tissue (BAT) and white adipose tissue (WAT). There are apparent differences between BAT and WAT, morphologically. One of these differences is related to the size of the fat storage drops. White adipose cells are placed in a large fat drop (unilocular), and brown fat cells are placed in several small fat cytoplasm drops (multilocular) [64]. Another apparent difference between these two types of tissue is the number of mitochondria in them. The number of mitochondria in BAT is much higher than in WAT but the number of mitochondria in the WAT is limited. Because WAT plays an important role in lipid metabolism processes, including beta-oxidation and the TCA cycle, maturation, and differentiation of adipocytes, the importance of mitochondrial function is highlighted [16]. In this way, the BAT can maintain body temperature, especially when it is cold. BAT owes this feature to the high number of mitochondria within it. Of course, this tissue also plays a very important role in lipid oxidation [65].

The mitochondria's inner membrane hosts a protein called uncoupling protein 1 (UCP1) that can generate heat by transferring protons to the mitochondrial matrix and separating oxidative phosphorylation and the electron transfer chain from ATP synthesis [60,66,67]. In fact, by consuming too many nutrients and being exposed to cold temperatures, UCP1 is expressed to protect the body's organisms against obesity and the cold. Sympathetic neurons stimulate UCP1 inside the BAT to lead to exotherm and energy loss. Thus, in both humans and rodents, the association between obesity and UCP1 expression is inverse [68]. Sesterins are among the vital proteins associated with obesity due to oxidative stress, which play an important role in regulating metabolic homeostasis, suppressing ROS accumulation, and regulating the AMP-activated protein kinase (AMPK)-mammalian target of rapamycin complex 1 (mTORC1) signaling pathway. However, reducing these proteins in the body is associated with obesity and other metabolic disorders [69]. In general, three different isoforms have been discovered for sestrin (sestrin 1–3), of which sestrin 2 is more expressed in liver and adipose tissue. It has been estimated that sestrin 2 can reduce fat accumulation in AT and improve metabolic homeostasis by suppressing ROS and mTORC1 [70].

The most important site for cold-induced exotherm in rodents is BAT. This tissue is also involved in the exotherm of fat tissue. The distribution of BAT in humans and rodents varies according to their age. In humans, there is a large amount of BAT in the body only in early infancy, and with age, the distribution of BAT in the body decreases. However, in the case of rodents, the opposite is true. Because BAT expands as rodents live longer, on the other hand, different types of BAT have been deposited in the body in a scattered manner; interscapular BAT (IBAT) is the most important and vital type [71]. The point here is that the calorific value of IBAT is consistent with the body's OS. The hypothalamus is responsible for controlling IBAT activity and is controlled primarily by the sympathetic nerves and body temperature control centers. Oxidation of FFA in IBAT provides the fuel needed for noradrenaline to activate lipolysis and heat production. However, since IBAT uses a lot of O_2 to generate heat, part of the oxygen molecules are converted to free radicals, especially $O_2^{\bullet-}$ by mitochondrial assemblies I and III [72]. However, WAT works more to maintain the balance of energy homeostasis and the source of fat storage and release. WAT content includes subcutaneous WAT (scWAT), visceral WAT (vWAT), and peripheral arteries, each containing fat such as the omental, gonadal, retroperitoneum, epicardial, mesenteric, and perineal. However, it should be noted that all of these reserves are responsible for regulating total energy homeostasis [16]. Most importantly, the endocrine properties of AT are embedded in WAT [66]. This means that there are many hormonal mediators, including cytokines (IL-1β, IL-6, and TNFα), adipokines (adiponectin, resistin, and leptin), and chemokines (macrophage inflammatory protein 1 (MIP1), monocyte chemoattractant protein-1 (MCP-1)), ROS and FFA from WAT [73]. However, various studies have shown that pro-inflammatory cytokines are much less expressed in scWAT than in vWAT [29].

On the other hand, different hormones affect the storage and release of WAT fat. For example, after a meal, TG storage is highly dependent on insulin action. While in fasting and when different body organs (skeletal muscle and liver) need energy, the TG stored in WAT is broken down by catecholamines into FFA and G [65]. Furthermore, beige or brown (brown-in-white) fat cells are a new type of exothermic fat cell within WAT. However, these cells have both white and brown fat cells at the same time. They are more similar to fat white cells in terms of growth and more similar to BAT in terms of function and morphology [64,74].

6. Metabolic Disorders in AT

One of the practical and important factors in maintaining animals' and humans' body weight is maintaining fat homeostasis in WAT. This depends a lot on the proper and adequate performance of the WAT-derived materials [75,76]. Excess fat due to the long-term balance of positive energy contributes to OS in adipocytes, obesity, and subsequent obesity-related metabolic disorders such as hyperglycemia, insulin resistance and cardiovascular disease. [55,65]. Thus, systemic OS is closely related to obesity. During obesity, the concentration of OS indicators such as high-sensitivity C-reactive protein (CRP) and oxidized LDL increases [33]. Obesity results from increasing the size and volume of body fat cells and has adverse effects on the health of living organisms [76,77]. On the other hand, the mismatch between height and body weight due to excessive fat accumulation is named obesity [78]. Today, the people of developed countries are facing obesity-related health problems, but the people of developing countries are grappling with this global dilemma [79].

It is estimated that by 2030, half of all retirees in the United States will suffer from chronic obesity and related diseases, especially cardiovascular disease. This has numerous negative effects on the global health system due to the costly and time-consuming treatment of obesity [80]. Countless BMI studies have accepted more than 30 as obesity, but several studies have shown that BMI is not a good indicator of obesity. These studies cited the inability of this criterion to measure lean mass. These studies have shown that the measure of waist-to-pelvis or waist circumference may be a more accurate and accurate indicator of BMI in calculating fat distribution [16,81]. According to statistics published by the WHO, the prevalence of this disease is likely to reach more than one billion people in the world by 2030, which is a worrying statistic [79,82]. On the other hand, there is a direct and positive relationship between BMI and indicators of oxidative damage to proteins (advanced oxidation protein products (AOPP), lipids (MDA or 8-iso-PGF2α), and DNA (8-OHdG) during obesity) [30].

Oxidative DNA damage due to prolonged exposure to OS impairs mitochondrial function and leads to excessive fat accumulation and subsequent insulin resistance [16]. In recent years, obesity has become widespread due to lifestyle changes such as increased consumption of nutrients, especially fast foods, and decreased physical activity (environmental factors). Of course, the role of genetics is also felt to some extent in the development of this disease, but its effect is not as tangible as environmental factors [83]. There should be a strong emphasis on body weight control as a practical way to prevent obesity-related diseases. This usually happens with lifestyle modifications and focuses on eating healthy and adequate nutrients and engaging in regular exercise [84]. Various studies have shown that obesity is associated with increased OS, decreased antioxidant activity, and insulin resistance in AT [85]. Numerous studies have shown that fat accumulation in obesity is closely related to an increase in ROS and subsequent OS. Therefore, obesity-induced OS plays an important role in disrupting adipokines regulation and amplifying inflammatory signals and even leads to changes in cellular composition and premature aging [84]. It is estimated that with excessive consumption of nutrients, WAT expands (size increases) by 10% per year, which is known as fatty remodeling and the penetration of immune cells into AT [21]. AT remodeling leads to the rapid spread of obesity, which is usually accompanied by changes in the size (hypertrophy) and number (hyperplasia) of fat cells [16,51].

Obesity is the cause of various other diseases such as T2D, dyslipidemia, cardiovascular disease, atherosclerosis, and hypertension [53]. Clinical studies have also shown that BMI is directly related to OS by-products such as protein carbonylation products or lipid peroxidation. By causing OS in AT, adipocytokines' secretion (IL-1β, IL-6, and TNFα) is disrupted and eventually leads to obesity and associated diseases [85]. Of course, the function of other cells and tissues in the body, including beta pancreatic cells, vascular endothelial cells, and myocytes, is affected by obesity due to adipose tissue [21].

7. Antioxidants

As mentioned earlier in this article, oxidative damage can be defined as an imbalance between ROS production and antioxidant defense, leading to overproduction of ROS. The result of this imbalance is a change in cellular redox status. In vivo, antioxidant defense systems play an important role in restoring cellular redox status, especially under normal and stress-free conditions [86]. To combat OS, the body uses enzymatic antioxidant systems (SOD, CAT, peroxidase (POD), peroxiredoxin (Prxs), and glutathione peroxidases (GPX)) and non-enzymatic (carotenoids, tocopherol, and ascorbic acid) [85]. When the body is under pressure using various stressors, especially fat accumulation in AT, these antioxidant defenses alone may not be sufficient and require the use of antioxidant supplements [87–91].

7.1. SOD

McCord and Fridovich, by discovering SOD, showed that this enzyme could defend cells exposed to O_2 as a defense mechanism [86]. To counteract $O_2^{\bullet-}$, SOD is the first enzyme to convert this free radical to H_2O_2 [92–94]. Based on specific cofactors and cell locations, there are three different isoforms of SOD. These isoforms include cytosolic (SOD1 or Cu/ZnSOD), mitochondrial SOD (SOD2 or MnSOD), and extracellular SOD (SOD3 or ecSOD) [16].

SOD1 consists of both copper and zinc ions, which are responsible for maintaining enzymatically active sites. SOD1 does this by working with the remaining imidazolate ligands of the histidine SOD1. On the other hand, zinc ions are responsible for stabilizing enzymes in different cells of the body. It has also been estimated that the nuclear part of mammalian cells, cytoplasm, peroxisomes, lysosomes, chloroplasts, and cytosols host SOD1. However, the highest SOD1 activity has been reported in the human liver [91]. The second cofactor of SOD is MnSOD, which has the most increased activity in the renal cortex, and mainly peroxisomes and mitochondrial matrix are the enzyme sites [95]. The third cofactor (EC-SOD) is also present in human lymphocytes and plasma. Zinc and copper are found in this enzyme and effectively remove $O_2^{\bullet-}$ from tissues [96].

7.2. GPx

GPx is usually in the mitochondria and cytosol of various cells and is mainly a glycoprotein containing selenocysteine residues. This antioxidant enzyme is skilled in converting H_2O_2 to water. This enzyme also participates in the catalysis cycle to reduce hydroperoxides to alcohol and ultimately involves the oxidation of GSSG induced by GSH [97]. There is a positive relationship between increased GPX concentration and anti-inflammatory activity of the cardiovascular system. On the other hand, lipid hydroperoxides such as cholesterol, free fatty acids, cholesterol esters, and phospholipids are rapidly neutralized by phospholipases and GPX. It is also noteworthy that the detoxification of lipid hydroperoxides is performed by the enzymes PRx, glutathione S-transferase (GST), and GPX [92]. To date, approximately five different isoforms of GPX have been identified. These isoforms include cytosolic or classical GPX (cGPX or GPx1), gastrointestinal GPX (GIGPX or GPX2), plasma GPX (PGPX or GPX3), phospholipid GPX (PHGPX or GPX4), and sperm nuclear GPx or GPx (Sn) [97].

7.3. CAT

The peroxisome part of many cells contains the enzyme CAT, which effectively reduces hydrogen peroxide to water. As mentioned earlier, both CAT and GPX are sensitive to H_2O_2. These two enzymes are exposed to high and low H_2O_2 concentrations, respectively. For this reason, the concentration of free radicals determines the importance of the two enzymes GPX and CAT [98].

Another way to measure ROS is to observe changes in the antioxidant defense system. Tools such as Total Antioxidant Status (TAS), Trolex Equivalent Antioxidant Capacity (TEAC), Total Radical Trapping Antioxidant Parameter (TRAP), Plasma Iron Reduction Capacity (FRAP), and Radical Oxygen Absorption Capacity (ORAC) can measure antioxidant capacity [41].

8. Obesity, OS, and Antioxidant Supplementation

Because obesity is more associated with physical inactivity and overeating, genetics play a very limited role in causing the disease. Therefore, for the treatment of obesity, special attention should be paid to the lifestyle because this disease can be prevented and even treated by lifestyle modification [73,99,100]. As mentioned, one of the most important treatment strategies and, of course, prevention of various diseases, especially obesity, is exercise. Multiple studies have shown that AT reserves are reduced by regulating exercise-induced lipase regulation, which ultimately leads to weight loss and obesity treatment. Another important role of exercise is to create antioxidant profiles, which can be a key solution to further reduce body fat due to OS [61]. The results of various studies show the depletion of both enzymatic and non-enzymatic antioxidant systems. However, the type of tissue and the degree of obesity is among the factors that play an important role in the rate of discharge of these systems [30].

These enzymes protect the body's cells by catalyzing free radicals into water. Various studies have shown that Prxs expression in humans and obese animals is closely related to OS induced by AT. The activity of this enzyme decreases with obesity. On the other hand, PRDXS in adipocytes can increase and decrease lipolytic and lipogenic gene expression, respectively [101]. Scientists have studied the effects of antioxidant supplements on the improvement of obesity caused by OS in various studies. These studies' results are contradictory, and further studies in this field are still required to reach a correct and logical conclusion. For example, vitamin E is one of the supplements for which the usefulness or harmfulness in treating obesity or other metabolic disorders remains unclear [30].

Simán et al. (1996) examined the effect of consuming an antioxidant diet containing butylated hydroxytoluene (BHT 0.5% and 1%) with or without vitamin E acetate (4%) for four weeks in 30 female Sprague Dawley rats. They concluded no change in the alpha-tocopherol concentration of abdominal AT with BHT supplementation [102]. In another study, Rodrigues et al. (2020) examined the effect of consuming an antioxidant fruit called chestnut at a dose of 1.1% in 18 FVB/Nn male 7-month-old mice. They concluded that this supplement reduced adipose tissue, serum cholesterol, and adipose tissue deposition [103].

Furthermore, Candiracci et al. (2014) investigated the effect of consuming an antioxidant source of rice bran enzymatic extract for 20 weeks in obese and lean Zucker rats. This study's results included the reduction of overproduction of IL-6, TNF-α, IL-1β, and NOS in abdominal and epidermal visceral AT. In addition, reducing the adipocyte size of abdominal and epidural visceral AT was another effect of this supplement on AT [29]. In a study, Valls et al. (2003) investigated the impact of eating a diet rich in corn oil with or without antioxidant supplementation of vitamin E (30 mg per day) on the antioxidant status and oxidative damage of AT in male Wistar rats. This study showed that the activity of the antioxidant enzymes CAT and SOD was reduced by taking a hyperlipidemia supplement along with vitamin E in AT [104].

In one study, Arias et al. (2014) examined the effect of quercetin (30 mg/kg body weight) in 28 male Wistar rats. This study shows that this supplement has no impact on reducing AT size and body weight. The activity of lipoprotein lipase and lipogenic enzymes remained unchanged with the use of this supplement [105]. Chen et al. (2020) investigated the effect of antioxidant supplementation of protease A-digested crude-chalaza hydrolysates (CCH-As) on Syrian male Golden Hamsters. They showed that adipose-perinatal/hepatic tissue size decreased as a result of consuming this antioxidant composition. Increased lipolysis (unpaired carnitine palmitoyltransferase 1, hormone-sensitive lipase, and protein 2) was also observed in these hamsters' AT [106]. Because mice, unlike humans, can endogenously synthesize vitamin C (ascorbate and ascorbic acid) and meet their daily needs, it is hypothesized that consuming extra amounts of vitamin C will counteract the anti-inflammatory effects. Therefore, in a study, researchers examined the effect of 4 weeks of vitamin C supplementation (low and high doses of 0.75 and 25 mg of ascorbic acid per kg of body weight, respectively) on male Wistar rats. Excessive consumption of this antioxidant supplement was able to strengthen antioxidant defenses (MnSOD, CuZnSOD, and CAT in AT [107]. Sung et al. (2012) investigated the effect of antioxidant supplementation of Polygonum aviculare L. (knotgrass) (PAE) in male C57BL/6J mice. They were given a high-fat diet or a high-fat diet with PAE antioxidant supplementation at a dose of 400 mg/kg body weight per day. In this article, the researchers found that adipose tissue weight, serum TG concentration, body weight, MDA and leptin concentrations, and fat cell area decreased as a result of taking this supplement [108]. Furthermore, Alcalá et al. (2015) examined the effect of taking antioxidant vitamin E supplementation (150 mg twice daily) in C57BL/6J mice. This study's results included a reduction in collagen deposition and OS in rat visceral AT. Consumption of this vitamin also led to increased storage capacity and fat cells' proliferation [30] (Table 1).

Table 1. The effect of antioxidant supplementation on obesity caused by oxidative stress (OS).

Reference	Subjects	Antioxidant Supplementation	Results
Simán et al. [102]	Sprague Dawley rats	BHT (0.5% and 1%) with or without vitamin E acetate (4%) for four weeks.	No change in the alpha-tocopherol concentration of abdominal AT with BHT supplementation.
Rodrigues et al. [103]	FVB/n male 7-month-old mice	Chestnut at a dose of 1.1%.	The reduction of serum cholesterol and AT deposition.
Candiracci et al. [29]	Obese and lean Zucker rats	Rice bran enzymatic extract (RBEE) for 20 weeks.	The reduction of overproduction of IL-6, TNF-α, IL-1β, and NOS in abdominal and epidermal visceral AT. Reducing the adipocyte size of abdominal and epidural visceral AT.
Valls et al. [104]	Male Wistar rats	Diet rich in corn oil with or without antioxidant supplementation of vitamin E (30 mg per day).	The reduction of activity of the antioxidant enzymes CAT and SOD.
Arias et al. [105]	Male Wistar rats	Quercetin (30 mg/kg body weight).	No impact on reducing AT size and body weight. No change in the activity of lipoprotein lipase and lipogenic enzymes.
Chen et al. [106]	Syrian male Golden Hamsters	Protease A-digested crude-chalaza hydrolysates (CCH-As).	The reduction adipose-perinatal/hepatic tissue size. The increase of lipolysis (unpaired carnitine palmitoyltransferase 1, hormone-sensitive lipase, and protein 2).
Djurasevic et al. [107]	Male Wistar rats	Vitamin C supplementation (low and high doses of 0.75 and 25 mg of ascorbic acid per kg of body weight, respectively) for 4 weeks.	Excessive consumption of this antioxidant supplement was able to strengthen antioxidant defenses (MnSOD, CuZnSOD, and CAT in AT
Sung et al. [108]	Male C57BL/6J mice	High-fat diet or a high-fat diet with PAE at a dose of 400 mg/kg body weight per day.	The reduction of AT weight, serum TG concentration, body weight, MDA and leptin concentrations, and fat cell area.
Alcalá et al. [30]	C57BL/6J mice	Vitamin E supplementation (150 mg twice daily).	The reduction in collagen deposition and OS in rat visceral AT. The increase of storage capacity and fat cells' proliferation.

9. AT, Coronavirus Disease 2019 (COVID-19), and Antioxidants

AT is one of the essential tissues that modulate innate and adaptive immune responses in the body. This tissue modulates these responses by secreting adipokines such as leptin and adiponectin. However, during obesity, the function of this tissue is impaired. This means that the secretion of leptin and adiponectin increases and decreases, respectively, and eventually, the immune system's role is impaired [109]. In such cases, the chest wall is also affected by fat accumulation and impairs the lungs' proper functioning [110]. One of the consequences of an impaired immune system is the induction of inflammatory cytokines and the development of viral infections such as COVID-19 due to reduced natural killer (NK) cell activity. This infectious disease is caused by SARS-COV-2 (Severe Acute Respiratory Syndrome Coronavirus 2) (Figure 2) [111–113].

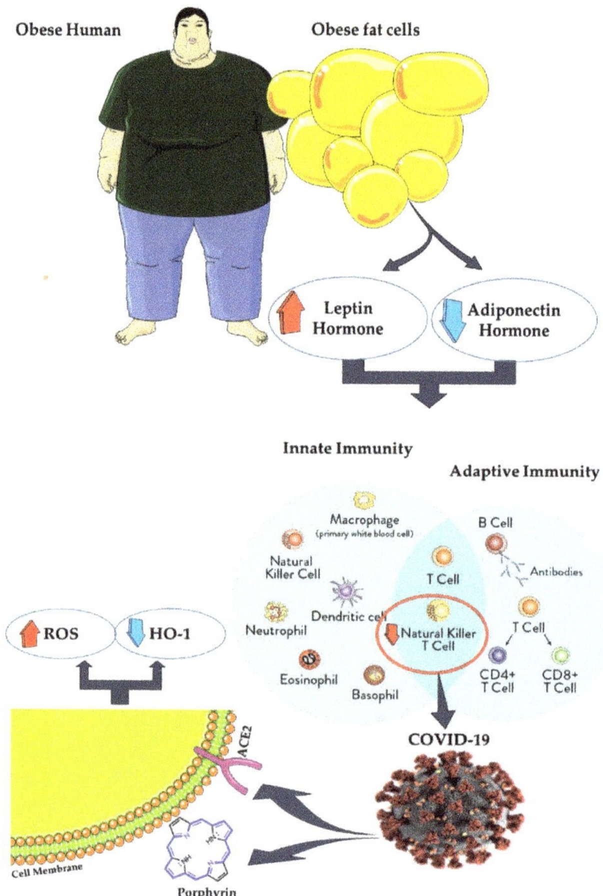

Figure 2. During obesity, adipose tissue (ATs) function is impaired, and secretion of leptin and adiponectin increases and decreases, respectively. Moreover, the immune system's function is impaired. One of the consequences of an impaired immune system is the induction of viral infections such as COVID-19 due to reduced natural killer (NK) cell activity. COVID-19 requires binding to the Angiotensin-Converting Enzyme 2 (ACE2) receptor and porphyrins on the cell surface to enter and then infect fat cells. Eventually, heme oxygenase-1 (HO-1) and ROS levels decrease and increase, respectively [109–113].

COVID-19 was first seen in December 2019 in Wuhan Province, China. Then, in January 2020, the disease's first cases were reported outside China (one in Japan and two in Thailand). Since then, the disease has spread rapidly to all countries of the world [114,115]. The condition was declared a pandemic on 11 March 2020, by WHO on 11 March 2020, and to date (22 February 2021), the total number of infected patients has reached 112,045,556, of which 2,479,625 people lost their lives (https://www.worldometers.info/coronavirus/, accessed on 20 February 2021). The virus requires binding to the Angiotensin-Converting Enzyme 2 (ACE2) receptor and porphyrins on the cell surface to enter and then infect fat cells. Eventually, heme oxygenase-1 enzymes (HO-1) and ROS levels decrease and increase, respectively [116–118].

Fatigue, headache, fever, and loss of taste and smell are symptoms associated with this disease, and most of these infected people recover without hospitalization. Various studies examining healthy people and people with underlying conditions have shown that people with cardiovascular disease, kidney damage, diabetes, and severe obesity (BMI $\geq$ 30 kg/m^2) are more susceptible to the virus [119]. The risk of developing COVID-19 does not depend on age, and the severity of the disease follows a different pattern at each age. According to Public Health England (PHE), the risk of COVID-19 death in people with a BMI between 35 and 40 kg/m^2 increases by 40%. However, this increase of risk in people with a BMI $\geq$ 40 kg reaches 90% [120]. As mentioned, obesity is directly related to COVID-19 disease and leads to increased inflammation, mitochondrial dysfunction, and increased ACE2 receptors. Numerous studies have shown that high BMI ($\geq$30 kg/m^2) and excess visceral fat (VF) are effective methods in diagnosing the severity of COVID-19, especially in obese patients [115,119,121].

For more than a year, the COVID-19 disease has affected human society in all aspects of life. The medical community has been able to develop effective vaccines against the disease. Furthermore, scientists in authoritative articles have suggested various drugs and nutrients reduce inflammation in the immune system, indirectly helping cure the disease. Among the various nutrients, antioxidants (vitamins C, D, and E, iron, and selenium) have always been at the forefront of strengthening the immune system and reducing inflammation in various body tissues, especially AT [122]. The recommended dose of vitamins C, D, and E in healthy individuals is 200 mg/day, 2000 IU/day (50 µg/day), and 15 mg/day, respectively. However, in patients who have inflammation in their immune system, it is better to increase the daily intake of vitamin C to 1-2 gr. It has also been suggested that the daily dose of vitamin D in these patients be increased to 10,000 IU in the first few weeks and then continued at a dose of 5000 IU. Also, the daily intake of vitamin E in these patients should be increased to 200 IU [123,124]. Consumption levels of another nutrient, iron, are usually about 8 mg daily in men, approximately 18 mg in women between the ages of 19 and 50, and around 8 mg in women over 51 years of age. However, if the person has inflammation in the immune system, 60 mg Fe should be consumed daily in both men and women and all age groups [114]. The daily intake of selenium in healthy men and women is 50 µg, respectively, but in inflammatory conditions, this amount increases to 200 µg per day [125].

10. NAFLD and OS

Various factors such as central obesity, IR, T2D, over nutrition, lack of exercise, and other metabolic syndrome parameters predispose multiple diseases such as non-alcoholic fatty liver disease (NAFLD) [126,127]. NAFLD is usually characterized by fat accumulation in the liver tissue, and oxidative stress plays a crucial role in its formation and development. It covers a wide range of liver-related diseases such as steatosis, steatohepatitis, liver fibrosis, liver cirrhosis, and even hepatocellular carcinoma [128,129]. For NAFLD, there are non-progressive forms (non-alcoholic fatty liver disease (NAFLD) or simple steatosis) and progressive and aggressive forms (non-alcoholic steatohepatitis (NASH)). Hepatocellular carcinoma (HCC) and cirrhosis are considered as consequences of NASH [130–132]. Macrophages and Kupffer cells can stimulate pro-inflammatory mechanisms and then satellite cell activity at the liver surface by secreting inflammatory cytokines such as IL-6, TNF-α, and IL-β. In such inflammatory conditions, conditions are provided for increasing the deterioration of insulin resistance and the development of liver fibrosis (Figure 3) [133]. In general, in patients with NAFLD, lipids' storage capacity in the liver tissue is so high that it leads to hepatocyte dysfunction and even death [134].

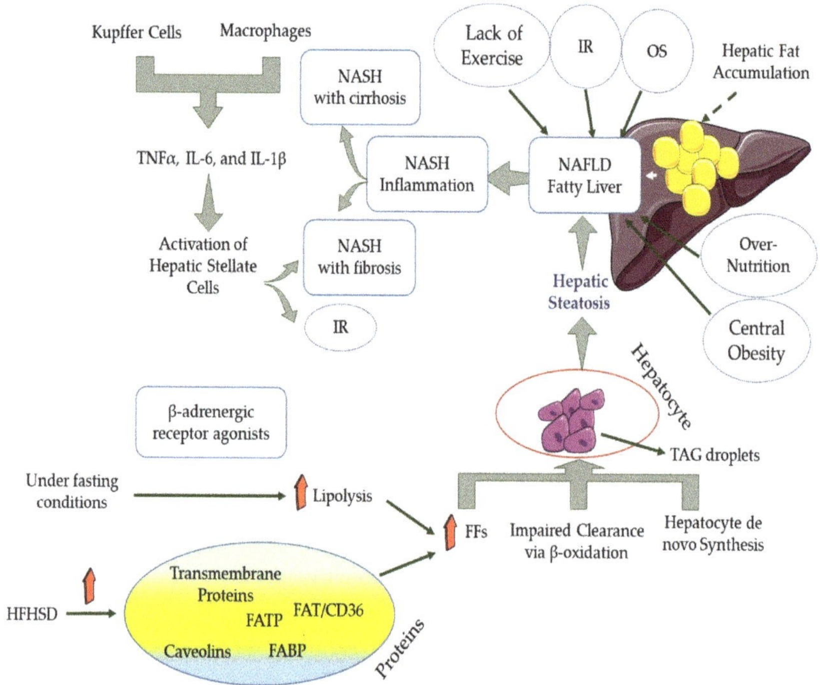

Figure 3. Various factors such as central obesity, insulin resistance (IR), type 2 diabetes (T2D), overnutrition, lack of exercise, and other metabolic syndrome parameters predispose multiple diseases such as NAFLD. Macrophages and Kupffer cells can stimulate pro-inflammatory mechanisms and then satellite cell activity at the liver surface by secreting inflammatory cytokines such as IL-6, TNF-α, and IL-β. In such inflammatory environments, conditions are provided for increasing the deterioration of IR and the development of liver fibrosis. It should be noted that the increased flow of FAs to the liver through the bloodstream, the synthesis of de novo hepatocytes, and impaired clearance through β-oxidation lead to the accumulation of TAG droplets in hepatocytes. Various proteins such as FATP, transmembrane proteins, FABP, caveolins, and FAT/CD36 can accelerate the absorption of FA by increasing the proliferation of FFA in blood vessels. It is noteworthy that these proteins' expression can improve by an HFHSD. On the other hand, in fasting conditions, FFAs are mainly produced during the lipolysis process by beta-adrenergic receptor agonists [126–133].

Numerous studies have shown that the mechanisms associated with the pathogenesis of obesity and NAFLD are the same [135]. A two-hit theory can usually explain the pathogenesis of NAFLD. The first theory is explained when triacylglycerol (TAG) droplets accumulate in hepatocytes and lead to simple hepatic steatosis development. In the second theory, NAFLD's pathogenesis is attributed to increased oxidative stress, IR, lipid peroxidation, and endoplasmic reticulum inflammation [136,137]. It should be noted that the increased flow of FAs to the liver through the bloodstream, the synthesis of de novo hepatocytes, and impaired clearance through β-oxidation lead to the accumulation of TAG droplets in hepatocytes. TG synthesis in the liver is mainly due to the lipids produced by de novo lipogenesis (DNL), dietary lipids, and carbohydrates. TG synthesis is dependent on the uptake of FFAs from the plasma by the liver [138]. DNL is the process by which exogenous energy sources or endogenous carbohydrates can synthesize lipids. Three steps are defined for this process. First, FAs can be synthesized through acetyl-CoA subunits produced during glycolysis and carbohydrate metabolism. Then, to form long-chain unsaturated FAs, FA elongation and desaturation must occur. Finally, the FAs formed from the previous step are assembled to convert to TG and very-low-density lipoproteins (VLDLs). When the balance between TG synthesis and degradation is lost, the conditions for NAFLD are created [130]. Various proteins such as fatty acid (FA) transporter protein (FATP), transmembrane proteins, FA binding protein (FABP), caveolins, FA translocase (FAT)/CD36 can accelerate the absorption of FA by increasing the proliferation of FFA in blood vessels. It is noteworthy that these proteins' expression can improve by a high-fat, high-sugar diet (HFHSD). On the other hand, in fasting conditions, FFAs are mainly produced during the lipolysis process by beta-adrenergic receptor agonists [139].

AT is severely affected by NAFLD because it is a source of FAs storage, and the secretion of adipokines is impaired. AT acts like a double-edged sword. This means that some hormones secreted by AT, such as adiponectin and visfatin, have protective effects against NAFLD; however, the hormones resistin and leptin contribute to hepatic development of steatosis and IR [140,141].

ROS production in hepatic mitochondria results from excessive oxidation of fatty acids, which ultimately causes OS in liver tissue. Proteins, DNA, and lipids are susceptible to OS and are easily damaged by activating pro-inflammatory cells such as Kupffer cells and stimulating the release of inflammatory cytokines. Furthermore, the expression and activity of antioxidant enzymes are usually inhibited by ROS overload, and thus, the liver's antioxidant capacity undergoes a declining trend. Finally, NAFLD occurs as a result of OS and chronic inflammation. Researchers should try to reduce OS to improve NAFLD disease [142].

Approximately 25% of adults worldwide are affected by this disease. This trend is increasing, and the number of these patients increases every year. According to a meta-analysis study, the global prevalence of NAFLD has risen to 25.2% in the last 20 years and has caused concern among the public [143]. On the other hand, no effective treatment for this disease has been achieved despite significant medical advances. Currently, the only treatment approach is lifestyle changes (diet and exercise) and bariatric surgery [144]. Because there are substantial differences between different communities in terms of lifestyle and diet, various studies have shown that the prevalence of NAFLD in Eastern societies is lower than in Western societies [145]. At the systemic level, there is impaired control of food intake resulting in hyperalimentation, intestinal dysbiosis leading to gastrointestinal hormone secretion, IR, gut dysfunction, abnormal adipokine, and activation of pro-inflammatory factors [146].

One of the effective strategies in the prevention and treatment of NAFLD is nuclear factor erythroid-derived 2-like 2 (Nrf2), which as a transcription factor consists of a highly protected basic region-leucine zipper (bZIP) structure and is mainly a member of the Cap "n" Collar (CNC) family. Activation of cellular antioxidant enzymes, regulation of lipid metabolism, and insulin sensitivity improvement are the essential cytoprotective effects of Nrf2. Hence, many researchers have tried to identify Nrf2 activators to improve NAFLD [147].

The Kelch-like-ECH-associated protein 1 (Keap1)-Nrf2-antioxidative response element (ARE) signaling pathway has been considered an essential antioxidant mechanism due to its effect on improving the oxidative stress response [148]. The Nrf2 gene includes six highly protected epichlorohydrin (EHC) domains (Nrf2-EHC homology, Neh) called Neh1-6. The C-terminus Neh1 subtends a protected bZIP DNA region that binds to musculoaponeurotic fibrosarcoma protein (Maf) to create a heterodimer. This heterodimer eventually binds to DNA and can detect ARE. On the other hand, Neh2 is composed of two vital regions, ETGE and DLG, which, by binding to KEAP1, can contribute to the strong binding of Nrf2 to the cytoplasm [129]. C-terminus is the site of another Nrf2 domain, Neh3, which participates in the transcriptional activity of ARE after binding with chromo-ATPase/helicase DNA-binding protein (CHD6). The other two Nrf2 domains, Neh4 and Neh5, initiate the transcription process when interacting with the cyclic adenosine monophosphate response element (CREB)-binding protein (CBP) [149]. Finally, the last Nrf2 domain, Neh6, and being rich in serine are used to Nrf2 decompose independent of KEAP1 [150]. The expression of Nrf2 in homeostatic conditions and combination with KEAP1 in the cytoplasm is considered a mediator for the degradation and ubiquitinoylation of Nrf2. However, when exposed to oxidative or electrophilic stress, KEAP1 modulates cysteine residues and ultimately releases Nrf2. On the other hand, the Nrf2 protein isolated from KEAP1 returns to the cell nucleus and is dimerized to bind to AREs, along with bZIP proteins such as Maf [151], and then promotes the expression of ARE-mediated downstream target genes containing antioxidant enzymes. The most critical antioxidant proteins targeting Nrf2 are HO-1, GSH, and NAD(P)H quinone oxidoreductase 1 (NQO1). It should also be noted that Nrf2 plays a pivotal role in suppressing the progression of NAFLD, maintaining cellular homeostasis, and protecting against oxidative or electrophilic stresses [152].

11. Roles of Nutraceuticals as an Antioxidant in Reducing Oxidative Stress

Nutraceuticals are not recognized traditionally as a nutrient but have physiological health benefits in the human body. Plant-derived nutraceuticals are well-known for their direct or indirect antioxidant activities, which relates to scavenging or eliminating free radicals during cellular metabolism. They can interact with the oxidized species at both cellular and molecular levels by regulating gene expression, epigenetic controls, and protein and DNA repair. Previously it was reported that nutraceuticals have potential properties in immunity modulation, gene expression, and various signaling process regulation [153–158]. The nutraceuticals can be prepared from the foods available in the local market, for example, ginger, garlic, avocado, and onion, in the form of polyphenols, carotenoids, sulforaphane and other isothiocyanates, glucosinolate, phytosterol, etc. [159]. They can increase the level of heme oxygenase (HO) 1, total glutathione, and other phase 2 enzymes by activating the transcription Nrf2. Moreover, treatment for a certain period with nutraceuticals may also improve the lipid profile and can reverse the harmful effects of obesity on blood lipids [160]. For example, curcumin is a potential nutraceutical, reduces macrophage infiltration in WAT, increase adiponectin in AT, decreases NF-κB activity, therefore reduces the expression of inflammatory markers and OS [161]. To date, the use of nutraceuticals, bioactive compounds or exercise could be an additional strategy in reducing obesity and related diseases [162,163].

12. Conclusions

OS affects various tissues, such as adipose tissue, skeletal muscle, and heart, in the body. In this study, we specifically examined adipose tissue response to OS. As mentioned in the text, this tissue is disrupted by various factors such as overconsumption of nutrients and sedentary lifestyle. This disorder eventually leads to lipid accumulation in adipose tissue and reduced energy expenditure. Of course, various treatments have been introduced for this disorder. However, most of them face limitations that are fully explained in the text. On the other hand, numerous studies have proven the effectiveness of diet, especially the use of antioxidant supplements, on the improvement of obesity caused by OS. The results of this treatment are inconsistent but have fewer side effects than other treatments such as medication and surgery. Further studies are needed because the results of the studies are contradictory. In future studies, researchers will investigate the effect of taking antioxidant supplements on heart and skeletal muscle tissues.

Author Contributions: S.T. conceived the review and drafted the manuscript. K.S. and R.T.R. made some additions to the text, revised the manuscript and approved the final version. All authors have read and agreed to the published version of the manuscript.

Funding: The publication was supported by the Scientific Research (A) (20H00574) from the Ministry of Education, Culture, Sports, Science, and Technology of Japan.

Acknowledgments: We gratefully appreciate our colleagues and laboratory team for research progress and discussion.

Conflicts of Interest: The authors declare no conflict of interest.

References

1. Tchang, B.G.; Saunders, K.H.; Igel, L.I. Best practices in the management of overweight and obesity. *Med. Clin. N. Am.* **2021**, *105*, 149–174. [CrossRef] [PubMed]
2. Curley, S.; Gall, J.; Byrne, R.; Yvan-Charvet, L.; McGillicuddy, F.C. Metabolic inflammation in obesity—At the crossroads between fatty acid and cholesterol metabolism. *Mol. Nutr. Food Res.* **2020**, *65*, 1900482. [CrossRef] [PubMed]
3. Goossens, G.H. The metabolic phenotype in obesity: Fat mass, body fat distribution, and adipose tissue function. *Obes. Facts.* **2017**, *10*, 207–215. [CrossRef]
4. Payab, M.; Abedi, M.; Foroughi, H.N.; Hadavandkhani, M.; Arabi, M.; Tayanloo-Beik, A.; Sheikh Hosseini, M.; Gerami, H.; Khatami, F.; Larijani, B.; et al. Brown adipose tissue transplantation as a novel alternative to obesity treatment: A systematic review. *Int. J. Obes.* **2021**, *45*, 109–121. [CrossRef]
5. Fernández-Sánchez, A.; Madrigal-Santillán, E.; Bautista, M.; Esquivel-Soto, J.; Morales-González, A.; Esquivel-Chirino, C.; Durante-Montiel, I.; Sánchez-Rivera, G.; Valadez-Vega, C.; Morales-González, J.A. Inflammation, oxidative stress, and obesity. *Int. J. Mol. Sci.* **2011**, *12*, 3117–3132. [CrossRef] [PubMed]
6. Manna, P.; Jain, S.K. Obesity, oxidative stress, adipose tissue dysfunction, and the associated health risks: Causes and therapeutic strategies. *Metab. Syndr. Relat. Disord.* **2015**, *13*, 423–444. [CrossRef] [PubMed]
7. Russo, L.; Lumeng, C.N. Properties and functions of adipose tissue macrophages in obesity. *Immunology* **2018**, *155*, 407–417. [CrossRef]
8. Kardinaal, A.F.; van 't Veer, P.; Brants, H.A.; van den Berg, H.; van Schoonhoven, J.; Hermus, R.J. Relations between antioxidant vitamins in adipose tissue, plasma, and diet. *Am. J. Epidemiol.* **1995**, *141*, 440–450. [CrossRef]
9. Boccellino, M.; D'Angelo, S. Anti-obesity effects of polyphenol intake: Current status and future possibilities. *Int. J. Mol. Sci.* **2020**, *21*, 5642. [CrossRef] [PubMed]
10. Jakubczyk, K.; Dec, K.; Kałduńska, J.; Kawczuga, D.; Kochman, J.; Janda, K. Reactive oxygen species—Sources, functions, oxidative damage. *Pol. Merkur. Lekarski.* **2020**, *48*, 124–127.
11. Bansal, M.; Kaushal, N. *Oxidative Stress Mechanisms and Their Modulation*; Springer: New Delhi, India, 2014. [CrossRef]
12. Lü, J.M.; Lin, P.H.; Yao, Q.; Chen, C. Chemical and molecular mechanisms of antioxidants: Experimental approaches and model systems. *J. Cell. Mol. Med.* **2010**, *14*, 840–860. [CrossRef]
13. Bloomer, R.J.; Goldfarb, A.H. Anaerobic exercise and oxidative stress: A review. *Can. J. Appl. Physiol.* **2004**, *29*, 245–263. [CrossRef]
14. Oh, J.; Jung, S.R.; Lee, Y.J.; Park, K.W.; Han, J. Antioxidant and antiobesity activities of seed extract from campbell early grape as a functional ingredient. *J. Food. Process. Pres.* **2013**, *37*, 291–298. [CrossRef]
15. Álvarez, E.; Rodiño-Janeiro, B.K.; Jerez, M.; Ucieda-Somoza, R.; Núñez, M.J.; González-Juanatey, J.R. Procyanidins from grape pomace are suitable inhibitors of human endothelial NADPH oxidase. *J. Cell. Biochem.* **2012**, *113*, 1386–1396. [CrossRef] [PubMed]
16. Lefranc, C.; Friederich-Persson, M.; Palacios-Ramirez, R.; Cat, A.N.D. Mitochondrial oxidative stress in obesity: Role of the mineralocorticoid receptor. *J. Endocrinol.* **2018**, *238*, R143–R159. [CrossRef] [PubMed]
17. Castro, J.P.; Grune, T.; Speckmann, B. The two faces of reactive oxygen species (ROS) in adipocyte function and dysfunction. *Biol. Chem.* **2016**, *397*, 709–724. [CrossRef]
18. García-Sánchez, A.; Miranda-Díaz, A.G.; Cardona-Muñoz, E.G. The role of oxidative stress in physiopathology and pharmacological treatment with pro- and antioxidant properties in chronic diseases. *Oxid. Med. Cell. Longev.* **2020**, *2020*, 2082145. [CrossRef]
19. Taherkhani, S.; Suzuki, K.; Castell, L. A short overview of changes in inflammatory cytokines and oxidative stress in response to physical activity and antioxidant supplementation. *Antioxidants* **2020**, *9*, 886. [CrossRef] [PubMed]
20. Sharif, A.; Akhtar, N.; Khan, M.S.; Menaa, A.; Menaa, B.; Khan, B.A.; Menaa, F. Formulation and evaluation on human skin of a water-in-oil emulsion containing Muscat hamburg black grape seed extract. *Int. J. Cosmet. Sci.* **2015**, *37*, 253–258. [CrossRef]
21. Le Lay, S.; Simard, G.; Martinez, M.C.; Andriantsitohaina, R. Oxidative stress and metabolic pathologies: From an adipocentric point of view. *Oxid. Med. Cell. Longev.* **2014**, *2014*, 908539. [CrossRef]
22. Fernández-Iglesias, A.; Pajuelo, D.; Quesada, H.; Díaz, S.; Bladé, C.; Arola, L.; Salvadó, M.J.; Mulero, M. Grape seed proanthocyanidin extract improves the hepatic glutathione metabolism in obese Zucker rats. *Mol. Nutr. Food. Res.* **2014**, *58*, 727–737. [CrossRef] [PubMed]
23. Santhakumar, A.B.; Bulmer, A.C.; Singh, I. A review of the mechanisms and effectiveness of dietary polyphenols in reducing oxidative stress and thrombotic risk. *J. Hum. Nutr. Diet.* **2014**, *27*, 1–21. [CrossRef] [PubMed]
24. Sano, A.; Tokutake, S.; Seo, A. Proanthocyanidin-rich grape seed extract reduces leg swelling in healthy women during prolonged sitting. *J. Sci. Food. Agric.* **2013**, *93*, 457–462. [CrossRef] [PubMed]
25. Saada, H.N.; Said, U.Z.; Meky, N.H.; Abd El Azime, A.S. Grape seed extract Vitis vinifera protects against radiation-induced oxidative damage and metabolic disorders in rats. *Phytother. Res.* **2009**, *23*, 434–438. [CrossRef]
26. Valko, M.; Leibfritz, D.; Moncol, J.; Cronin, M.T.; Mazur, M.; Telser, J. Free radicals and antioxidants in normal physiological functions and human disease. *Int. J. Biochem. Cell. Biol.* **2007**, *39*, 44–84. [CrossRef]
27. Friedenreich, C.M.; Pialoux, V.; Wang, Q.; Shaw, E.; Brenner, D.R.; Waltz, X.; Conroy, S.M.; Johnson, R.; Woolcott, C.G.; Poulin, M.J.; et al. Effects of exercise on markers of oxidative stress: An ancillary analysis of the Alberta physical activity and breast cancer prevention trial. *BMJ. Open. Sport. Exerc. Med.* **2016**, *2*, e000171. [CrossRef]

28. Kurata, A.; Nishizawa, H.; Kihara, S.; Maeda, N.; Sonoda, M.; Okada, T.; Ohashi, K.; Hibuse, T.; Fujita, K.; Yasui, A.; et al. Blockade of Angiotensin II type-1 receptor reduces oxidative stress in adipose tissue and ameliorates adipocytokine dysregulation. *Kidney Int.* **2006**, *70*, 1717–1724. [CrossRef]

29. Candiracci, M.; Justo, M.L.; Castaño, A.; Rodriguez-Rodriguez, R.; Herrera, M.D. Rice bran enzymatic extract-supplemented diets modulate adipose tissue inflammation markers in Zucker rats. *Nutrition* **2014**, *30*, 466–472. [CrossRef] [PubMed]

30. Alcalá, M.; Sánchez-Vera, I.; Sevillano, J.; Herrero, L.; Serra, D.; Ramos, M.P.; Viana, M. Vitamin E reduces adipose tissue fibrosis, inflammation, and oxidative stress and improves metabolic profile in obesity. *Obesity* **2015**, *23*, 1598–1606. [CrossRef]

31. Preiser, J.C. Oxidative stress. *JPEN. J. Parenter. Enteral. Nutr.* **2012**, *36*, 147–154. [CrossRef]

32. Vilchis-Landeros, M.M.; Matuz-Mares, D.; Vázquez-Meza, H. Regulation of metabolic processes by hydrogen peroxide generated by NADPH oxidases. *Processes* **2020**, *8*, 1424. [CrossRef]

33. Sakurai, T.; Ogasawara, J.; Shirato, K.; Izawa, T.; Oh-Ishi, S.; Ishibashi, Y.; Radák, Z.; Ohno, H.; Kizaki, T. Exercise training attenuates the dysregulated expression of adipokines and oxidative stress in white adipose tissue. *Oxid. Med. Cell. Longev.* **2017**, *2017*, 9410954. [CrossRef]

34. DeVallance, E.; Li, Y.; Jurczak, M.J.; Cifuentes-Pagano, E.; Pagano, P.J. The role of NADPH oxidases in the etiology of obesity and metabolic syndrome: Contribution of individual isoforms and cell biology. *Antioxid. Redox. Signal.* **2019**, *31*, 687–709. [CrossRef] [PubMed]

35. Paredes, F.; Suster, I.; Martin, A.S. Poldip2 takes a central role in metabolic reprograming. *Oncoscience* **2018**, *5*, 130–131. [CrossRef] [PubMed]

36. Tosetti, P.; Dunlap, K. Assays of RGS3 activation and modulation. *Methods. Enzymol.* **2004**, *390*, 99–119.

37. Chazin, W.J. Relating form and function of EF-hand calcium binding proteins. *Acc. Chem. Res.* **2011**, *44*, 171–179. [CrossRef]

38. Brownlee, M. The pathobiology of diabetic complications: A unifying mechanism. *Diabetes* **2005**, *54*, 1615–1625. [CrossRef]

39. Matsuda, M.; Shimomura, I. Increased oxidative stress in obesity: Implications for metabolic syndrome, diabetes, hypertension, dyslipidemia, atherosclerosis, and cancer. *Obes. Res. Clin. Pract.* **2013**, *7*, e330–e341. [CrossRef]

40. Dillard, C.J.; Litov, R.E.; Savin, W.M.; Dumelin, E.E.; Tappel, A.L. Effects of exercise, vitamin E, and ozone on pulmonary function and lipid peroxidation. *J. Appl. Physiol.* **1978**, *45*, 927–932. [CrossRef]

41. Fisher-Wellman, K.; Bloomer, R.J. Acute exercise and oxidative stress: A 30 year history. *Dyn. Med.* **2009**, *8*, 1. [CrossRef]

42. Halliwell, B.; Whiteman, M. Measuring reactive species and oxidative damage in vivo and in cell culture: How should you do it and what do the results mean? *Br. J. Pharmacol.* **2004**, *142*, 231–255. [CrossRef]

43. Ohshima, H.; Sawa, T.; Akaike, T. 8-nitroguanine, a product of nitrative DNA damage caused by reactive nitrogen species: Formation, occurrence, and implications in inflammation and carcinogenesis. *Antioxid. Redox. Signal.* **2006**, *8*, 1033–1045. [CrossRef]

44. Kilhovd, B.K.; Juutilainen, A.; Lehto, S.; Rönnemaa, T.; Torjesen, P.A.; Hanssen, K.F.; Laakso, M. Increased serum levels of advanced glycation endproducts predict total, cardiovascular and coronary mortality in women with type 2 diabetes: A population-based 18 year follow-up study. *Diabetologia* **2007**, *50*, 1409–1417. [CrossRef]

45. Fang, Y.Z.; Yang, S.; Wu, G. Free radicals, antioxidants, and nutrition. *Nutrition* **2002**, *18*, 872–879. [CrossRef]

46. Jones, D.P. Redox potential of GSH/GSSG couple: Assay and biological significance. *Methods. Enzymol.* **2002**, *348*, 93–112.

47. Smas, C.M.; Sul, H.S. Control of adipocyte differentiation. *Biochem. J.* **1995**, *309*, 697–710. [CrossRef] [PubMed]

48. Berry, R.; Church, C.D.; Gericke, M.T.; Jeffery, E.; Colman, L.; Rodeheffer, M.S. Imaging of adipose tissue. *Methods Enzymol.* **2014**, *537*, 47–73. [PubMed]

49. Nakajima, I.; Yamaguchi, T.; Ozutsumi, K.; Aso, H. Adipose tissue extracellular matrix: Newly organized by adipocytes during differentiation. *Differentiation* **1998**, *63*, 193–200. [CrossRef] [PubMed]

50. Luo, W.; Cao, J.; Li, J.; He, W. Adipose tissue-specific PPARgamma deficiency increases resistance to oxidative stress. *Exp. Gerontol.* **2008**, *43*, 154–163. [CrossRef]

51. Sun, K.; Kusminski, C.M.; Scherer, P.E. Adipose tissue remodeling and obesity. *J. Clin. Investig.* **2011**, *121*, 2094–2101. [CrossRef]

52. Trayhurn, P.; Beattie, J.H. Physiological role of adipose tissue: White adipose tissue as an endocrine and secretory organ. *Proc. Nutr. Soc.* **2001**, *60*, 329–339. [CrossRef]

53. Choe, S.S.; Huh, J.Y.; Hwang, I.J.; Kim, J.I.; Kim, J.B. Adipose tissue remodeling: Its role in energy metabolism and metabolic disorders. *Front. Endocrinol.* **2016**, *7*, 30. [CrossRef]

54. Siiteri, P.K. Adipose tissue as a source of hormones. *Am. J. Clin. Nutr.* **1987**, *45*, 277–282. [CrossRef]

55. Kershaw, E.E.; Flier, J.S. Adipose tissue as an endocrine organ. *J. Clin. Endocrinol. Metab.* **2004**, *89*, 2548–2556. [CrossRef]

56. Cook, K.S.; Min, H.Y.; Johnson, D.; Chaplinsky, R.J.; Flier, J.S.; Hunt, C.R.; Spiegelman, B.M. Adipsin: A circulating serine protease homolog secreted by adipose tissue and sciatic nerve. *Science* **1987**, *237*, 402–405. [CrossRef] [PubMed]

57. Ryu, K.Y.; Jeon, E.J.; Leem, J.; Park, J.H.; Cho, H. Regulation of adipsin expression by endoplasmic reticulum stress in adipocytes. *Biomolecules* **2020**, *10*, 314. [CrossRef] [PubMed]

58. Lo, J.C.; Ljubicic, S.; Leibiger, B.; Kern, M.; Leibiger, I.B.; Moede, T.; Kelly, M.E.; Bhowmick, D.C.; Murano, I.; Cohen, P.; et al. M. Adipsin is an adipokine that improves β cell function in diabetes. *Cell* **2014**, *158*, 41–53. [CrossRef] [PubMed]

59. Liu, J.; DeYoung, S.M.; Zhang, M.; Zhang, M.; Cheng, A.; Saltiel, A.R. Changes in integrin expression during adipocyte differentiation. *Cell. Metab.* **2005**, *2*, 165–177. [CrossRef]

60. Scheja, L.; Heeren, J. The endocrine function of adipose tissues in health and cardiometabolic disease. *Nat. Rev. Endocrinol.* **2019**, *15*, 507–524. [CrossRef] [PubMed]

61. De Farias, J.M.; Bom, K.F.; Tromm, C.B.; Luciano, T.F.; Marques, S.O.; Tuon, T.; Silva, L.A.; Lira, F.S.; de Souza, C.T.; Pinho, R.A. Effect of physical training on the adipose tissue of diet-induced obesity mice: Interaction between reactive oxygen species and lipolysis. *Horm. Metab. Res.* **2013**, *45*, 190–196. [CrossRef]

62. Esteve Ràfols, M. Adipose tissue: Cell heterogeneity and functional diversity. *Endocrinol. Nutr.* **2014**, *61*, 100–112. [CrossRef] [PubMed]

63. Lee, M.J.; Wu, Y.; Fried, S.K. Adipose tissue remodeling in pathophysiology of obesity. *Curr. Opin. Clin. Nutr. Metab. Care* **2010**, *13*, 371–376. [CrossRef] [PubMed]

64. Zoico, E.; Rubele, S.; De Caro, A.; Nori, N.; Mazzali, G.; Fantin, F.; Rossi, A.; Zamboni, M. Brown and beige adipose tissue and aging. *Front. Endocrinol.* **2019**, *10*, 368. [CrossRef] [PubMed]

65. Vegiopoulos, A.; Rohm, M.; Herzig, S. Adipose tissue: Between the extremes. *EMBO J.* **2017**, *36*, 1999–2017. [CrossRef]

66. Berry, D.C.; Stenesen, D.; Zeve, D.; Graff, J.M. The developmental origins of adipose tissue. *Development* **2013**, *140*, 3939–3949. [CrossRef] [PubMed]

67. Colaianni, G.; Colucci, S.; Grano, M. *Anatomy and Physiology of Adipose Tissue*; Springer: Cham, Switzerland, 2014. [CrossRef]

68. Ro, S.H.; Nam, M.; Jang, I.; Park, H.W.; Park, H.; Semple, I.A.; Kim, M.; Kim, J.S.; Park, H.; Einat, P.; et al. Sestrin2 inhibits uncoupling protein 1 expression through suppressing reactive oxygen species. *Proc. Natl. Acad. Sci. USA* **2014**, *111*, 7849–7854. [CrossRef] [PubMed]

69. Lee, J.H.; Budanov, A.V.; Talukdar, S.; Park, E.J.; Park, H.L.; Park, H.W.; Bandyopadhyay, G.; Li, N.; Aghajan, M.; Jang, I.; et al. Maintenance of metabolic homeostasis by Sestrin2 and Sestrin3. *Cell. Metab.* **2012**, *16*, 311–321. [CrossRef]

70. Bae, S.H.; Sung, S.H.; Oh, S.Y.; Lim, J.M.; Lee, S.K.; Park, Y.N.; Lee, H.E.; Kang, D.; Rhee, S.G. Sestrins activate Nrf2 by promoting p62-dependent autophagic degradation of Keap1 and prevent oxidative liver damage. *Cell. Metab.* **2013**, *17*, 73–84. [CrossRef]

71. Rothwell, N.J.; Stock, M.J. A role for brown adipose tissue in diet-induced thermogenesis. *Obes. Res.* **1997**, *5*, 650–656. [CrossRef] [PubMed]

72. Turrens, J.F. Superoxide production by the mitochondrial respiratory chain. *Biosci. Rep.* **1997**, *17*, 3–8. [CrossRef]

73. Krüger, K.; Mooren, F.C.; Eder, K.; Ringseis, R. Immune and inflammatory signaling pathways in exercise and obesity. *Am. J. Lifestyle. Med.* **2014**, *10*, 268–279. [CrossRef] [PubMed]

74. Kaisanlahti, A.; Glumoff, T. Browning of white fat: Agents and implications for beige adipose tissue to type 2 diabetes. *J. Physiol. Biochem.* **2019**, *75*, 1–10. [CrossRef] [PubMed]

75. Fain, J.N. Release of interleukins and other inflammatory cytokines by human adipose tissue is enhanced in obesity and primarily due to the nonfat cells. *Vitam. Horm.* **2006**, *74*, 443–477. [PubMed]

76. Skurk, T.; Alberti-Huber, C.; Herder, C.; Hauner, H. Relationship between adipocyte size and adipokine expression and secretion. *J. Clin. Endocrinol. Metab.* **2007**, *92*, 1023–1033. [CrossRef] [PubMed]

77. Zhao, J.; Zhai, L.; Liu, Z.; Wu, S.; Xu, L. Leptin level and oxidative stress contribute to obesity-induced low testosterone in murine testicular tissue. *Oxid. Med. Cell. Longev.* **2014**, *2014*, 190945. [CrossRef]

78. Alexandre, E.C.; Calmasini, F.B.; Sponton, A.C.D.S.; de Oliveira, M.G.; André, D.M.; Silva, F.H.; Delbin, M.A.; Mónica, F.Z.; Antunes, E. Influence of the periprostatic adipose tissue in obesity-associated mouse urethral dysfunction and oxidative stress: Effect of resveratrol treatment. *Eur. J. Pharmacol.* **2018**, *836*, 25–33. [CrossRef] [PubMed]

79. Ruhee, R.T.; Suzuki, K. Dietary fiber and its effect on obesity: A review article. *Adv. Med. Res.* **2018**, *1*, 1–13. [CrossRef]

80. Bailey-Downs, L.C.; Tucsek, Z.; Toth, P.; Sosnowska, D.; Gautam, T.; Sonntag, W.E.; Csiszar, A.; Ungvari, Z. Aging exacerbates obesity-induced oxidative stress and inflammation in perivascular adipose tissue in mice: A paracrine mechanism contributing to vascular redox dysregulation and inflammation. *J. Gerontol. A. Biol. Sci. Med. Sci.* **2013**, *68*, 780–792. [CrossRef]

81. Abdali, D.; Samson, S.E.; Grover, A.K. How effective are antioxidant supplements in obesity and diabetes? *Med. Princ. Pract.* **2015**, *24*, 201–215. [CrossRef]

82. Bjørklund, G.; Chirumbolo, S. Role of oxidative stress and antioxidants in daily nutrition and human health. *Nutrition* **2017**, *33*, 311–321. [CrossRef]

83. Ruiz-Ojeda, F.J.; Méndez-Gutiérrez, A.; Aguilera, C.M.; Plaza-Díaz, J. Extracellular matrix remodeling of adipose tissue in obesity and metabolic diseases. *Int. J. Mol. Sci.* **2019**, *20*, 4888. [CrossRef]

84. Timmers, S.; de Vogel-van den Bosch, J.; Towler, M.C.; Schaart, G.; Moonen-Kornips, E.; Mensink, R.P.; Hesselink, M.K.; Hardie, D.G.; Schrauwen, P. Prevention of high-fat diet-induced muscular lipid accumulation in rats by alpha lipoic acid is not mediated by AMPK activation. *J. Lipid. Res.* **2010**, *51*, 352–359. [CrossRef]

85. Martínez-Fernández, L.; Fernández-Galilea, M.; Felix-Soriano, E.; Escoté, X.; González-Muniesa, P.; Moreno-Aliaga, M.J. Inflammation and oxidative stress in adipose tissue: Nutritional regulation. In *Obesity: Oxidative Stress and Dietary Antioxidants*; del Moral, A.M., Garcia, C.M.A., Eds.; Elsevier: Amsterdam, The Netherlands, 2018; Chapter 4, pp. 63–92.

86. Aldini, G.; Yeum, K.J.; Niki, E.; Russell, R.M. *Biomarkers for Antioxidant Defense and Oxidative Damage: Principles and Practical Applications*; Blackwell: Oxford, UK, 2010. [CrossRef]

87. Brooker, S.; Martin, S.; Pearson, A.; Bagchi, D.; Earl, J.; Gothard, L.; Hall, E.; Porter, L.; Yarnold, J. Double-blind, placebo-controlled, randomised phase II trial of IH636 grape seed proanthocyanidin extract (GSPE) in patients with radiation-induced breast induration. *Radiother. Oncol.* **2006**, *79*, 45–51. [CrossRef] [PubMed]

88. Han, Y.; Song, S.; Wu, H.; Zhang, J.; Ma, E. Antioxidant enzymes and their role in phoxim and carbaryl stress in Caenorhabditis elegans. *Pestic. Biochem. Physiol.* **2017**, *138*, 43–50. [CrossRef] [PubMed]
89. Keong, C.C.; Singh, H.J.; Singh, R. Effects of palm vitamin e supplementation on exercise-induced oxidative stress and endurance performance in the heat. *J. Sports Sci. Med.* **2006**, *5*, 629–639.
90. Leong, P.K.; Chen, J.; Chan, W.M.; Leung, H.Y.; Chan, L.; Ko, K.M. Acute pre-/post-treatment with 8th day SOD-like supreme (a free radical scavenging health product) protects against oxidant-induced injury in cultured cardiomyocytes and hepatocytes in vitro as well as in mouse myocardium and liver in vivo. *Antioxidants* **2017**, *6*, 28. [CrossRef] [PubMed]
91. Li, Z.; Han, X.; Song, X.; Zhang, Y.; Jiang, J.; Han, Q.; Liu, M.; Qiao, G.; Zhuo, R. Overexpressing the sedum alfredii Cu/Zn superoxide dismutase increased resistance to oxidative stress in transgenic arabidopsis. *Front. Plant. Sci.* **2017**, *8*, 1010. [CrossRef]
92. Staerck, C.; Vandeputte, P.; Gastebois, A.; Calenda, A.; Giraud, S.; Papon, N.; Bouchara, J.P.; Fleury, M.J.J. Enzymatic mechanisms involved in evasion of fungi to the oxidative stress: Focus on scedosporium apiospermum. *Mycopathologia* **2018**, *183*, 227–239. [CrossRef]
93. Berwal, M.K.; Padmanabhan, S.; Vittal, N.; Hebbar, K.B. Variability in superoxide dismutase isoforms in tall and dwarf cultivars of coconut (*Cocos nucifera* L.) Leaves. *Ind. J. Agric. Biochem.* **2016**, *29*, 184. [CrossRef]
94. Morrow, J.D.; Roberts, L.J. Mass spectrometric quantification of F2-isoprostanes in biological fluids and tissues as measure of oxidant stress. *Methods. Enzymol.* **1999**, *300*, 3–12.
95. Kang, S.W. Superoxide dismutase 2 gene and cancer risk: Evidence from an updated meta-analysis. *Int. J. Clin. Exp. Med.* **2015**, *8*, 14647–14655.
96. Halliwell, B.; Gutteridge, J.M.C. *Free Radicals in Biology and Medicine*; Oxford Scholarship Online: Oxford, UK, 2015. [CrossRef]
97. Margis, R.; Dunand, C.; Teixeira, F.K.; Margis-Pinheiro, M. Glutathione peroxidase family—An evolutionary overview. *FEBS J.* **2008**, *275*, 3959–3970. [CrossRef] [PubMed]
98. Chance, B.; Greenstein, D.S.; Roughton, F.J.W. The mechanism of catalase action. I. Steady-state analysis. *Arch. Biochem. Biophys.* **1952**, *37*, 301–321. [CrossRef]
99. Kahn, S.E.; Hull, R.L.; Utzschneider, K.M. Mechanisms linking obesity to insulin resistance and type 2 diabetes. *Nature* **2006**, *444*, 840–846. [CrossRef] [PubMed]
100. Suzuki, K. Chronic inflammation as an immunological abnormality and effectiveness of exercise. *Biomolecules* **2019**, *9*, 223. [CrossRef] [PubMed]
101. Masschelin, P.M.; Cox, A.R.; Chernis, N.; Hartig, S.M. The impact of oxidative stress on adipose tissue energy balance. *Front. Physiol.* **2020**, *10*, 1638. [CrossRef] [PubMed]
102. Simán, C.M.; Eriksson, U.J. Effect of butylated hydroxytoluene on alpha-tocopherol content in liver and adipose tissue of rats. *Toxicol. Lett.* **1996**, *87*, 103–138. [CrossRef]
103. Rodrigues, P.; Ferreira, T.; Nascimento-Gonçalves, E.; Seixas, F.; Gil da Costa, R.M.; Martins, T.; Neuparth, M.J.; Pires, M.J.; Lanzarin, G.; Félix, L.; et al. Dietary supplementation with chestnut (*Castanea sativa*) reduces abdominal adiposity in FVB/n mice: A preliminary study. *Biomedicines* **2020**, *8*, 75. [CrossRef]
104. Valls, V.; Goicoechea, M.; Muñiz, P.; Cabo, S.J.R. Effect of corn oil and vitamin E on the oxidative status of adipose tissues and liver in rat. *Food. Chem.* **2003**, *81*, 281–286. [CrossRef]
105. Arias, N.; Macarulla, M.T.; Aguirre, L.; Martínez-Castaño, M.G.; Portillo, M.P. Quercetin can reduce insulin resistance without decreasing adipose tissue and skeletal muscle fat accumulation. *Genes. Nutr.* **2014**, *9*, 361. [CrossRef]
106. Chen, J.W.; Lin, Y.L.; Chou, C.H.; Wu, Y.H.S.; Wang, S.Y.; Chen, Y.C. Antiobesity and hypolipidemic effects of protease A-digested crude-chalaza hydrolysates in a high-fat diet. *J. Funct. Foods.* **2020**, *66*, 103788. [CrossRef]
107. Djurasevic, S.F.; Cvijic, G.; Djordjevic, J.; Davidovic, V. The influence of vitamin C supplementation on the oxidative status of rat interscapular brown adipose tissue. *J. Therm. Biol.* **2008**, *33*, 238–243. [CrossRef]
108. Sung, Y.Y.; Yoon, T.; Yang, W.K.; Kim, S.J.; Kim, D.S.; Kim, H.K. The antiobesity effect of polygonum aviculare L. ethanol extract in high-fat diet-induced obese mice. *Evid. Based. Complement. Altern. Med.* **2013**, *2013*, 626397. [CrossRef]
109. Maurya, R.; Bhattacharya, P.; Dey, R.; Nakhasi, H.L. Leptin functions in infectious diseases. *Front. Immunol.* **2018**, *9*, 2741. [CrossRef] [PubMed]
110. Soeroto, A.Y.; Soetedjo, N.N.; Purwiga, A.; Santoso, P.; Kulsum, I.D.; Suryadinata, H.; Ferdian, F. Effect of increased BMI and obesity on the outcome of COVID-19 adult patients: A systematic review and meta-analysis. Diabetes. *Metab. Syndr.* **2020**, *14*, 1897–1904. [CrossRef]
111. O'Shea, D.; Hogan, A.E. Dysregulation of natural killer cells in obesity. *Cancers* **2019**, *11*, 573. [CrossRef] [PubMed]
112. Hinchliffe, N.; Bullen, V.; Haslam, D.; Feenie, J. COVID-19 and obesity. *Pract. Diabetes* **2020**, *37*, 149–151. [CrossRef]
113. Khoramipour, K.; Basereh, A.; Hekmatikar, A.A.; Castell, L.; Ruhee, R.T.; Suzuki, K. Physical activity and nutrition guidelines to help with the fight against COVID-19. *J. Sports Sci.* **2021**, *39*, 101–107. [CrossRef]
114. Fernández-Quintela, A.; Milton-Laskibar, I.; Trepiana, J.; Gómez-Zorita, S.; Kajarabille, N.; Léniz, A.; González, M.; Portillo, M.P. Key aspects in nutritional management of COVID-19 patients. *J. Clin. Med.* **2020**, *9*, 2589. [CrossRef]
115. Mohammad, S.; Aziz, R.; Al Mahri, S.; Malik, S.S.; Haji, E.; Khan, A.H.; Khatlani, T.S.; Bouchama, A. Obesity and COVID-19: What makes obese host so vulnerable? *Immun. Ageing* **2021**, *18*, 1. [CrossRef]
116. Fakhouri, E.W.; Peterson, S.J.; Kothari, J.; Alex, R.; Shapiro, J.I.; Abraham, N.G. Genetic polymorphisms complicate COVID-19 therapy: Pivotal role of HO-1 in cytokine storm. *Antioxidants* **2020**, *9*, 636. [CrossRef] [PubMed]

117. Favre, G.; Legueult, K.; Pradier, C.; Raffaelli, C.; Ichai, C.; Iannelli, A.; Redheuil, A.; Lucidarme, O.; Esnault, V. Visceral fat is associated to the severity of COVID-19. *Metabolism* **2021**, *115*, 154440. [CrossRef] [PubMed]
118. Gheblawi, M.; Wang, K.; Viveiros, A.; Nguyen, Q.; Zhong, J.C.; Turner, A.J.; Raizada, M.K.; Grant, M.B.; Oudit, G.Y. Angiotensin-Converting enzyme 2: SARS-CoV-2 receptor and regulator of the Renin-Angiotensin System: Celebrating the 20th anniversary of the discovery of ACE2. *Circ. Res.* **2020**, *126*, 1456–1474. [CrossRef]
119. Xie, Z.J.; Novograd, J.; Itzkowitz, Y.; Sher, A.; Buchen, Y.D.; Sodhi, K.; Abraham, N.G.; Shapiro, J.I. The pivotal role of adipocyte-Na K peptide in reversing systemic inflammation in obesity and COVID-19 in the development of heart failure. *Antioxidants* **2020**, *9*, 1129. [CrossRef]
120. Li, S.; Cao, Z.; Yang, H.; Zhang, Y.; Xu, F.; Wang, Y. Metabolic healthy obesity, vitamin D status, and risk of COVID-19. *Aging. Dis.* **2021**, *12*, 61–71. [CrossRef] [PubMed]
121. Mohseni, H.; Amini, S.; Abiri, B.; Kalantar, M. Do body mass index (BMI) and history of nutritional supplementation play a role in the severity of COVID-19? A retrospective study. *Nutr. Food. Sci.* **2021**. preprint. [CrossRef]
122. Maggini, S.; Pierre, A.; Calder, P.C. Immune function and micronutrient requirements change over the life course. *Nutrients* **2018**, *10*, 1531. [CrossRef]
123. Calder, P.C.; Carr, A.C.; Gombart, A.F.; Eggersdorfer, M. Optimal nutritional status for a well-functioning immune system is an important factor to protect against viral infections. *Nutrients* **2020**, *12*, 1181. [CrossRef]
124. Grant, W.B.; Lahore, H.; McDonnell, S.L.; Baggerly, C.A.; French, C.B.; Aliano, J.L.; Bhattoa, H.P. Evidence that vitamin D supplementation could reduce risk of influenza and COVID-19 infections and deaths. *Nutrients* **2020**, *12*, 988. [CrossRef] [PubMed]
125. Zhang, J.; Taylor, E.W.; Bennett, K.; Saad, R.; Rayman, M.P. Association between regional selenium status and reported outcome of COVID-19 cases in China. *Am. J. Clin. Nutr.* **2020**, *111*, 1297–1299. [CrossRef] [PubMed]
126. Dornas, W.; Schuppan, D. Mitochondrial oxidative injury: A key player in nonalcoholic fatty liver disease. *Am. J. Physiol. Gastrointest. Liver Physiol.* **2020**, *319*, G400–G411. [CrossRef]
127. Zhang, C.; Wang, P.; Li, Y.; Huang, C.; Ni, W.; Chen, Y.; Shi, J.; Chen, G.; Hu, X.; Ye, M.; et al. Role of MicroRNAs in the development of hepatocellular carcinoma in nonalcoholic fatty liver disease. *Anat. Rec.* **2019**, *302*, 193–200. [CrossRef] [PubMed]
128. Rolo, A.P.; Teodoro, J.S.; Palmeira, C.M. Role of oxidative stress in the pathogenesis of nonalcoholic steatohepatitis. *Free. Radic. Biol. Med.* **2012**, *52*, 59–69. [CrossRef] [PubMed]
129. He, Y.; Jiang, J.; He, B.; Shi, Z. Chemical activators of the Nrf2 signaling pathway in nonalcoholic fatty liver disease. *Nat. Prod. Commun.* **2021**, *16*, 1–9.
130. Wang, J.; He, W.; Tsai, P.J.; Chen, P.H.; Ye, M.; Guo, J.; Su, Z.H. Mutual interaction between endoplasmic reticulum and mitochondria in nonalcoholic fatty liver disease. *Lipids. Health Dis.* **2020**, *19*, 72. [CrossRef] [PubMed]
131. Kawanishi, N.; Mizokami, T.; Yada, K.; Suzuki, K. Exercise training suppresses scavenger receptor CD36 expression in kupffer cells of nonalcoholic steatohepatitis model mice. *Physiol. Rep.* **2018**, *6*, e13902. [CrossRef]
132. Kawanishi, N.; Yano, H.; Mizokami, T.; Takahashi, M.; Oyanagi, E.; Suzuki, K. Exercise training attenuates hepatic inflammation, fibrosis and macrophage infiltration during diet induced-obesity in mice. *Brain. Behav. Immun.* **2012**, *26*, 931–941. [CrossRef]
133. Simoes, I.C.M.; Janikiewicz, J.; Bauer, J.; Karkucinska-Wieckowska, A.; Kalinowski, P.; Dobrzyń, A.; Wolski, A.; Pronicki, M.; Zieniewicz, K.; Dobrzyń, P.; et al. Fat and sugar-A dangerous duet. A comparative review on metabolic remodeling in rodent models of nonalcoholic fatty liver disease. *Nutrients* **2019**, *11*, 2871. [CrossRef]
134. Videla, L.A.; Rodrigo, R.; Orellana, M.; Fernandez, V.; Tapia, G.; Quiñones, L.; Varela, N.; Contreras, J.; Lazarte, R.; Csendes, A.; et al. Oxidative stress-related parameters in the liver of non-alcoholic fatty liver disease patients. *Clin. Sci.* **2004**, *106*, 261–268. [CrossRef] [PubMed]
135. Yki-Järvinen, H. Non-alcoholic fatty liver disease as a cause and a consequence of metabolic syndrome. *Lancet Diabetes Endocrinol.* **2014**, *2*, 901–910. [CrossRef]
136. Buzzetti, E.; Pinzani, M.; Tsochatzis, E.A. The multiple-hit pathogenesis of non-alcoholic fatty liver disease (NAFLD). *Metabolism* **2016**, *65*, 1038–1048. [CrossRef] [PubMed]
137. Tarantino, G.; Citro, V.; Capone, D. Nonalcoholic fatty liver disease: A challenge from mechanisms to therapy. *J. Clin. Med.* **2019**, *9*, 15. [CrossRef] [PubMed]
138. Chiu, S.; Mulligan, K.; Schwarz, J.M. Dietary carbohydrates and fatty liver disease: De novo lipogenesis. *Curr. Opin. Clin. Nutr. Metab. Care.* **2018**, *21*, 277–282. [CrossRef]
139. Drygalski, K.; Berk, K.; Charytoniuk, T.; Iłowska, N.; Łukaszuk, B.; Chabowski, A.; Konstantynowicz-Nowicka, K. Does the enterolactone (ENL) affect fatty acid transporters and lipid metabolism in liver? *Nutr. Metab* **2020**, *14*, 69. [CrossRef]
140. Pagano, C.; Soardo, G.; Esposito, W.; Fallo, F.; Basan, L.; Donnini, D.; Federspil, G.; Sechi, L.A.; Vettor, R. Plasma adiponectin is decreased in nonalcoholic fatty liver disease. *Eur. J. Endocrinol.* **2005**, *152*, 113–118. [CrossRef]
141. Khoramipour, K.; Chamari, K.; Hekmatikar, A.A.; Ziyaiyan, A.; Taherkhani, S.; Elguindy, N.M.; Bragazzi, N.L. Adiponectin: Structure, physiological functions, role in diseases, and effects of nutrition. *Nutrients* **2021**, *13*, 1180. [CrossRef]
142. Liu, W.; Baker, S.S.; Baker, R.D.; Zhu, L. Antioxidant mechanisms in nonalcoholic fatty liver disease. *Curr. Drug Targets* **2015**, *16*, 1301–1314. [CrossRef]
143. Vanni, E.; Bugianesi, E.; Kotronen, A.; De Minicis, S.; Yki-Järvinen, H.; Svegliati-Baroni, G. From the metabolic syndrome to NAFLD or vice versa? *Dig. Liver Dis.* **2010**, *42*, 320–330. [CrossRef] [PubMed]

144. Younossi, Z.M.; Loomba, R.; Rinella, M.E.; Bugianesi, E.; Marchesini, G.; Neuschwander-Tetri, B.A.; Serfaty, L.; Negro, F.; Caldwell, S.H.; Ratziu, V.; et al. Current and future therapeutic regimens for nonalcoholic fatty liver disease and nonalcoholic steatohepatitis. *Hepatology* **2018**, *68*, 361–371. [CrossRef] [PubMed]
145. Brunt, E.M.; Wong, V.W.; Nobili, V.; Day, C.P.; Sookoian, S.; Maher, J.J.; Bugianesi, E.; Sirlin, C.B.; Neuschwander-Tetri, B.A.; Rinella, M.E. Nonalcoholic fatty liver disease. *Nat. Rev. Dis. Primers* **2015**, *1*, 15080. [CrossRef]
146. Alisi, A.; Carpino, G.; Oliveira, F.L.; Panera, N.; Nobili, V.; Gaudio, E. The role of tissue macrophage-mediated inflammation on NAFLD pathogenesis and its clinical implications. *Mediators. Inflamm.* **2017**, *2017*, 8162421. [CrossRef] [PubMed]
147. Maher, J.; Yamamoto, M. The rise of antioxidant signaling—The evolution and hormetic actions of Nrf2. *Toxicol. Appl. Pharmacol.* **2010**, *244*, 4–15. [CrossRef] [PubMed]
148. Kobayashi, M.; Yamamoto, M. Molecular mechanisms activating the Nrf2-Keap1 pathway of antioxidant gene regulation. *Antioxid. Redox. Signal.* **2005**, *7*, 385–394. [CrossRef]
149. Michalopoulos, G.K. Liver regeneration. *J. Cell. Physiol.* **2007**, *213*, 286–300. [CrossRef]
150. McMahon, M.; Thomas, N.; Itoh, K.; Yamamoto, M.; Hayes, J.D. Redox-regulated turnover of Nrf2 is determined by at least two separate protein domains, the redox-sensitive Neh2 degron and the redox-insensitive Neh6 degron. *J. Biol. Chem.* **2004**, *279*, 31556–31567. [CrossRef]
151. Yamamoto, T.; Suzuki, T.; Kobayashi, A.; Wakabayashi, J.; Maher, J.; Motohashi, H.; Yamamoto, M. Physiological significance of reactive cysteine residues of Keap1 in determining Nrf2 activity. *Mol. Cell. Biol.* **2008**, *28*, 2758–2770. [CrossRef] [PubMed]
152. Sharma, R.S.; Harrison, D.J.; Kisielewski, D.; Cassidy, D.M.; McNeilly, A.D.; Gallagher, J.R.; Walsh, S.V.; Honda, T.; McCrimmon, R.J.; Dinkova-Kostova, A.T.; et al. Experimental nonalcoholic steatohepatitis and liver fibrosis are ameliorated by pharmacologic activation of Nrf2 (NF-E2 p45-related factor 2). *Cell. Mol. Gastroenterol. Hepatol.* **2017**, *5*, 367–398. [CrossRef]
153. Surh, Y.J. Cancer chemoprevention with dietary phytochemicals. *Nat. Rev. Cancer* **2003**, *3*, 768–780. [CrossRef] [PubMed]
154. Agrawal, D.K.; Mishra, P.K. Curcumin and its analogues: Potential anticancer agents. *Med. Res. Rev.* **2010**, *30*, 818–860. [CrossRef]
155. Ma, S.; Yada, K.; Lee, H.; Fukuda, Y.; Iida, A.; Suzuki, K. Taheebo polyphenols attenuate free fatty acid-induced inflammation in murine and human macrophage cell lines as inhibitor of cyclooxygenase-2. *Front. Nutr.* **2017**, *4*, 63. [CrossRef]
156. Yada, K.; Suzuki, K.; Oginome, N.; Ma, S.; Fukuda, Y.; Iida, A.; Radak, Z. Single dose administration of taheebo polyphenol enhances endurance capacity in mice. *Sci. Rep.* **2018**, *8*, 1–12. [CrossRef] [PubMed]
157. Ruhee, R.T.; Ma, S.; Suzuki, K. Sulforaphane protects cells against lipopolysaccharide-stimulated inflammation in murine macrophages. *Antioxidants* **2019**, *8*, 577. [CrossRef] [PubMed]
158. Ruhee, R.T.; Ma, S.; Suzuki, K. Protective effects of sulforaphane on exercise-induced organ damage via inducing antioxidant defense responses. *Antioxidants* **2020**, *9*, 136. [CrossRef] [PubMed]
159. Espín, J.C.; García-Conesa, M.T.; Tomás-Barberán, F.A. Nutraceuticals: Facts and fiction. *Phytochemistry* **2007**, *68*, 2986–3008. [CrossRef] [PubMed]
160. Tramontin, N.d.S.; Luciano, T.F.; Marques, S.d.O.; de Souza, C.T.; Muller, A.P. Ginger and avocado as nutraceuticals for obesity and its comorbidities. *Phytother. Res.* **2020**, *34*, 1282–1290. [CrossRef]
161. Kota, S.; Jammula, S.; Kota, S.; Satya Krishna, S.; Meher, L.; Rao, E.; Modi, K. Nutraceuticals in pathogenic obesity; striking the right balance between energy imbalance and inflammation. *J. Med. Nutr. Nutraceuticals* **2012**, *1*, 63–76. [CrossRef]
162. Suzuki, K.; Tominaga, T.; Ruhee, R.T.; Ma, S. Characterization and Modulation of Systemic Inflammatory Response to Exhaustive Exercise in Relation to Oxidative Stress. *Antioxidants* **2020**, *9*, 401. [CrossRef]
163. Suzuki, K.; Hayashida, H. Effect of exercise intensity on cell-mediated immunity. *Sports* **2021**, *9*, 8. [CrossRef]

Review

Circadian Clocks, Redox Homeostasis, and Exercise: Time to Connect the Dots?

Conor McClean * and Gareth W. Davison

Sport and Exercise Sciences Research Institute, Ulster University,
Newtownabbey BT37 0QB, Northern Ireland, UK; gw.davison@ulster.ac.uk
* Correspondence: cm.mcclean@ulster.ac.uk

Abstract: Compelling research has documented how the circadian system is essential for the maintenance of several key biological processes including homeostasis, cardiovascular control, and glucose metabolism. Circadian clock disruptions, or losses of rhythmicity, have been implicated in the development of several diseases, premature ageing, and are regarded as health risks. Redox reactions involving reactive oxygen and nitrogen species (RONS) regulate several physiological functions such as cell signalling and the immune response. However, oxidative stress is associated with the pathological effects of RONS, resulting in a loss of cell signalling and damaging modifications to important molecules such as DNA. Direct connections have been established between circadian rhythms and oxidative stress on the basis that disruptions to circadian rhythms can affect redox biology, and vice versa, in a bi-directional relationship. For instance, the expression and activity of several key antioxidant enzymes (SOD, GPx, and CAT) appear to follow circadian patterns. Consequently, the ability to unravel these interactions has opened an exciting area of redox biology. Exercise exerts numerous benefits to health and, as a potent environmental cue, has the capacity to adjust disrupted circadian systems. In fact, the response to a given exercise stimulus may also exhibit circadian variation. At the same time, the relationship between exercise, RONS, and oxidative stress has also been scrutinised, whereby it is clear that exercise-induced RONS can elicit both helpful and potentially harmful health effects that are dependent on the type, intensity, and duration of exercise. To date, it appears that the emerging interface between circadian rhythmicity and oxidative stress/redox metabolism has not been explored in relation to exercise. This review aims to summarise the evidence supporting the conceptual link between the circadian clock, oxidative stress/redox homeostasis, and exercise stimuli. We believe carefully designed investigations of this nexus are required, which could be harnessed to tackle theories concerned with, for example, the existence of an optimal time to exercise to accrue physiological benefits.

Keywords: circadian rhythms; reactive oxygen and nitrogen species (RONS); exercise training; antioxidant

Citation: McClean, C.; Davison, G.W. Circadian Clocks, Redox Homeostasis, and Exercise: Time to Connect the Dots? *Antioxidants* **2022**, *11*, 256. https://doi.org/10.3390/antiox11020256

Academic Editor: Stanley Omaye

Received: 6 December 2021
Accepted: 18 January 2022
Published: 28 January 2022

Publisher's Note: MDPI stays neutral with regard to jurisdictional claims in published maps and institutional affiliations.

1. Introduction

A diverse array of physiological functions, human behaviours, and social connections are determined by the complex interactions between the environment (light/dark cycles, temperature variations, and seasonal food opportunities) and endogenous biological, or circadian drivers. Such an intricate controlling mechanism is susceptible to disruption and, in current modern society, misalignment between the circadian system and environmental cues is a frequent occurrence associated with negative health consequences [1]. Modern chronic diseases are linked to changes in human lifestyle compared to our hunter-gatherer ancestors [2] involving, but not exclusively limited to, low levels of physical activity (PA) and exercise; prolonged sitting; regular access to highly palatable and energy-dense foods; inadequate/disrupted sleep quality/duration; shift work; and social jetlag [3]. These changes in contemporary work and domestic habits have outpaced genome adaption and the underlying circadian rhythms are thus exposed to dysregulation (or shifts) that can predispose to chronic disease [4].

Oxidative stress has long been implicated in the pathogenesis of several chronic lifestyle-related diseases (e.g., type 2 diabetes mellitus) [5] and the interplay between circadian clocks and oxidative stress is evident whereby: (i) disruption to circadian rhythms can alter redox homeostasis leading to oxidative stress and (ii) elevated production of reactive oxygen and nitrogen species (RONS) may induce circadian oscillations [6]. While still in its relative infancy, research on the effects of exercise on circadian rhythmicity seems to be encouraging; certainly not discouraging [7–9]. At the same time, certain exercise stimuli may evoke oxidative stress, but to the contrary, many exercise-mediated adaptations seem to occur via contractile-induced RONS signalling, acting in a manner that can be explained by the concept of hormesis and other multi-dimensional models [10]. Hormesis, in this context, is when a potentially harmful agent (e.g., exercise) provokes an adaptation to a damaging agent (such as RONS-disrupted signalling) to up-regulate, for example, enzymatic antioxidant capacity [11]. Hence, exercise appears to be a potent activator of oxidative stress, the circadian clock, and, at the correct dose (intensity and volume), elicits desirable health outcomes. As a clear bi-directional link has already been established between the circadian clock and redox homeostasis/metabolism, and that exercise appears to be a potent stimulus of both, we propose that further exploration of exercise, circadian rhythms, and redox biology combined may reveal further important insights in this intriguing field of investigation. In this perspective, the exercise paradigm may help unravel some emerging concepts such as the potential existence of an optimal time to exercise for health benefits. The positive effects of exercise on health may be partially mediated via changes in tissue molecular clocks and/or the outcomes may be modified depending on the timing (and intensity) when exercise is performed [12], but the role, whether directly or indirectly, that RONS and oxidative stress/redox signalling play in such responses remains largely unexplored. Therefore, given the emerging nexus of circadian and redox biology for health and disease prevention, this narrative review will examine the multifactorial interrelationship between circadian clocks and oxidative stress relevant to the important biological stressor, exercise.

2. Circadian Rhythms and Molecular Clock Control

Apparent in virtually all forms of life, circadian rhythms are endogenous 24-h oscillations in behaviour and biological processes enabling organisms to physiologically adjust to the transitions between light and dark (day and night). The circadian clock drives fluctuations in a diverse set of biological processes including sleep, locomotor activity, blood pressure, body temperature, and blood hormone levels [13]. Driven by cellular clocks distributed across the body, these rhythms control mammalian adjustment by preparing the brain and other tissues to perform biologically appropriate functions relative to the anticipated day or night pattern [14,15]. The principal pacemaker of the circadian clock in humans, the so-called central clock, is located within the suprachiasmatic nucleus (SCN) region of the hypothalamus, functioning as an autonomous timekeeper that coordinates the activity of downstream peripheral tissue clocks [15]. Neurons of the SCN receive input from the retina, via the retino-hypothalamic tract, to ensure its molecular clocks are synchronised to light-dark cycles, known as photic entrainment [16]. The SCN can be stimulated by cues other than light, especially by serotonin and melatonin, although these are thought to be internal feedback regulators as opposed to primary circadian rhythm initiators [6]. Using neuro-endocrine actions involving melatonin, insulin, and glucocorticoids (via the HPA axis), as well as the autonomic nervous system [6,17], the SCN relays its coordinating prompts to peripheral clocks [12]. In turn, these peripheral clocks oversee localised, temporal gene and protein expression patterns required for several physiological processes and functions across the ~24-h period [15,17]. Beyond the autonomous cues from the SCN, mammals can also synchronise their inherent timing systems in response to fluctuations in other environmental cues (termed *zeitgebers*) such as changes in temperature and food availability [18], but light remains the strongest zeitgeber [6]—illustrated in Figure 1. While this innate flexibility to varying environmental stimuli is unquestionably important for adaptation and survival, it may also expose the host to negative consequences, such as the onset of sleeping and metabolic disorders, when the rhythms become misaligned or dysregulated [19].

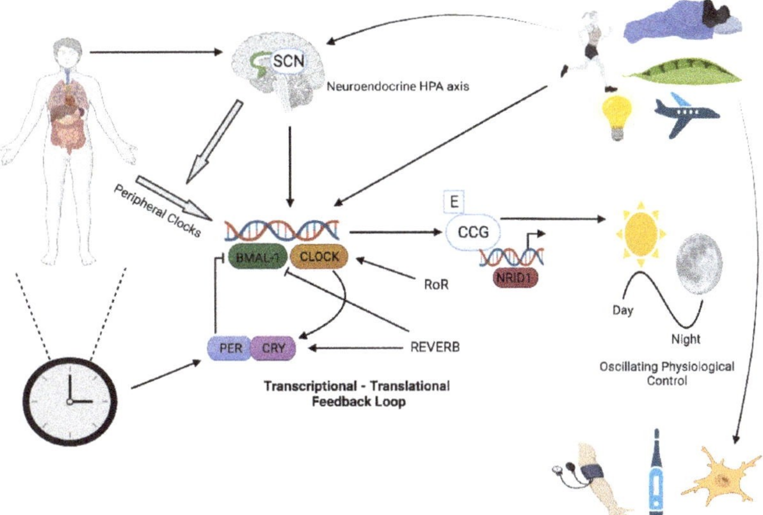

Figure 1. Circadian Rhythms in Humans. Crucial biological and physiological processes such as blood pressure control, antioxidant expression, body temperature, and immune function normally fluctuate in a circadian pattern (~24 h). These oscillations are coordinated by circadian clocks distributed in virtually every cell. The core molecular clock functions to direct a daily program of gene transcription and protein expression. The molecular clock involves a transcriptional–translational feedback loop (TTFL) of core clock genes (positive limb: *BMAL-1* and *CLOCK*; negative limb: *PERs* and *CRYs*) that act to modulate the gene expression of clock-controlled genes (CCG) which generate tissue-specific circadian rhythms in transcription and cellular function across the day, even in the absence of external cues. The principal pacemaker of the circadian clock in humans is located within the suprachiasmatic nucleus (SCN) region of the hypothalamus. This master clock coordinates the activity of downstream peripheral tissue clocks in response to several stimuli such as the day-night cycle, food ingestion, exercise, and sleep. Figure key = SCN: suprachiasmatic nucleus; HPA: hypothalamic-pituitary-adrenal axis; CCG: clock-controlled genes; E: E-box motif.

The core molecular clock is a self-sustaining, transcriptional–translational feedback loop (TTFL) that exists in virtually every human cell and functions to direct a daily program of gene transcription and protein formation, permitting rhythmic adjustments in response to specific entrainment signals [12,13]. This molecular clock is comprised of a positive transcriptional limb and a negative feedback limb [16]; interrogation of this regulatory mechanism was recognized in 2017 when Jeffery Hall, Michael Rosbash, and Michael Young were awarded the Nobel Prize in Physiology or Medicine [13]. Their work, initially in the *Drosophila* model, identified some of the most important clock genes involved in the negative limb of the TTFL such as *Period* (*PER 1-3*) and *Timeless* (*Tim*; or *Cryptochrome* (*CRY*) as it is known in mammals). These were later followed by the identification of the positive limb transcription factors, CLOCK and BMAL-1 [20]. The heterodimerization of BMAL-1 and CLOCK, or its homolog NPAS2, initiates the circadian cycle. Once dimerized, BMAL-1 and CLOCK bind to E-box motifs in the promoters of target genes, initiating the intracellular transcriptional processes of various rhythmic proteins [6,16]. Two groups of these transcriptional targets are the PERs and the CRYs. The PER and CRY proteins, through the formation of a second heterodimer, translocate back into the nucleus, and interfere with the activity of BMAL1 and CLOCK at promoter sites, thus completing the cycle by inhibiting CLOCK/BMAL-1 transcriptional activity [6]. This core clock is refined to a 24-h period by the combined actions of several post-translational mechanisms (e.g., covalent histone modifications) carried out by a network of secondary clock proteins involving the kinase family [16,21]. An influential secondary loop driven by BMAL-1 activity involving the nuclear hormone receptors REVERB and ROR has also been identified. These proteins can inhibit and activate the transcription of the BMAL-1 gene (ARNTL), respectively, to ensure its rhythmic expression, and thus they can exert influence on circadian regulation [22]. In this loop, the expression of REVERB proteins serves to repress transcription in the promoter and enhancer regions of target genes, including ARNTL, whereas ROR competes to enhance ARNTL expression. A further CLOCK/BMAL1-driven sub-loop contains the PAR-bZip factors DBP, TEF, and HLF [23–34]. These compete with the repressor NFIL3 (or E4BP4), driven by the REV-ERB/ROR loop, to drive the expression of clock genes from D-box-containing promoters [22].

In addition to its timekeeping function via the TTFLs discussed above, the core molecular clock directly modulates the expression of over 4000 genes in 24-h cycles of transcription with diverse expression phases, resting with the combination of cis-elements (E-box, RORE, D-box) as the promoters and enhancers of specific clock-controlled genes (CCGs) [12,23,24]. These CCGs are expressed in virtually every cell and can generate tissue-specific circadian rhythms in transcription and cellular function, even in the absence of external cues. The circadian clock regulates between 10 and 50% of all transcripts in a cell, depending on tissue type, and influences critical processes such as cell cycle, redox homeostasis, inflammation, and metabolism [13]. The fact that nearly half of the mammalian protein-coding genome expresses tissue-specific circadian rhythmicity [13] underlines the importance of maintaining these oscillations and the seemingly critical role played by circadian rhythms in optimising cell and tissue function for normal health. In fact, cardiovascular function, energy metabolism, fluid balance, inflammation/immune function, cognition/neurological responses, and some of their biological drivers (CCGs and endocrine hormones) all appear to exhibit, either directly or indirectly, some degree of rhythmicity [25,26]. Beyond the central or core factors that directly control the core molecular clock, external factors can regulate the stability, phase, or function of core molecular clock proteins. Hypoxia (and fluctuating oxygen levels) acts on circadian rhythms (body temperature, metabolic rate, cortisol, and melatonin release in humans) through a number of mechanisms involving HIF-1α, with mounting evidence showing cross-talk between the HIF pathway and the circadian clock [27].

Disruption of the circadian system has direct consequences for human health by the uncoupling of multiple physiological processes and the disturbance of normal homeostasis, thus increasing the risk of several disorders and diseases [28]. In addition to genetic disruption to the clock circuit (via specific mutations), humans encounter several potent zeitgebers, often in simultaneous exposures, which hinder circadian regulation such as shift work, excessive artificial/night-time light exposure, disrupted eating patterns (often as a result of shift work), and international travel across several time zones (inducing jet lag).

The current understanding of the effects of circadian disruption has relied mainly on observational studies exploring occupational practices. Jet lagged cabin crew demonstrated temporary cognitive defects and structural brain changes following trans-meridian flights (across at least seven times zones) with five-day recovery intervals when compared to colleagues with fourteen-day recovery intervals [29]. Shift work and social jetlag also disrupt the circadian system and are associated with an increased risk of multiple diseases such as neurological disorders, diabetes, cancer, and cardiovascular disorders [30–32]. In addition to occupational requirements, recreational activities may also affect normal circadian rhythms. For example, the common use of light-emitting (via so-called 'blue light') electronic devices for reading, communication, and entertainment inhibits melatonin production which poses an increased risk for several circadian-related disorders when such devices are used in the evening [33]. As a consequence of modern living per se, many people are no longer exposed to the natural light/dark cycle of our ancestors. Humans are now inclined to be more active later into the night-time hours, often accompanied by eating and alcohol consumption, leading to a delayed bedtime but rising earlier than our natural rhythm would instigate. This is arguably leading to a chronic state of circadian disruption and associated risks [34].

3. Oxidative Stress and Redox Homeostasis

Humans are continually exposed to RONS from both endogenous (e.g., from aerobic metabolism and the immune response) and exogenous sources (e.g., tobacco smoke, UV light, etc.) and thus, antioxidants function to help prevent/curb high levels of oxidative stress and maintain redox homeostasis. Accordingly, aerobic organisms possess an intricate antioxidant defence system [35] that comprises an orchestrated synergism between several endogenous and exogenous antioxidants attempting to control the RONS produced in cells and tissues [36,37] which arise as a consequence of everyday activities such as the food we eat [38]. Oxidative stress is defined as 'an imbalance between oxidants and antioxidants in favor (sic) of the oxidants, leading to a disruption of redox signaling (sic) and control' [39]. Thus, the biological (physiological or pathological) effects of RONS depend critically on the amounts and location of production, relative to local antioxidant defences. For instance, when formed in low/moderate amounts, they can act as crucial mediators of signal transduction pathways e.g., in the growth of vascular smooth muscle cells (VSMC) and fibroblasts [40] and for exosome control and myokine release [41]. Yet, excessive RONS may cause widespread cellular toxicity [6,42]. Indeed, work from several labs, including our own, has demonstrated how RONS instigate oxidative damage by altering susceptible lipids, proteins, and DNA in human volunteers [43–45]. This can lead to impaired transcription and translational processes, altered protein function, and production of secondary by-products and metabolites which can further RONS production and/or cellular damage [46]. Although still an area of some discourse, RONS and other redox-active species have increasingly been acknowledged as fundamental regulators of genes, proteins, and the associated molecular signalling pathways integral to the regulation of many biological processes and functions [46,47].

3.1. Oxidative Stress, Redox Homeostasis, and Exercise

Multiple studies have now established a link between exercise (especially strenuous and/or exhaustive exercise) and oxidative stress [10,48]. The production of RONS in response to exercise, especially acutely, has long interested researchers and offers an intriguing model to examine the dynamic role of RONS from both the physiological and pathological perspectives—as depicted in Figure 2. The latter is important to note as evidence suggests that exhaustive (long duration) and/or strenuous exercise (high-intensity maximal exercise, marathons, triathlons, and overtraining) can induce detrimental, oxidative DNA alterations if left unrepaired [49,50]. Yet, during low or moderate-intensity exercise, the generated RONS may serve to act as signalling molecules responsible for the initiation of exercise and skeletal muscle adaptation [50–52], including accentuation in antioxidant enzymes and drivers of mitochondrial biogenesis [53]. In fact, the observation that low doses of stressors can exert beneficial effects, but when in excess can be toxic, forms the fundamental basis of the hitherto mentioned hormesis theory. This paradoxical effect has led to the extension of the hormesis theory to exercise induced RONS formation: transient increases in RONS can alter signalling pathways and/or cause molecular damage that induces adaptive responses that protect against subsequent stronger stress [54] such as that evident in human volunteers for the repeated bout effect following eccentric muscle contractions [55]. Moreover, when a single bout of exhaustive exercise is performed by well-trained human volunteers, a large elevation in oxidative damage is not observed [56]. Mounting evidence suggests that exercise-induced RONS are essential upstream signals for the activation of redox-sensitive transcription factors (e.g., Nrf2, AP1) and the induction of gene expression associated with exercise. As such, redox processes are increasingly recognized as an integral part of normal human biology and especially exercise (for further reading see the review by Margaritelis et al. [57]).

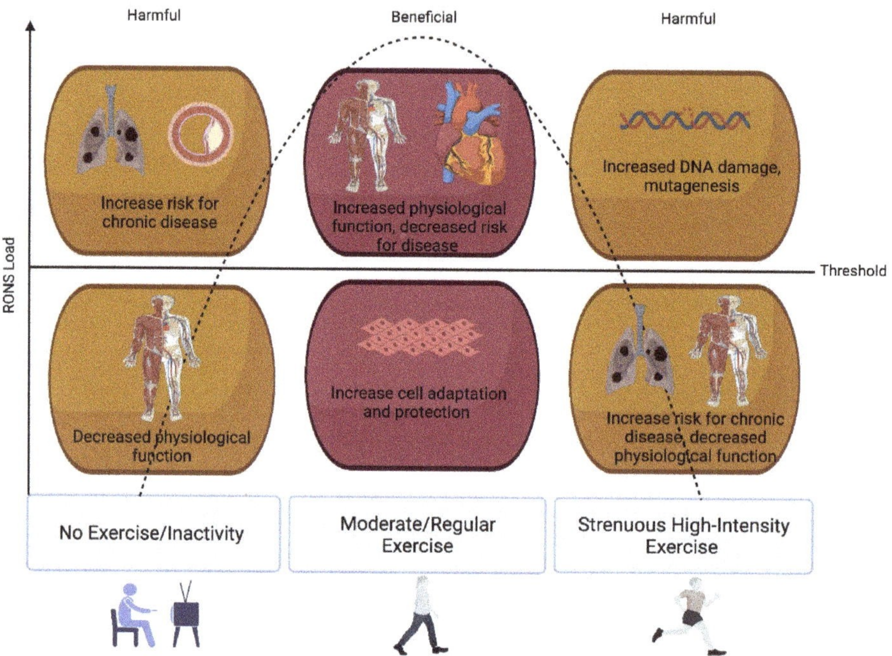

Figure 2. Exercise-induced RONS, Oxidative Stress, and Hormesis. The fundamental basis of the hormesis theory can be applied to exercise-induced RONS formation: low or transient increases in RONS, from moderate-intensity and regular exercise, activate signalling pathways that induce adaptive and protective responses. Whereas inactivity and/or sporadic strenuous/high-intensity exercise can lead to a RONS load that overwhelms antioxidant defences leading to oxidative stress, impaired physiological function, and an increased risk for chronic disease.

The adaptive potential of exercise and the extent of RONS production seems to be related to the exercise intensity of the bout or training stimulus [58]. High-intensity exercise produces higher concentrations of RONS than exercise at low/moderate-intensity aerobic exercise at <50% VO_{2max} [59]. While we have recently reported that acute intense/prolonged exercise can induce oxidative DNA damage in human volunteers, this damage does not persist indefinitely and is most likely repaired, at least in trained/experienced athletes [10]. This damage might act as a trigger for repair to prime the cell for further, subsequent damage and thus underpin the adaptive process and promote longevity. For instance, master endurance athletes are shown to have a longer telomere length (TL), a marker of biological age, than non-athlete, age-matched controls [60,61]. In addition, three weeks of high-intensity interval training (HIIT) in humans improves plasma antioxidant capacity [62]. Furthermore, recent evidence in animal models has reported reductions in maximal exercise capacity, several exercise-responsive proteins, and mitochondrial network adaptations in mice lacking a functional NOX2 complex following a HIIT programme [62]. Thus, the fine balance between redox signalling and oxidative stress in the response to exercise requires further elucidation. A growing appreciation for the role of redox homeostasis has allowed researchers to understand that RONS can offer distinct signalling roles depending on where they originate. The following is a summary of the main sources of RONS production in contracting muscle and the corresponding antioxidants that converge to support redox balance.

3.2. Skeletal Muscle as a Biological Source of RONS

Several RONS are generated continuously in, or very close to, skeletal muscle both at rest and during exercising contractions [63]. Of these, $O_2^{\bullet-}$ is the primary species. Early work revealed mitochondria as a source of free radical production at rest [64], but it appears mitochondria are not a major source of RONS generation in exercising muscle [65]. A number of other potential sources for primary and subsequent RONS (e.g., H_2O_2) have been identified and include NADPH oxidases (NOX; which is active across several cellular organelles), xanthine oxidases (XO), and phospholipase A_2 (PLA_2) [49]. It should also be noted that nitric oxide ($NO^{\bullet}$) production occurs both at rest and during/following exercise in muscle. Interestingly, via the convergence of NOX4 and two forms of neuronal NOS in mouse models, the resultant $NO^{\bullet}$ yields peroxynitrite ($ONOO^-$) proposed to be involved in the signal transduction pathways for mTORC1 integral to muscle hypertrophy associated with strength/resistance training [66–69]. $NO^{\bullet}$ itself has also been postulated to confer an intrinsic effect on the regulation of insulin secretion, and glucose transport and metabolism, which are also accentuated via exercise [70,71].

The NOX enzymes are viewed as an important source of RONS in skeletal muscle, particularly the NOX2 and NOX4 isoforms [72]. NOX2 is located within the sarcolemma and T-tubule, whereas NOX4 is located in both the sarcoplasmic reticulum and the mitochondria [73]. NOX4 is constitutively expressed and postulated to be involved in basal RONS production, whereas NOX2, sensitive to several stimuli, is believed to be the primary source of NOX-mediated RONS production in contracting muscle [11,72]. Whether or not NOX enzymes are the major source of muscular RONS is debatable as complexities remain in measuring their activity [49]; although recent advances using genetically encoded redox probes offer attractive possibilities in this respect [11,63], further scrutiny is beyond the scope of this narrative.

3.3. Antioxidants

As RONS and redox reactions appear to be pivotal for cellular health and function, mammalian cells require an inherent regulatory control mechanism to sustain redox homeostasis and prevent damage from oxidative stress. The antioxidant network of enzymatic and non-enzymatic antioxidants maintains RONS at physiological levels, thereby maintaining redox control. The principal antioxidant enzymes include superoxide dismutase (SOD), catalase (CAT), and glutathione peroxidase (GPx) [74]. Important non-enzymatic or dietary antioxidants include ascorbic acid (vitamin C), α-tocopherol (vitamin E), carotenoids, and flavonoids. In addition to the three main groups of enzymatic antioxidants (SODs, GPx, and Catalase), other accessory antioxidant enzymes help maintain redox balance. The Thioredoxin (Trx) antioxidant system comprising of Trx and the enzyme Trx reductase, serves as an electron donor to drive antioxidant systems and regulate proteins (in dithiol-disulfide exchange reactions) in response to a changing redox environment. This antioxidant assists in general metabolism (including DNA synthesis) and the prevention of deleterious levels of oxidative stress [75,76]. Peroxiredoxins (PRDXs) are a group of ubiquitous enzymes (PRDX1-6) that have emerged as important and widespread peroxide and $ONOO^-$ scavenging enzymes [77,78]. A growing body of evidence has begun to recognise the role of H_2O_2 in mediating redox signalling pathways during exercise, and thus it has been suggested that PRDX may be a salient regulator of exercise induced H_2O_2 levels (see review by Wadley et al. [79]).

4. Circadian Rhythms, Oxidative Stress, and Redox Homeostasis

Much is known about the cellular and molecular effects of RONS and circadian rhythms, respectively, but relatively less attention has been paid to the cross-talk and integration between the two and if/how these complex interactions affect physiological processes in health and disease. Current investigations have thus begun to focus on the molecular mechanisms linking RONS/oxidative stress and dysregulated circadian rhythms, sustaining the basis of a seemingly vital biological connection [80,81]—please see Figure 3. For instance, evidence now depicts a circadian influence on the immune system and this extends to the regulation of inflammatory and oxidative stress responses (especially in the ageing context—see review by [16]). It is now known that macrophages possess a molecular clock potentially able to impose temporal fluctuations in immune function [82]. RONS are integral to the proper functioning of the immune system and, interestingly, their production may be regulated by BMAL-1. In an elegant study by Early and colleagues [83] conducted on mice, ROS accumulation (detected by fluorescent probes) was increased in *BMAL-1$^{-/-}$* macrophages, and related to decreased activity of Nrf2 in cells, resulting in a diminished antioxidant response (including a reduced synthesis of glutathione) and increased production of the proinflammatory cytokine, IL-1β. Conversely, the redox milieu may also modulate core clock regulation as the DNA binding activity of the CLOCK: BMAL1 heterodimer is dependent on the prevailing cellular redox status (as evidenced in the NADH/NAD$^+$ and NADPH/NADP ratios), whereby binding is enhanced under reducing conditions [84]. Moreover, oxidative activation of Nrf2 via H_2O_2 has been shown to regulate core clock function by altering clock gene expression (*Per3*, *Nr1d1*, *Nr1d2*, *Dbp*, and *Tef*) and circadian function in mice in a comparable manner to overexpression or elimination of Nrf2 conditions [85].

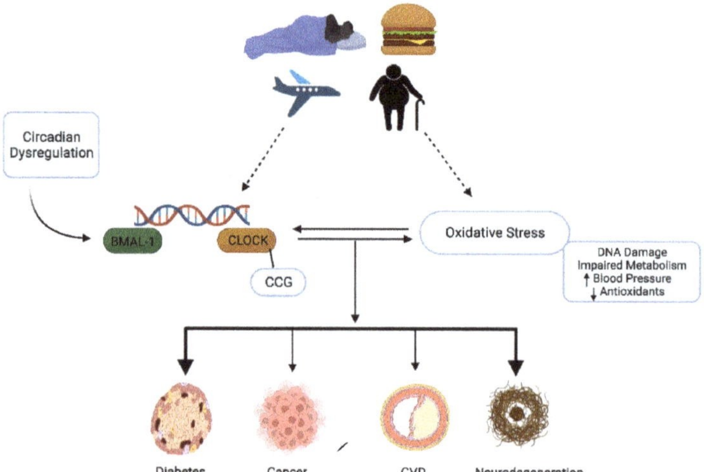

Figure 3. The interplay between Circadian and Redox Biology: Oxidative Stress and Disrupted Circadian Rhythms. A proposed bidirectional relationship exists between oxidative stress and circadian dysregulation whereby external factors and behaviours that disrupt circadian rhythms (jet lag, the ageing process, and regular consumption of certain foods) may also induce oxidative stress and, likely vice versa. The result is a deterioration of normal physiological functions (DNA damage, increased blood pressure, and insulin resistance) and control that increases the risk for chronic diseases.

4.1. Antioxidant Regulation and Control

Insight into the cross-talk between the circadian clock and oxidative stress has been gleaned from studies illustrating temporal expression patterns for antioxidant enzymes and compounds. For instance, O'Neill and Reddy [86] observed a circadian rhythm in mouse PRDXs that was altered in Cry1/2 double knockout cells, suggesting it may be regulated by the molecular clock. Using human RBCs, they also showed a robust ∼24-h redox cycle for PRDXs to directly show rhythms persist even without active transcription [86]. Furthermore, deletion of core clock components leads to increased thiol oxidation and protein carbonylation in flies and mice [87], and Per1-knockout flies exposed to oxidative stress display a shorter lifespan and increased oxidative damage compared to controls [88] which may, in part, be due to decreased activity of RONS scavenging enzymes. Given that RONS production can oscillate over the course of the day often due to normal, habitual behaviours, i.e., the by-products of metabolism, circadian control of antioxidants seems intuitive to maintain redox control [89]. Several studies have reported differences in DNA damage [90], lipid peroxidation [91], and protein oxidation [92] at different times of the day. These oscillations directly reflect the daily rhythm of antioxidant expression and protective enzyme activity levels. Those that peak in the morning includes: GPx; CAT; SOD; and PRDXs [6]. On the other hand, melatonin, plasma thiols, and ascorbic acid peak in the evening which also corresponds to the reported peaks for PERs1 and 2 and the CRYs (see [6]). Daily rhythmicity in SOD activity was first reported by Diaz-Munoz and colleagues in 1985. In contrast to humans, they found that in the rat cerebral cortex, SOD activity peaked in the dark phase, coinciding with the peak level of malondialdehyde (MDA), a marker of lipid peroxidation [93]. Thus, circadian oscillations in antioxidant expression seem to vary depending on the species investigated (i.e., nocturnal vs. diurnal), and this is most likely due to variations in sleep/wake cycles and the corresponding differences in their respective feeding patterns and light exposure [6].

The master regulator of antioxidant defence appears to be the transcription factor Nrf2, which translocates to the nucleus where it drives the expression of several antioxidant enzymes and is directly regulated by the expression of NRF through E box elements in the promoter [89]. As antioxidants are important for governing intracellular/local RONS levels, which, in turn, have been documented to impinge on the expression of clock genes, it appears that further appreciation and understanding of this cross-talk may yield purposeful applications for those interested in redox and chronobiology given how circadian rhythmicity and oxidative stress are central to many disease pathologies. For example, life-long administration of the antioxidant, N-acetyl-L-cysteine, has been shown to delay the onset of premature ageing induced by chronic oxidative stress in *BMAL-1*$^{-/-}$ mice [94].

4.2. Circadian and Oxidative Influence on Cardiovascular Physiology and Disease

Despite considerable progress in understanding, preventing, and treating cardiovascular diseases (CVD), ischaemic heart disease and strokes remain a major source of global morbidity and mortality [95]. Clock mechanisms are integral to normal cardiovascular function by coordinating rhythms in blood pressure, heart rate, and cardiac muscle contractility. However, circadian dysregulation increases cardiovascular risk with strong data suggesting adverse cardiovascular events such as sudden cardiac death [96] and myocardial infarction [97] are more likely to occur in the morning after awakening. Notably, a recent meta-analysis reported how the risk of acute myocardial infarction (AMI) increases after daylight saving transitions to emphasise the precarity of the circadian system [98]. The complete mechanism(s) surrounding such phenomena are unclear but may be linked to disruptions to redox control as RONS and free radical biology have been implicated in both the pathophysiology and treatment of heart disease [25]. For instance, we have reported increased peripheral arterial stiffness in healthy volunteers following the ingestion of a high-fat meal which was associated with augmented lipid hydroperoxides, while decreases in SOD were also observed [38]. We believe the high-fat meal increased $O_2^{\bullet-}$ production that was subsequently able to react with endothelium-derived $NO^{\bullet}$ (via eNOS) to impair blood vessel function. The consequences of this (and similar reactions that elevate vascular RONS) may be detrimental to cardiovascular function by reducing $NO^{\bullet}$ bioactivity and increasing the formation of the $ONOO^-$ [37], possibly leading to endothelial dysfunction and a pro-atherogenic environment [5]. $ONOO^-$ interacts with lipids, DNA, and proteins via direct or indirect radical mechanisms, and its generation has been cited in the pathogenesis of stroke, myocardial infarction, atherosclerosis, circulatory shock, and chronic inflammatory diseases [99,100]. These RONS-instigated reactions trigger cellular responses ranging from subtle modulations of cell signalling to oxidative injury, committing cells to necrosis and apoptosis [100]. Of note, Man et al. [101] reported how the peripheral circadian clock can regulate eNOS and NO$^\bullet$ production (which is lower during the morning) and that lipid metabolism also displays circadian oscillations. Combined, the misalignment of the circadian clock with these parameters could lead to the development/progression of atherosclerosis, which may be heightened with frequent exposure to conditions that amplify RONS production such as high-fat meal ingestion, as discussed above. Support for a circadian-redox mechanism is further evident from a study in middle-aged adults where vascular endothelial function, as measured by flow-mediated dilation (FMD), was impaired across the night and into the morning period and was accompanied by a pronounced rise in plasma MDA and a concomitant augmentation in the vasoconstrictor, endothelin-1 (ET-1) [102]. Such interactions are clearly pertinent to those already predisposed to increased risk for cardiovascular events, and other diseases like cancer, such as shift workers.

4.3. Shift Work: DNA Damage and Repair

In a recent cross-sectional study, higher levels of H_2O_2 and lower SOD and catalase were observed in night workers when compared to day workers [103]. In fact, the circadian clock was first implicated as a factor in various diseases as epidemiological studies reported an increased incidence of cancers in long-term shift workers [19,104,105]. Those working night shifts also appear to have increased DNA damage that is linked to lower melatonin induced by night working [106]. In strict laboratory conditions, when compared to simulated day shifts for three consecutive days, simulated night shifts of the same duration also caused circadian dysregulation of genes involved in key DNA repair pathways. Moreover, the percentage of cells with BRCA1 and γH2AX foci (representing DNA damage biomarkers using immunofluorescent microscopy) was significantly higher in the night shift condition, whereas the effectiveness of the processes to repair leukocyte DNA damage from both endogenous and exogenous sources were compromised in samples from the night shift volunteers [107]. Given that it exerts antioxidant and DNA repair properties (via the nucleotide excision repair (*NER*) pathways), melatonin supplementation, acting as a so-called chronobiotic [32], has been suggested to reduce the potentially damaging effects of shift work [108]. The studies and data summarised in this section are undoubtedly useful for helping tie the molecular and biochemical connections between shift workers and elevated cancer risk, but much remains open for further discovery.

4.4. Diabetes, Obesity, and Metabolic Control

Normal CLOCK and BMAL-1 activity play a role in defending the body against metabolic disturbances, but CLOCK gene mutations are associated with hyperphagia, hyperlipidemia, hyperinsulinemia, hyperglycaemia, and sleep disorders—all of which are common in metabolic diseases such as diabetes and obesity [109]. Diabetes, especially Type 2 Diabetes (T2D), is connected to disruptions in normal circadian rhythms with shift-work, light pollution, jet lag, and increased screen time, all acting as potential contributory factors [33,110,111]. Emerging evidence has also identified melatonin, already known for its roles in circadian and redox physiology, respectively, as a potential mediator of glucose levels and insulin production [109] adding further credence to the existence of an important cross-talk between circadian rhythms and redox pathways in the regulation of metabolic control. Underpinning all forms of diabetes is either a decrease in β-cell mass or β-cell function [112–114]. Mechanistically, oxidative stress, potentially stemming from increased mitochondrial O_2 production following excessive caloric intake and low activity levels, has also been implicated in diabetes as a key mediator of β-cell dysfunction and insulin resistance [5,115]. This appears to be secondary to circadian dysregulation as β-cells contain critical antioxidant genes that are targets not only of Nrf2, but also BMAL-1 which are vulnerable to circadian disruption in *BMAL-1$^{-/-}$* mice [116]. Circadian disruption to these key antioxidants can thus lead to augmented β-cell mitochondrial RONS production in cells that already have relatively less antioxidant capacity, culminating in the impairment in β-cell function, insulin resistance, and diabetes [111,117].

Alterations to normal circadian control and oxidative stress are understood to be instrumental in the pathogenesis of other metabolic conditions associated with T2D, such as obesity [118]. Adipose tissue is a critical modulator of metabolic health, and oxidative stress can cause adipose tissue dysfunction by stimulating preadipocyte proliferation, adipogenesis, and chronic inflammation, which leads to obesity [119]. The circadian clock controls energy homeostasis by regulating circadian expression and/or activity of enzymes, hormones, and transport systems involved in metabolism with evidence showing that knockout and mutations in clock genes induce disruptions in adipose tissue function, differentiation, and metabolism (see review by Froy and Garaulet [120]). Disruptions to circadian control and oxidative stress may also affect other important metabolic organs. For instance, a relatively recent study found that night shift workers have a higher serum concentration of alanine aminotransferase (ALT) than daytime workers, indicating a potential association between circadian disruption and liver function [121] that could be important in the development of conditions such as non-alcoholic fatty liver disease (NAFLD). Moreover, such conditions proposed to be related to dietary factors (see review by Arrigo et al. [122]), are often accompanied by oxidative stress and seem to favourably respond to antioxidant and circadian therapy in the form of melatonin administration [123]. As alluded to, dietary factors and feeding patterns undoubtedly represent an important metabolic focus for their ability to modulate circadian function and redox control in health and disease, though further scrutiny is beyond the scope of this review.

Far from being fully determined, it does appear that a bidirectional relationship exists between RONS production and circadian rhythms: core clock function appears to be sensitive to changes in redox status and redox homeostasis may conversely be governed by clock machinery [124]. When out of sync, circadian dysregulations, oxidative stress, or both, can manifest. This is an intriguing frontier of research emerging, and the ability to harness these insights, combined with other important drivers of circadian and redox biology, will help to provide more focussed health and lifestyle prescriptions.

5. Exercise—Zeitgeber and Modulator of Redox Homeostasis

Regular exercise is often regarded as one of the 'best buys' for public health given its multiple health-promoting benefits [125], with the Academy of Medical Royal Colleges [126] describing 'the miracle cure' of performing 30 min of moderate exercise, five times a week, as more powerful than many drugs administered for chronic disease prevention and management. Aerobic exercise training improves risk factors of metabolic syndrome such as glucose intolerance, hyperlipidemia, high blood pressure, low high-density lipoprotein content, and visceral obesity. Improved aerobic fitness increases neurogenesis, enhances memory, and may prevent brain atrophy [127]. Exercise is widely regarded for its substantial health benefits and there is now ample evidence from both observational studies and randomised trials to support that regular exercise is a contributing factor in the prevention of cardiovascular disease, cancer, diabetes, and other chronic conditions, as well as reducing the risk of all-cause mortality [128,129]. As more is understood regarding the molecular pathways and mechanisms through which the beneficial effects of exercise are transmitted, interest continues to grow in strategies to optimise the benefits of such effects and how these can be translated for optimal health and performance purposes (see Figure 4).

The circadian clock can be synchronised by photic and non-photic stimuli (temperature, physical activity, and food intake). Given that exercise represents a major challenge to whole-body homeostasis, provoking widespread perturbations in cells, tissues, and organs [130], modern theories have begun to probe the connections between exercise and circadian rhythms for both performance and health purposes. An extensive body of literature has established that exercise can influence the circadian system in rodents (see [1]) and emerging human evidence shows exercise can elicit phase-shifting effects which may be dependent on chronotype [34]. Exercise appears to be a potent entrainment factor for central as well as peripheral clocks, including those in muscle. For instance, the average core-clock gene expression (*BMAL1, ROR-α, CRY1, PER2, PER1,* and *NR1D1*) in male rugby players is significantly higher compared to sedentary males [131]. Therefore, the possibility of (a) exercise being able to attenuate the negative health effects of circadian misalignment and (b) the existence of an optimal time to exercise to maximise its therapeutic effects have become increasingly attractive to researchers and clinicians.

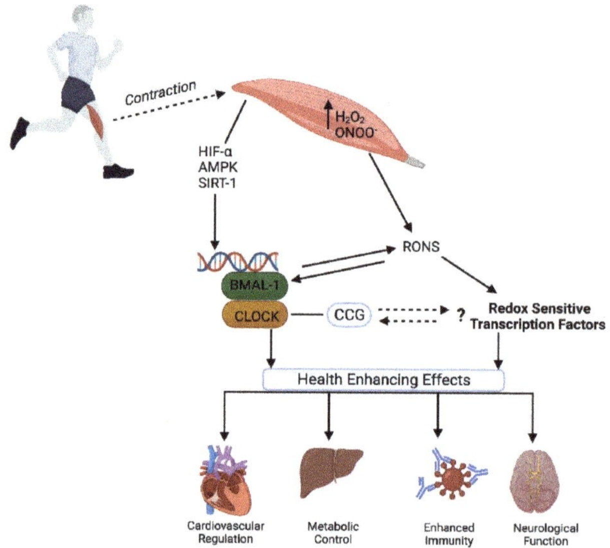

Figure 4. The interplay between Circadian and Redox Biology: Exercise as a zeitgeber and source of RONS. Exercise is a potent stimulus to entrain dysregulated circadian systems and positively affect the core molecular clock and subsequent expression of CCGs throughout the body. At the same time, exercise-induced RONS are integral to several recognised physiological responses and adaptations via the activation of redox-sensitive transcription factors (e.g., up-regulation of immunity, antioxidant enzymatic activity, etc.). It is possible that both circadian and redox signalling are inter-connected and operate synergistically to confer protective health effects following regular exercise.

Of all the peripheral tissues, skeletal muscle represents a major downstream target for clock activity. Among other functions, elegant studies have shown how the molecular clock governs glucose metabolism in skeletal muscle. In one such study, *BMAL-1* deletion in mouse skeletal muscle manifested in impaired glucose uptake, reduced GLUT-4, and disrupted the activity of key glycolytic enzymes [132]. Thus, impairments in the muscle molecular clock appear to have important implications for the development of metabolic diseases such as T2D. As exercise is recommended for the prevention and treatment of T2D [9], it now appears that the beneficial metabolic effects exercise confers are, at least partly, achieved through actions on the muscle molecular clock to restore local circadian regulation [7]. For instance, skeletal muscle gene and protein expression of *BMAL-1* and *PER2* were increased in adults with obesity and pre-diabetes following 12 weeks of exercise training and this was accompanied by improvements in body composition, peripheral insulin sensitivity (glucose disposal rate), and maximal oxygen consumption. Specifically, *BMAL-1* gene expression correlated with glucose disposal rate [133].

While exercise undoubtedly elicits favourable, modulating effects in skeletal muscle metabolism, relatively little is known about the potency of these effects at different times of the day. This is important as the biological clock seems to drive patent rhythms in human skeletal muscle metabolism whereby mitochondrial oxidative capacity follows a day-night rhythm, peaking in the late evening and being lowest in the early afternoon [134]. As oxidative capacity is a vital determinant of exercise performance, it is unsurprising that studies report clear time of day effects (~10%) for exercise performance, capacity [8], and/or strength measures in human volunteers, which in one recent paper were correlated with *PER2* daily profiles [135]. The underlying mechanisms for such phenomena are complex but are thought to involve circadian fluctuations in core temperature, endocrine hormones, neuromuscular function, and metabolic flux [8,9]. One of the key contributing factors in these performance fluctuations is body temperature which can exert a myriad of regulatory effects on neuromuscular and metabolic activity and appears to peak in the late afternoon. As thermoregulation during exercise itself appears to follow a circadian rhythm, this may also explain the variability of fatigue onset when the same activity is performed at different times across the daily cycle, especially in longer, endurance-type exercise bouts [136]. Variations in human exercise efficiency (improved in Late vs. Early) have been ascribed to clock-driven fluctuations in metabolic control such as carbohydrate metabolism that requires lower oxygen consumption, a lower heart rate, and a lower rate of perceived exertion [8]. In competitive and elite sporting settings, this connection between exercise capacity and the molecular clock may be useful, to some extent, when planning to optimise training and competition schedules, but this is not an entirely new concept (see [137]).

Syncing exercise bouts, alongside other interventions not scrutinised in the current review (e.g., time-restricted eating), to align with circadian rhythms is an appealing paradigm to consider when trying to maximise the metabolic and health effects of exercise. In a crossover study involving 2 weeks of high-intensity interval training (HIIT), afternoon HIIT was more efficacious than morning HIIT at improving blood glucose in $n = 11$ men with T2D [138]. Such data indicate that the timing of exercise should be considered when prescribed for the management of T2D. However, in a larger study investigating the effects of exercise timing on glycaemic control in those with and without T2D, no distinct glycaemic benefits or alterations in circadian rhythm (as assessed by skin temperature) were detected between morning versus evening exercise [139].

Regular physical activity and exercise programmes have been shown to reduce the symptoms of patients with established CVD; additionally, prospective epidemiological studies of occupational and leisure-time physical activity have consistently documented a reduced incidence of CVD in the more physically active and fit individuals [140–142]. Consequently, exercise is regarded as an important intervention in tackling the burden imposed by CVD. Yet, the precise mechanisms by which exercise exhibits its ameliorating effects on the cardiovascular system still require elucidation. For instance, relatively little is known about the potential for exercise to entrain central and peripheral clocks and how this might influence cardiovascular health. It is conceivable that exercise may regulate circadian factors to partially influence cardiovascular health through the documented effects on skeletal muscle—given the connections between metabolic and cardiovascular health [5,7]. Studies have begun to explore this theme as well as the possible existence of diurnal effects of exercise on cardiovascular risk markers. Elevated arterial blood pressure, or hypertension, is a traditional cardiovascular risk factor that also forms one of the main constituents of metabolic syndrome [143,144]. Acute exercise can induce a transient reduction in blood pressure or post-exercise hypotension (PEH) [145]. This reduction is largely thought to be a result of increased blood flow to vascular beds and the subsequent decrease in the total peripheral resistance associated with exercise. The prolonged hypotensive effects of regular aerobic exercise may thus be due to repeated instances of PEH [146]. Despite this, BP reductions after aerobic training vary across studies and some factors, such as higher initial BP, moderate to high training intensities, and concomitant diet-induced weight loss, have been identified as promoters of a greater BP decrease [147]. Moreover, alterations to dietary composition (reduced fat and increased fibre) alongside exercise training have also been reported to decrease BP and oxidative stress markers in the absence of weight loss [148]. It is also possible that the time of day when aerobic training is performed may influence the extent of BP reductions after training. As most aspects of cardiovascular regulation demonstrate a circadian or diurnal pattern [149], it is plausible that the mechanisms driving sustained PEH are impacted by time of day. Some studies have tested this hypothesis with mixed results. In one study, the acute hypotensive effects following 30 min of steady state exercise were less marked in the morning versus the afternoon [150], but this may be masked due to the circadian effect on morning blood pressure [7]. While de Brito et al. [151] demonstrated that aerobic training performed in the evening decreased clinic and ambulatory BP when compared to morning training in hypertensive men, this finding may be explained by the use of antihypertensive medications in the morning group. In contrast, while aerobic exercise performed both in the morning and the afternoon/evening contributed to PEH when circadian influences of morning blood pressure were considered, PEH was greater following morning exercise rather than evening exercise in a separate study [152]. The influence of exercise intensity and other zeitgebers like food ingestion, light exposure, and sleeping patterns are complex confounders in such studies, but exercise may induce re-alignment of the circadian clock and better cardiovascular outcomes by modulating hormonal responses and heart rate (see review by [7]).

It is well established that aerobic exercise can improve endothelial function and flow-mediated dilation (FMD) responses [153], particularly in those with cardiovascular disease and related risk factors [154]. Despite this link, the diurnal effects of exercise on FMD have received relatively little attention, even though FMD has been shown previously to vary with time of day [155]. A handful of studies have nevertheless provided useful insights to illustrate the apparent diurnal effects of exercise on FMD in some [156,157] but not all instances [152]. Although beyond the scope of this paper, the equivocal findings in such studies may be partially explained by variations in the exercise stimuli used e.g., whole body versus isolated limbs.

Endurance exercise training elicits a number of benefits such as increased skeletal muscle mitochondrial number and volume density leading to improved oxidative capacity [158]. An important development in unraveling the cellular events that promote mitochondrial biogenesis and other beneficial effects such as angiogenesis, improved antioxidant defences, and enhanced fat metabolism [11] was the discovery of the transcriptional coactivator PGC-1α, thought to be the master regulator of the process. Recently, this exercise-stimulated pathway has been identified as being downstream of the molecular clock, providing a molecular mechanism through which circadian timing can influence exercise responses [12,159]. In recent years, the understanding of this mechanism has grown to the extent that the pathway relies on redox signals, whereby exercise-induced RONS appear to potentiate PGC-1α and NF-kB [48]. Evidence to support this supposition comes from studies such as that from Ristow and Colleagues [53] who showed the use of a vitamin C and vitamin E antioxidant regime blunted mRNA responses in several markers of mitochondrial biogenesis, including PGC-1α, in healthy young men completing a training intervention. Mechanistically, PGC-1α regulation at rest and during exercise depends on several factors with evidence suggesting that AMPK, Nrf2, and p38 serve as the principal intermediate molecules connecting RONS with PGC-1α activation and/or mitochondrial biogenesis markers [57]. Redox activation of Nrf2 has already been identified in the priming of the antioxidant network [83] and in the regulation of core clock function through its activation of clock machinery [85]. Moreover, exercise-induced RONS and redox-sensitive genes and transcription factors, including AMPK, HIF-1α, and PGC-1α, may influence the expression of core molecular clock genes [11,12]. Given the cross-talk between HIF pathways and circadian clock control [27], and that RONS are thought to be instrumental in the exercise-induced increases in HIF-1α [57], HIF-1α may be a crucial mediator connecting circadian control, redox homeostasis, and exercise.

While the majority of redox-focussed exercise studies have examined aerobic/endurance training, a growing appreciation for the role of RONS signalling in resistance training-mediated adaptations have been outlined. In particular, the mechanistic target of rapamycin complex 1 (mTORC1) activation, which stimulates protein synthesis via increased translation of contractile protein mRNA central to muscular hypertrophy, has shown to be redox-sensitive. For example, and as previously highlighted, $ONOO^-$ has been implicated in the signal transduction pathways for mTORC1 that lead to hypertrophy [67]. As mTORC1 appears to exhibit diurnal oscillations—current evidence has highlighted the suppressive effect of PER2 on mTORC1 [160], the extent to which these are influenced by RONS signalling and if/how this might be integral to diurnal exercise responses remains uncovered and ripe for investigation.

6. Future Directions

Contemporary research has identified circadian disruptions to augment oxidative stress and subsequently aberrate adipose tissue function and metabolism. Therefore, circadian machinery in the adipose tissue may be a novel therapeutic target for the prevention and treatment of metabolic and cardiovascular diseases [119]. We believe that exploratory exercise studies connecting circadian principles to adipose tissue and redox metabolism may represent one such approach. Based on the existing literature, there appears to be scope for mechanistic and maybe even clinical progress, but it is imperative that future investigations are carefully designed to control for confounding factors including participant chronotype and the influence of other zeitgebers [34] and testing of insufficient numbers of volunteers in a narrow range of times across the 24 h day [1]. Importantly, new studies could also benefit from the careful selection of appropriate methods to measure and assess the impact of potential interventions on circadian rhythms [161]. Exercise stimuli and protocols such as intensity, mode, etc. should also be carefully chosen when considering the potential implications for biological rhythms and RONS. Some of the equivocal findings on the health/physiological effects of exercise timing described in the review may be due to the heterogeneity of the volunteers. More parallel investigations are warranted for this potentially promising realm of exercise prescription research, such as those that incorporate larger sample sizes, crossover study designs, and, crucially, chronic exercise interventions that assay a range of clinical and mechanistic redox markers. Of course, as alterations in feeding patterns and diet can affect circadian rhythms and redox markers [119,162], further mechanistic investigations are also welcome that seek to simultaneously explore dietary manipulations with exercise to aid in the development of targeted treatment strategies to improve health in those with/or at risk from chronic disease.

7. Conclusions

Roenneberg and Merrow [19] offer a plan on how to implement chronobiological principles into medicine through the identification of a mechanism that determines the circadian clocks of an individual and their ability to be synchronised. This knowledge could inform the prescription of specific exposures of zeitgebers to target the circadian system and personalise the therapeutic schedules for any individual [109]. Targeting circadian health would be especially pertinent to those at risk from circadian misalignment from shift work, frequent bright light exposure, sleeping disorders, and frequent travel across multiple time zones. We believe that exercise might represent one such zeitgeber given its ability to potentially entrain the human circadian system; additionally, careful interrogation of how it may modulate the cross-talk between the molecular clock and redox biology could yield further insights into the salient mechanistic pathways such as those that control antioxidant defences and DNA repair. At present, and despite the promise of studies showing diurnal variations in exercise responses/adaptations, much more work is needed to comprehensively identify whether an optimal exercise time exists. As it stands, exercise prescription should be personalised based on several variables such as the individual's health and fitness status (i.e., the absence or presence of CVD risk factors), chronotype, work and domestic patterns (including mealtimes and work schedules), and preferences. Given the recent advances in redox and molecular biology, informed by advancements in analytical techniques, exercise studies positioned at the emerging nexus between circadian rhythms and the redox milieu represent a fascinating area for exploration, especially for those that advocate, or wish to scrutinise, exercise prescription for health.

Funding: This research received no external funding.

Acknowledgments: All figures contained within this manuscript were created with BioRender.com.

Conflicts of Interest: The authors declare no conflict of interest.

References

1. Youngstedt, S.D.; Elliott, J.A.; Kripke, D.F. Human circadian phase-response curves for exercise. *J. Physiol.* **2019**, *597*, 2253–2268. [CrossRef]
2. Booth, F.W.; Chakravarthy, M.V.; Spangenburg, E.E. Exercise and gene expression: Physiological regulation of the human genome through physical activity. *J. Physiol.* **2002**, *543*, 399–411. [CrossRef]
3. Parr, E.B.; Heilbronn, L.K.; Hawley, J.A. A Time to Eat and a Time to Exercise. *Exerc. Sport Sci. Rev.* **2020**, *48*, 4–10. [CrossRef]
4. Zimmet, P.; Thomas, C.R. Genotype, obesity and cardiovascular disease-has technical and social advancement outstripped evolution? *J. Intern. Med.* **2003**, *254*, 114–125. [CrossRef]
5. Ceriello, A.; Motz, E. Is oxidative stress the pathogenic mechanism underlying insulin resistance, diabetes, and cardiovascular disease? The common soil hypothesis revisited. *Arter. Thromb. Vasc. Biol.* **2004**, *24*, 816–823. [CrossRef]
6. Wilking, M.; Ndiaye, M.; Mukhtar, H.; Ahmad, N. Circadian rhythm connections to oxidative stress: Implications for human health. *Antioxid. Redox Signal.* **2013**, *19*, 192–208. [CrossRef]
7. Hower, I.M.; Harper, S.A.; Buford, T.W. Circadian Rhythms, Exercise, and Cardiovascular Health. *J. Circadian Rhythm.* **2018**, *16*, 7. [CrossRef]
8. Ezagouri, S.; Zwighaft, Z.; Sobel, J.; Baillieul, S.; Doutreleau, S.; Ladeuix, B.; Golik, M.; Verges, S.; Asher, G. Physiological and Molecular Dissection of Daily Variance in Exercise Capacity. *Cell Metab.* **2019**, *30*, 78–91.e4. [CrossRef]
9. Gabriel, B.M.; Zierath, J.R. Circadian rhythms and exercise—Re-setting the clock in metabolic disease. *Nat. Rev. Endocrinol.* **2019**, *15*, 197–206. [CrossRef]
10. Tryfidou, D.V.; McClean, C.; Nikolaidis, M.G.; Davison, G.W. DNA Damage Following Acute Aerobic Exercise: A Systematic Review and Meta-analysis. *Sports Med.* **2020**, *50*, 103–127. [CrossRef]
11. Bouviere, J.; Fortunato, R.S.; Dupuy, C.; Werneck-de-Castro, J.P.; Carvalho, D.P.; Louzada, R.A. Exercise-Stimulated ROS Sensitive Signaling Pathways in Skeletal Muscle. *Antioxidants* **2021**, *10*, 537. [CrossRef]
12. Wolff, C.A.; Esser, K.A. Exercise Timing and Circadian Rhythms. *Curr. Opin. Physiol.* **2019**, *10*, 64–69. [CrossRef]
13. Zhang, R.; Lahens, N.F.; Ballance, H.I.; Hughes, M.E.; Hogenesch, J.B. A circadian gene expression atlas in mammals: Implications for biology and medicine. *Proc. Natl. Acad. Sci. USA* **2014**, *111*, 16219–16224. [CrossRef]
14. Reppert, S.M.; Weaver, D.R. Coordination of circadian timing in mammals. *Nature* **2002**, *418*, 935–941. [CrossRef]
15. Hastings, M.H.; Maywood, E.S.; Brancaccio, M. Generation of circadian rhythms in the suprachiasmatic nucleus. *Nat. Rev. Neurosci.* **2018**, *19*, 453–469. [CrossRef]
16. Lananna, B.V.; Musiek, E.S. The wrinkling of time: Aging, inflammation, oxidative stress, and the circadian clock in neurodegeneration. *Neurobiol. Dis.* **2020**, *139*, 104832. [CrossRef]
17. Tahara, Y.; Shibata, S. Entrainment of the mouse circadian clock: Effects of stress, exercise, and nutrition. *Free Radic. Biol. Med.* **2018**, *119*, 129–138. [CrossRef]
18. Bae, S.A.; Fang, M.Z.; Rustgi, V.; Zarbl, H.; Androulakis, I.P. At the Interface of Lifestyle, Behavior, and Circadian Rhythms: Metabolic Implications. *Front. Nutr.* **2019**, *6*, 132. [CrossRef]
19. Roenneberg, T.; Merrow, M. The Circadian Clock and Human Health. *Curr. Biol.* **2016**, *26*, R432–R443. [CrossRef]
20. Huang, R.C. The discoveries of molecular mechanisms for the circadian rhythm: The 2017 Nobel Prize in Physiology or Medicine. *Biomed. J.* **2018**, *41*, 5–8. [CrossRef]
21. Wang, J.; Symul, L.; Yeung, J.; Gobet, C.; Sobel, J.; Luck, S.; Westermark, P.O.; Molina, N.; Naef, F. Circadian clock-dependent and -independent posttranscriptional regulation underlies temporal mRNA accumulation in mouse liver. *Proc. Natl. Acad. Sci. USA* **2018**, *115*, E1916–E1925. [CrossRef]
22. Patke, A.; Young, M.W.; Axelrod, S. Molecular mechanisms and physiological importance of circadian rhythms. *Nat. Rev. Mol. Cell Biol.* **2020**, *21*, 67–84. [CrossRef]
23. Koike, N.; Yoo, S.H.; Huang, H.C.; Kumar, V.; Lee, C.; Kim, T.K.; Takahashi, J.S. Transcriptional architecture and chromatin landscape of the core circadian clock in mammals. *Science* **2012**, *338*, 349–354. [CrossRef]
24. Xie, Y.; Tang, Q.; Chen, G.; Xie, M.; Yu, S.; Zhao, J.; Chen, L. New Insights into the Circadian Rhythm and Its Related Diseases. *Front. Physiol.* **2019**, *10*, 682. [CrossRef]
25. Khaper, N.; Bailey, C.D.C.; Ghugre, N.R.; Reitz, C.; Awosanmi, Z.; Waines, R.; Martino, T.A. Implications of disturbances in circadian rhythms for cardiovascular health: A new frontier in free radical biology. *Free Radic. Biol. Med.* **2018**, *119*, 85–92. [CrossRef]
26. Ayyar, V.S.; Sukumaran, S. Circadian rhythms: Influence on physiology, pharmacology, and therapeutic interventions. *J. Pharmacokinet. Pharmacodyn.* **2021**, *48*, 321–338. [CrossRef]

27. O'Connell, E.J.; Martinez, C.A.; Liang, Y.G.; Cistulli, P.A.; Cook, K.M. Out of breath, out of time: Interactions between HIF and circadian rhythms. *Am. J. Physiol. Cell Physiol.* **2020**, *319*, C533–C540. [CrossRef]

28. Kondratov, R.V.; Gorbacheva, V.Y.; Antoch, M.P. The role of mammalian circadian proteins in normal physiology and genotoxic stress responses. *Curr. Top. Dev. Biol.* **2007**, *78*, 173–216.

29. Cho, K. Chronic 'jet lag' produces temporal lobe atrophy and spatial cognitive deficits. *Nat. Neurosci.* **2001**, *4*, 567–568. [CrossRef]

30. Roenneberg, T.; Allebrandt, K.V.; Merrow, M.; Vetter, C. Social jetlag and obesity. *Curr. Biol.* **2012**, *22*, 939–943. [CrossRef]

31. James, S.M.; Honn, K.A.; Gaddameedhi, S.; Van Dongen, H.P.A. Shift Work: Disrupted Circadian Rhythms and Sleep-Implications for Health and Well-Being. *Curr. Sleep Med. Rep.* **2017**, *3*, 104–112. [CrossRef]

32. Potter, G.D.M.; Wood, T.R. The Future of Shift Work: Circadian Biology Meets Personalised Medicine and Behavioural Science. *Front. Nutr.* **2020**, *7*, 116. [CrossRef]

33. Chang, A.M.; Aeschbach, D.; Duffy, J.F.; Czeisler, C.A. Evening use of light-emitting eReaders negatively affects sleep, circadian timing, and next-morning alertness. *Proc. Natl. Acad. Sci. USA* **2015**, *112*, 1232–1237. [CrossRef]

34. Thomas, J.M.; Kern, P.A.; Bush, H.M.; McQuerry, K.J.; Black, W.S.; Clasey, J.L.; Pendergast, J.S. Circadian rhythm phase shifts caused by timed exercise vary with chronotype. *JCI Insight* **2020**, *5*, e134270. [CrossRef]

35. Finkel, T.; Holbrook, N.J. Oxidants, oxidative stress and the biology of ageing. *Nature* **2000**, *408*, 239–247. [CrossRef]

36. Sen, C.K.; Packer, L. Thiol homeostasis and supplements in physical exercise. *Am. J. Clin. Nutr.* **2000**, *72*, 653S–669S. [CrossRef]

37. Fukai, T.; Folz, R.J.; Landmesser, U.; Harrison, D.G. Extracellular superoxide dismutase and cardiovascular disease. *Cardiovasc. Res.* **2002**, *55*, 239–249. [CrossRef]

38. McClean, C.M.; McLaughlin, J.; Burke, G.; Murphy, M.H.; Trinick, T.; Duly, E.; Davison, G.W. The effect of acute aerobic exercise on pulse wave velocity and oxidative stress following postprandial hypertriglyceridemia in healthy men. *Eur. J. Appl. Physiol.* **2007**, *100*, 225–234. [CrossRef]

39. Jones, D.P. Redefining oxidative stress. *Antioxid. Redox Signal.* **2006**, *8*, 1865–1879. [CrossRef]

40. Dröge, W. Free radicals in the physiological control of cell function. *Physiol. Rev.* **2002**, *82*, 47–95. [CrossRef]

41. Louzada, R.A.; Bouviere, J.; Matta, L.P.; Werneck-de-Castro, J.P.; Dupuy, C.; Carvalho, D.P.; Fortunato, R.S. Redox Signaling in Widespread Health Benefits of Exercise. *Antioxid. Redox Signal.* **2020**, *33*, 745–760. [CrossRef]

42. Stocker, R.; Keaney, J.F., Jr. Role of oxidative modifications in atherosclerosis. *Physiol. Rev.* **2004**, *84*, 1381–1478. [CrossRef]

43. Davison, G.W.; Hughes, C.M.; Bell, R.A. Exercise and mononuclear cell DNA damage: The effects of antioxidant supplementation. *Int. J. Sport Nutr. Exerc. Metab.* **2005**, *15*, 480–492. [CrossRef]

44. Fogarty, M.C.; Devito, G.; Hughes, C.M.; Burke, G.; Brown, J.C.; McEneny, J.; Brown, D.; McClean, C.; Davison, G.W. Effects of alpha-lipoic acid on mtDNA damage after isolated muscle contractions. *Med. Sci. Sports Exerc.* **2013**, *45*, 1469–1477. [CrossRef]

45. Brown, M.; McClean, C.M.; Davison, G.W.; Brown, J.C.W.; Murphy, M.H. Preceding exercise and postprandial hypertriglyceridemia: Effects on lymphocyte cell DNA damage and vascular inflammation. *Lipids Health Dis.* **2019**, *18*, 125. [CrossRef]

46. McKeegan, K.; Mason, S.A.; Trewin, A.J.; Keske, M.A.; Wadley, G.D.; Della Gatta, P.A.; Nikolaidis, M.G.; Parker, L. Reactive oxygen species in exercise and insulin resistance: Working towards personalized antioxidant treatment. *Redox Biol.* **2021**, *44*, 102005. [CrossRef]

47. Gomez-Cabrera, M.C.; Domenech, E.; Vina, J. Moderate exercise is an antioxidant: Upregulation of antioxidant genes by training. *Free Radic. Biol. Med.* **2008**, *44*, 126–131. [CrossRef]

48. Powers, S.K.; Deminice, R.; Ozdemir, M.; Yoshihara, T.; Bomkamp, M.P.; Hyatt, H. Exercise-induced oxidative stress: Friend or foe? *J. Sport Health Sci.* **2020**, *9*, 415–425. [CrossRef]

49. Radak, Z.; Chung, H.Y.; Koltai, E.; Taylor, A.W.; Goto, S. Exercise, oxidative stress and hormesis. *Ageing Res. Rev.* **2008**, *7*, 34–42. [CrossRef]

50. Powers, S.K.; Duarte, J.; Kavazis, A.N.; Talbert, E.E. Reactive oxygen species are signalling molecules for skeletal muscle adaptation. *Exp. Physiol.* **2010**, *95*, 1–9. [CrossRef]

51. Irrcher, I.; Ljubicic, V.; Hood, D.A. Interactions between ROS and AMP kinase activity in the regulation of PGC-1alpha transcription in skeletal muscle cells. *Am. J. Physiol.-Cell Physiol.* **2009**, *296*, C116–C123. [CrossRef]

52. Nemes, R.; Koltai, E.; Taylor, A.W.; Suzuki, K.; Gyori, F.; Radak, Z. Reactive Oxygen and Nitrogen Species Regulate Key Metabolic, Anabolic, and Catabolic Pathways in Skeletal Muscle. *Antioxidants* **2018**, *7*, 85. [CrossRef] [PubMed]

53. Ristow, M.; Zarse, K.; Oberbach, A.; Kloting, N.; Birringer, M.; Kiehntopf, M.; Stumvoll, M.; Kahn, C.R.; Bluher, M. Antioxidants prevent health-promoting effects of physical exercise in humans. *Proc. Natl. Acad. Sci. USA* **2009**, *106*, 8665–8670. [CrossRef] [PubMed]

54. Radak, Z.; Chung, H.Y.; Goto, S. Exercise and hormesis: Oxidative stress-related adaptation for successful aging. *Biogerontology* **2005**, *6*, 71–75. [CrossRef] [PubMed]

55. Byrnes, W.C.; Clarkson, P.M.; White, J.S.; Hsieh, S.S.; Frykman, P.N.; Maughan, R.J. Delayed onset muscle soreness following repeated bouts of downhill running. *J. Appl. Physiol.* **1985**, *59*, 710–715. [CrossRef]

56. Radak, Z.; Taylor, A.W.; Ohno, H.; Goto, S. Adaptation to exercise-induced oxidative stress: From muscle to brain. *Exerc. Immunol. Rev.* **2001**, *7*, 90–107.

57. Margaritelis, N.V.; Paschalis, V.; Theodorou, A.A.; Kyparos, A.; Nikolaidis, M.G. Redox basis of exercise physiology. *Redox Biol.* **2020**, *35*, 101499. [CrossRef]

58. Radak, Z.; Ishihara, K.; Tekus, E.; Varga, C.; Posa, A.; Balogh, L.; Boldogh, I.; Koltai, E. Exercise, oxidants, and antioxidants change the shape of the bell-shaped hormesis curve. *Redox Biol.* **2017**, *12*, 285–290. [CrossRef]

59. Lovlin, R.; Cottle, W.; Pyke, I.; Kavanagh, M.; Belcastro, A.N. Are indices of free radical damage related to exercise intensity. *Eur. J. Appl. Physiol. Occup. Physiol.* **1987**, *56*, 313–316. [CrossRef]

60. Corbett, N.; Alda, M. On telomeres long and short. *J. Psychiatry Neurosci.* **2015**, *40*, 3–4. [CrossRef]

61. Sousa, C.V.; Aguiar, S.S.; Santos, P.A.; Barbosa, L.P.; Knechtle, B.; Nikolaidis, P.T.; Deus, L.A.; Sales, M.M.; Rosa, E.C.C.C.; Rosa, T.S.; et al. Telomere length and redox balance in master endurance runners: The role of nitric oxide. *Exp. Gerontol.* **2019**, *117*, 113–118. [CrossRef] [PubMed]

62. Bogdanis, G.C.; Stavrinou, P.; Fatouros, I.G.; Philippou, A.; Chatzinikolaou, A.; Draganidis, D.; Ermidis, G.; Maridaki, M. Short-term high-intensity interval exercise training attenuates oxidative stress responses and improves antioxidant status in healthy humans. *Food Chem. Toxicol. Int. J. Publ. Br. Ind. Biol. Res. Assoc.* **2013**, *61*, 171–177. [CrossRef] [PubMed]

63. Henriquez-Olguin, C.; Renani, L.B.; Arab-Ceschia, L.; Raun, S.H.; Bhatia, A.; Li, Z.; Knudsen, J.R.; Holmdahl, R.; Jensen, T.E. Adaptations to high-intensity interval training in skeletal muscle require NADPH oxidase 2. *Redox Biol.* **2019**, *24*, 101188. [CrossRef] [PubMed]

64. Chance, B.; Sies, H.; Boveris, A. Hydroperoxide metabolism in mammalian organs. *Physiol. Rev.* **1979**, *59*, 527–605. [CrossRef]

65. Gomez-Cabrera, M.C.; Salvador-Pascual, A.; Cabo, H.; Ferrando, B.; Vina, J. Redox modulation of mitochondriogenesis in exercise. Does antioxidant supplementation blunt the benefits of exercise training? *Free Radic. Biol. Med.* **2015**, *86*, 37–46. [CrossRef]

66. Henriquez-Olguin, C.; Meneses-Valdes, R.; Jensen, T.E. Compartmentalized muscle redox signals controlling exercise metabolism— Current state, future challenges. *Redox Biol.* **2020**, *35*, 101473. [CrossRef]

67. Ito, N.; Ruegg, U.T.; Kudo, A.; Miyagoe-Suzuki, Y.; Takeda, S. Activation of calcium signaling through Trpv1 by nNOS and peroxynitrite as a key trigger of skeletal muscle hypertrophy. *Nat. Med.* **2013**, *19*, 101–106. [CrossRef]

68. Ito, N.; Ruegg, U.T.; Kudo, A.; Miyagoe-Suzuki, Y.; Takeda, S. Capsaicin mimics mechanical load-induced intracellular signaling events: Involvement of TRPV1-mediated calcium signaling in induction of skeletal muscle hypertrophy. *Channels* **2013**, *7*, 221–224. [CrossRef]

69. Ito, N.; Ruegg, U.T.; Takeda, S. ATP-Induced Increase in Intracellular Calcium Levels and Subsequent Activation of mTOR as Regulators of Skeletal Muscle Hypertrophy. *Int. J. Mol. Sci.* **2018**, *19*, 2804. [CrossRef]

70. McClean, C.M.; McNeilly, A.M.; Trinick, T.R.; Murphy, M.H.; Duly, E.; McLaughlin, J.; McEneny, J.; Burke, G.; Davison, G.W. Acute exercise and impaired glucose tolerance in obese humans. *J. Clin. Lipidol.* **2009**, *3*, 262–268. [CrossRef]

71. McConell, G.K.; Rattigan, S.; Lee-Young, R.S.; Wadley, G.D.; Merry, T.L. Skeletal muscle nitric oxide signaling and exercise: A focus on glucose metabolism. *Am. J. Physiol. Endocrinol. Metab.* **2012**, *303*, E301–E307. [CrossRef] [PubMed]

72. Ward, C.W.; Prosser, B.L.; Lederer, W.J. Mechanical stretch-induced activation of ROS/RNS signaling in striated muscle. *Antioxid. Redox Signal.* **2014**, *20*, 929–936. [CrossRef] [PubMed]

73. Ferreira, L.F.; Laitano, O. Regulation of NADPH oxidases in skeletal muscle. *Free Radic. Biol. Med.* **2016**, *98*, 18–28. [CrossRef] [PubMed]

74. Valko, M.; Leibfritz, D.; Moncol, J.; Cronin, M.T.; Mazur, M.; Telser, J. Free radicals and antioxidants in normal physiological functions and human disease. *Int. J. Biochem. Cell Biol.* **2007**, *39*, 44–84. [CrossRef]

75. Nordberg, J.; Arner, E.S. Reactive oxygen species, antioxidants, and the mammalian thioredoxin system. *Free Radic. Biol. Med.* **2001**, *31*, 1287–1312. [CrossRef]

76. Balsera, M.; Buchanan, B.B. Evolution of the thioredoxin system as a step enabling adaptation to oxidative stress. *Free Radic. Biol. Med.* **2019**, *140*, 28–35. [CrossRef]

77. Perkins, A.; Nelson, K.J.; Parsonage, D.; Poole, L.B.; Karplus, P.A. Peroxiredoxins: Guardians against oxidative stress and modulators of peroxide signaling. *Trends Biochem. Sci.* **2015**, *40*, 435–445. [CrossRef]

78. Cobley, J.N.; Husi, H. Immunological Techniques to Assess Protein Thiol Redox State: Opportunities, Challenges and Solutions. *Antioxidants* **2020**, *9*, 315. [CrossRef]

79. Wadley, A.J.; Aldred, S.; Coles, S.J. An unexplored role for Peroxiredoxin in exercise-induced redox signalling? *Redox Biol.* **2016**, *8*, 51–58. [CrossRef]

80. Fanjul-Moles, M.L.; Lopez-Riquelme, G.O. Relationship between Oxidative Stress, Circadian Rhythms, and AMD. *Oxidative Med. Cell. Longev.* **2016**, *2016*, 7420637. [CrossRef]

81. Yang, Z.; Kim, H.; Ali, A.; Zheng, Z.; Zhang, K. Interaction between stress responses and circadian metabolism in metabolic disease. *Liver Res.* **2017**, *1*, 156–162. [CrossRef] [PubMed]

82. Timmons, G.A.; O'Siorain, J.R.; Kennedy, O.D.; Curtis, A.M.; Early, J.O. Innate Rhythms: Clocks at the Center of Monocyte and Macrophage Function. *Front. Immunol.* **2020**, *11*, 1743. [CrossRef] [PubMed]

83. Early, J.O.; Menon, D.; Wyse, C.A.; Cervantes-Silva, M.P.; Zaslona, Z.; Carroll, R.G.; Palsson-McDermott, E.M.; Angiari, S.; Ryan, D.G.; Corcoran, S.E.; et al. Circadian clock protein BMAL1 regulates IL-1beta in macrophages via NRF2. *Proc. Natl. Acad. Sci. USA* **2018**, *115*, E8460–E8468. [CrossRef] [PubMed]

84. Rutter, J.; Reick, M.; Wu, L.C.; McKnight, S.L. Regulation of clock and NPAS2 DNA binding by the redox state of NAD cofactors. *Science* **2001**, *293*, 510–514. [CrossRef] [PubMed]

85. Wible, R.S.; Ramanathan, C.; Sutter, C.H.; Olesen, K.M.; Kensler, T.W.; Liu, A.C.; Sutter, T.R. NRF2 regulates core and stabilizing circadian clock loops, coupling redox and timekeeping in Mus musculus. *Elife* **2018**, *7*, e31656. [CrossRef]

86. O'Neill, J.S.; Reddy, A.B. Circadian clocks in human red blood cells. *Nature* **2011**, *469*, 498–503. [CrossRef]
87. Sato, T.; Greco, C.M. Expanding the link between circadian rhythms and redox metabolism of epigenetic control. *Free Radic. Biol. Med.* **2021**, *170*, 50–58. [CrossRef]
88. Krishnan, N.; Kretzschmar, D.; Rakshit, K.; Chow, E.; Giebultowicz, J.M. The circadian clock gene period extends healthspan in aging Drosophila melanogaster. *Aging* **2009**, *1*, 937–948. [CrossRef]
89. Patel, S.A.; Velingkaar, N.S.; Kondratov, R.V. Transcriptional control of antioxidant defense by the circadian clock. *Antioxid. Redox Signal.* **2014**, *20*, 2997–3006. [CrossRef]
90. Ashok Kumar, P.V.; Dakup, P.P.; Sarkar, S.; Modasia, J.B.; Motzner, M.S.; Gaddameedhi, S. It's About Time: Advances in Understanding the Circadian Regulation of DNA Damage and Repair in Carcinogenesis and Cancer Treatment Outcomes. *Yale J. Biol. Med.* **2019**, *92*, 305–316.
91. Lapenna, D.; De Gioia, S.; Mezzetti, A.; Porreca, E.; Ciofani, G.; Marzio, L.; Capani, F.; Di Ilio, C.; Cuccurullo, F. Circadian variations in antioxidant defences and lipid peroxidation in the rat heart. *Free Radic. Res. Commun.* **1992**, *17*, 187–194. [CrossRef]
92. Tomas-Zapico, C.; Coto-Montes, A.; Martinez-Fraga, J.; Rodriguez-Colunga, M.J.; Tolivia, D. Effects of continuous light exposure on antioxidant enzymes, porphyric enzymes and cellular damage in the Harderian gland of the Syrian hamster. *J. Pineal Res.* **2003**, *34*, 60–68. [CrossRef]
93. Diaz-Munoz, M.; Hernandez-Munoz, R.; Suarez, J.; Chagoya de Sanchez, V. Day-night cycle of lipid peroxidation in rat cerebral cortex and their relationship to the glutathione cycle and superoxide dismutase activity. *Neuroscience* **1985**, *16*, 859–863. [CrossRef]
94. Kondratov, R.V.; Vykhovanets, O.; Kondratova, A.A.; Antoch, M.P. Antioxidant N-acetyl-L-cysteine ameliorates symptoms of premature aging associated with the deficiency of the circadian protein BMAL1. *Aging* **2009**, *1*, 979–987. [CrossRef] [PubMed]
95. GBD 2019 Diseases and Injuries Collaborators. Global burden of 369 diseases and injuries in 204 countries and territories, 1990–2019: A systematic analysis for the Global Burden of Disease Study 2019. *Lancet* **2020**, *396*, 1204–1222. [CrossRef]
96. Willich, S.N.; Goldberg, R.J.; Maclure, M.; Perriello, L.; Muller, J.E. Increased onset of sudden cardiac death in the first three hours after awakening. *Am. J. Cardiol.* **1992**, *70*, 65–68. [CrossRef]
97. Takeda, N.; Maemura, K. Circadian clock and the onset of cardiovascular events. *Hypertens. Res. Off. J. Jpn. Soc. Hypertens.* **2016**, *39*, 383–390. [CrossRef] [PubMed]
98. Manfredini, R.; Fabbian, F.; Cappadona, R.; De Giorgi, A.; Bravi, F.; Carradori, T.; Flacco, M.E.; Manzoli, L. Daylight Saving Time and Acute Myocardial Infarction: A Meta-Analysis. *J. Clin. Med.* **2019**, *8*, 404. [CrossRef] [PubMed]
99. Gewaltig, M.T.; Kojda, G. Vasoprotection by nitric oxide: Mechanisms and therapeutic potential. *Cardiovasc. Res.* **2002**, *55*, 250–260. [CrossRef]
100. Pacher, P.; Beckman, J.S.; Liaudet, L. Nitric oxide and peroxynitrite in health and disease. *Physiol. Rev.* **2007**, *87*, 315–424. [CrossRef]
101. Man, A.W.C.; Li, H.; Xia, N. Circadian Rhythm: Potential Therapeutic Target for Atherosclerosis and Thrombosis. *Int. J. Mol. Sci.* **2021**, *22*, 676. [CrossRef] [PubMed]
102. Thosar, S.S.; Berman, A.M.; Herzig, M.X.; McHill, A.W.; Bowles, N.P.; Swanson, C.M.; Clemons, N.A.; Butler, M.P.; Clemons, A.A.; Emens, J.S.; et al. Circadian Rhythm of Vascular Function in Midlife Adults. *Arterioscler. Thromb. Vasc. Biol.* **2019**, *39*, 1203–1211. [CrossRef]
103. Teixeira, K.R.C.; Dos Santos, C.P.; de Medeiros, L.A.; Mendes, J.A.; Cunha, T.M.; De Angelis, K.; Penha-Silva, N.; de Oliveira, E.P.; Crispim, C.A. Night workers have lower levels of antioxidant defenses and higher levels of oxidative stress damage when compared to day workers. *Sci. Rep.* **2019**, *9*, 1–11. [CrossRef] [PubMed]
104. Tynes, T.; Hannevik, M.; Andersen, A.; Vistnes, A.I.; Haldorsen, T. Incidence of breast cancer in Norwegian female radio and telegraph operators. *Cancer Causes Control* **1996**, *7*, 197–204. [CrossRef]
105. Schernhammer, E.S.; Laden, F.; Speizer, F.E.; Willett, W.C.; Hunter, D.J.; Kawachi, I.; Fuchs, C.S.; Colditz, G.A. Night-shift work and risk of colorectal cancer in the nurses' health study. *J. Natl. Cancer Inst.* **2003**, *95*, 825–828. [CrossRef] [PubMed]
106. Bhatti, P.; Mirick, D.K.; Randolph, T.W.; Gong, J.; Buchanan, D.T.; Zhang, J.J.; Davis, S. Oxidative DNA damage during sleep periods among nightshift workers. *Occup. Environ. Med.* **2016**, *73*, 537–544. [CrossRef]
107. Koritala, B.S.C.; Porter, K.I.; Arshad, O.A.; Gajula, R.P.; Mitchell, H.D.; Arman, T.; Manjanatha, M.G.; Teeguarden, J.; Van Dongen, H.P.A.; McDermott, J.E.; et al. Night shift schedule causes circadian dysregulation of DNA repair genes and elevated DNA damage in humans. *J. Pineal Res.* **2021**, *70*, e12726. [CrossRef]
108. Bhatti, P.; Mirick, D.K.; Randolph, T.W.; Gong, J.; Buchanan, D.T.; Zhang, J.J.; Davis, S. Oxidative DNA damage during night shift work. *Occup. Environ. Med.* **2017**, *74*, 680–683. [CrossRef]
109. Hudec, M.; Dankova, P.; Solc, R.; Bettazova, N.; Cerna, M. Epigenetic Regulation of Circadian Rhythm and Its Possible Role in Diabetes Mellitus. *Int. J. Mol. Sci.* **2020**, *21*, 3005. [CrossRef]
110. Pan, A.; Schernhammer, E.S.; Sun, Q.; Hu, F.B. Rotating night shift work and risk of type 2 diabetes: Two prospective cohort studies in women. *PLoS Med.* **2011**, *8*, e1001141. [CrossRef]
111. Lee, J.; Ma, K.; Moulik, M.; Yechoor, V. Untimely oxidative stress in beta-cells leads to diabetes—Role of circadian clock in beta-cell function. *Free Radic. Biol. Med.* **2018**, *119*, 69–74. [CrossRef] [PubMed]
112. Bell, G.I.; Polonsky, K.S. Diabetes mellitus and genetically programmed defects in beta-cell function. *Nature* **2001**, *414*, 788–791. [CrossRef] [PubMed]

113. Meier, J.J.; Bonadonna, R.C. Role of reduced beta-cell mass versus impaired beta-cell function in the pathogenesis of type 2 diabetes. *Diabetes Care* **2013**, *36* (Suppl. 2), S113–S119. [CrossRef]
114. Burrack, A.L.; Martinov, T.; Fife, B.T. T Cell-Mediated Beta Cell Destruction: Autoimmunity and Alloimmunity in the Context of Type 1 Diabetes. *Front. Endocrinol.* **2017**, *8*, 343. [CrossRef] [PubMed]
115. Eguchi, N.; Vaziri, N.D.; Dafoe, D.C.; Ichii, H. The Role of Oxidative Stress in Pancreatic beta Cell Dysfunction in Diabetes. *Int. J. Mol. Sci.* **2021**, *22*, 1509. [CrossRef] [PubMed]
116. Lee, J.; Moulik, M.; Fang, Z.; Saha, P.; Zou, F.; Xu, Y.; Nelson, D.L.; Ma, K.; Moore, D.D.; Yechoor, V.K. Bmal1 and beta-cell clock are required for adaptation to circadian disruption, and their loss of function leads to oxidative stress-induced beta-cell failure in mice. *Mol. Cell. Biol.* **2013**, *33*, 2327–2338. [CrossRef]
117. Patti, M.E.; Corvera, S. The role of mitochondria in the pathogenesis of type 2 diabetes. *Endocr. Rev.* **2010**, *31*, 364–395. [CrossRef]
118. Man, A.W.C.; Xia, N.; Li, H. Circadian Rhythm in Adipose Tissue: Novel Antioxidant Target for Metabolic and Cardiovascular Diseases. *Antioxidants* **2020**, *9*, 968. [CrossRef]
119. Furukawa, S.; Fujita, T.; Shimabukuro, M.; Iwaki, M.; Yamada, Y.; Nakajima, Y.; Nakayama, O.; Makishima, M.; Matsuda, M.; Shimomura, I. Increased oxidative stress in obesity and its impact on metabolic syndrome. *J. Clin. Investig.* **2004**, *114*, 1752–1761. [CrossRef]
120. Froy, O.; Garaulet, M. The Circadian Clock in White and Brown Adipose Tissue: Mechanistic, Endocrine, and Clinical Aspects. *Endocr. Rev.* **2018**, *39*, 261–273. [CrossRef]
121. Wang, F.; Zhang, L.; Wu, S.; Li, W.; Sun, M.; Feng, W.; Ding, D.; Yeung-Shan Wong, S.; Zhu, P.; Evans, G.J.; et al. Night shift work and abnormal liver function: Is non-alcohol fatty liver a necessary mediator? *Occup. Environ. Med.* **2019**, *76*, 83–89. [CrossRef] [PubMed]
122. Arrigo, T.; Leonardi, S.; Cuppari, C.; Manti, S.; Lanzafame, A.; D'Angelo, G.; Gitto, E.; Marseglia, L.; Salpietro, C. Role of the diet as a link between oxidative stress and liver diseases. *World J. Gastroenterol.* **2015**, *21*, 384–395. [CrossRef] [PubMed]
123. Sato, K.; Meng, F.; Francis, H.; Wu, N.; Chen, L.; Kennedy, L.; Zhou, T.; Franchitto, A.; Onori, P.; Gaudio, E.; et al. Melatonin and circadian rhythms in liver diseases: Functional roles and potential therapies. *J. Pineal Res.* **2020**, *68*, e12639. [CrossRef] [PubMed]
124. Stangherlin, A.; Reddy, A.B. Regulation of circadian clocks by redox homeostasis. *J. Biol. Chem.* **2013**, *288*, 26505–26511. [CrossRef] [PubMed]
125. Haskell, W.L.; Lee, I.M.; Pate, R.R.; Powell, K.E.; Blair, S.N.; Franklin, B.A.; Macera, C.A.; Heath, G.W.; Thompson, P.D.; Bauman, A. Physical activity and public health: Updated recommendation for adults from the American College of Sports Medicine and the American Heart Association. *Circulation* **2007**, *116*, 1081–1093. [CrossRef] [PubMed]
126. Academy of Medical Royal Colleges. Exercise: The Miracle Cure and the Role of the Doctor in Promoting It. 2015. Available online: https://www.aomrc.org.uk/reports-guidance/exercise-the-miracle-cure-0215/ (accessed on 26 November 2021).
127. Zilberter, T.; Paoli, A. Editorial: Metabolic Shifting: Nutrition, Exercise, and Timing. *Front. Nutr.* **2020**, *7*, 592863. [CrossRef]
128. Finaud, J.; Lac, G.; Filaire, E. Oxidative stress: Relationship with exercise and training. *Sports Med.* **2006**, *36*, 327–358. [CrossRef]
129. Warburton, D.E.; Nicol, C.W.; Bredin, S.S. Prescribing exercise as preventive therapy. *CMAJ Can. Med. Assoc. J.* **2006**, *174*, 961–974. [CrossRef]
130. Hawley, J.A.; Hargreaves, M.; Joyner, M.J.; Zierath, J.R. Integrative biology of exercise. *Cell* **2014**, *159*, 738–749. [CrossRef]
131. Song, Y.; Choi, G.; Laeguen, J.; Kim, S.-W.; Jung, K.-H.; Park, H. Circadian rhythm gene expression and daily melatonin levels vary in athletes and sedentary males. *Biol. Rhythm Res.* **2018**, *49*, 237–245. [CrossRef]
132. Harfmann, B.D.; Schroder, E.A.; Kachman, M.T.; Hodge, B.A.; Zhang, X.; Esser, K.A. Muscle-specific loss of Bmal1 leads to disrupted tissue glucose metabolism and systemic glucose homeostasis. *Skelet. Muscle* **2016**, *6*, 12. [CrossRef]
133. Erickson, M.L.; Zhang, H.; Mey, J.T.; Kirwan, J.P. Exercise Training Impacts Skeletal Muscle Clock Machinery in Prediabetes. *Med. Sci. Sports Exerc.* **2020**, *52*, 2078–2085. [CrossRef] [PubMed]
134. Van Moorsel, D.; Hansen, J.; Havekes, B.; Scheer, F.A.J.L.; Jorgensen, J.A.; Hoeks, J.; Schrauwen-Hinderling, V.B.; Duez, H.; Lefebvre, P.; Schaper, N.C.; et al. Demonstration of a day-night rhythm in human skeletal muscle oxidative capacity. *Mol. Metab.* **2016**, *5*, 635–645. [CrossRef] [PubMed]
135. Basti, A.; Yalcin, M.; Herms, D.; Hesse, J.; Aboumanify, O.; Li, Y.; Aretz, Z.; Garmshausen, J.; El-Athman, R.; Hastermann, M.; et al. Diurnal variations in the expression of core-clock genes correlate with resting muscle properties and predict fluctuations in exercise performance across the day. *BMJ Open Sport Exerc. Med.* **2021**, *7*, e000876. [CrossRef] [PubMed]
136. Waterhouse, J.; Drust, B.; Weinert, D.; Edwards, B.; Gregson, W.; Atkinson, G.; Kao, S.; Aizawa, S.; Reilly, T. The circadian rhythm of core temperature: Origin and some implications for exercise performance. *Chronobiol. Int.* **2005**, *22*, 207–225. [CrossRef] [PubMed]
137. Atkinson, G.; Reilly, T. Circadian variation in sports performance. *Sports Med.* **1996**, *21*, 292–312. [CrossRef]
138. Savikj, M.; Gabriel, B.M.; Alm, P.S.; Smith, J.; Caidahl, K.; Bjornholm, M.; Fritz, T.; Krook, A.; Zierath, J.R.; Wallberg-Henriksson, H. Afternoon exercise is more efficacious than morning exercise at improving blood glucose levels in individuals with type 2 diabetes: A randomised crossover trial. *Diabetologia* **2019**, *62*, 233–237. [CrossRef]
139. Teo, S.Y.M.; Kanaley, J.A.; Guelfi, K.J.; Marston, K.J.; Fairchild, T.J. The Effect of Exercise Timing on Glycemic Control: A Randomized Clinical Trial. *Med. Sci. Sports Exerc.* **2020**, *52*, 323–334. [CrossRef]
140. Morris, J.N.; Heady, J.A.; Raffle, P.A.; Roberts, C.G.; Parks, J.W. Coronary heart-disease and physical activity of work. *Lancet* **1953**, *265*, 1053–1057. [CrossRef]

141. Morris, J.N.; Heady, J.A.; Raffle, P.A.; Roberts, C.G.; Parks, J.W. Coronary heart-disease and physical activity of work. *Lancet* **1953**, *265*, 1111–1120. [CrossRef]

142. Thompson, P.D.; Buchner, D.; Pina, I.L.; Balady, G.J.; Williams, M.A.; Marcus, B.H.; Berra, K.; Blair, S.N.; Costa, F.; Franklin, B.; et al. Exercise and physical activity in the prevention and treatment of atherosclerotic cardiovascular disease: A statement from the Council on Clinical Cardiology (Subcommittee on Exercise, Rehabilitation, and Prevention) and the Council on Nutrition, Physical Activity, and Metabolism (Subcommittee on Physical Activity). *Circulation* **2003**, *107*, 3109–3116. [PubMed]

143. Safar, M.E.; Czernichow, S.; Blacher, J. Obesity, arterial stiffness, and cardiovascular risk. *J. Am. Soc. Nephrol.* **2006**, *17*, S109–S111. [CrossRef] [PubMed]

144. Van Gaal, L.F.; Mertens, I.L.; De Block, C.E. Mechanisms linking obesity with cardiovascular disease. *Nature* **2006**, *444*, 875–880. [CrossRef]

145. MacDonald, J.R.; Hogben, C.D.; Tarnopolsky, M.A.; MacDougall, J.D. Post exercise hypotension is sustained during subsequent bouts of mild exercise and simulated activities of daily living. *J. Hum. Hypertens.* **2001**, *15*, 567–571. [CrossRef] [PubMed]

146. Jones, H.; George, K.; Edwards, B.; Atkinson, G. Is the magnitude of acute post-exercise hypotension mediated by exercise intensity or total work done? *Eur. J. Appl. Physiol.* **2007**, *102*, 33–40. [CrossRef]

147. De Brito, L.C.; Ely, M.R.; Sieck, D.C.; Mangum, J.E.; Larson, E.A.; Minson, C.T.; Forjaz, C.L.M.; Halliwill, J.R. Effect of Time of Day on Sustained Postexercise Vasodilation Following Small Muscle-Mass Exercise in Humans. *Front. Physiol.* **2019**, *10*, 762. [CrossRef]

148. Roberts, C.K.; Vaziri, N.D.; Barnard, R.J. Effect of diet and exercise intervention on blood pressure, insulin, oxidative stress, and nitric oxide availability. *Circulation* **2002**, *106*, 2530–2532. [CrossRef]

149. Portaluppi, F.; Hermida, R.C. Circadian rhythms in cardiac arrhythmias and opportunities for their chronotherapy. *Adv. Drug Deliv. Rev.* **2007**, *59*, 940–951. [CrossRef]

150. Jones, H.; Pritchard, C.; George, K.; Edwards, B.; Atkinson, G. The acute post-exercise response of blood pressure varies with time of day. *Eur. J. Appl. Physiol.* **2008**, *104*, 481–489. [CrossRef]

151. De Brito, L.C.; Pecanha, T.; Fecchio, R.Y.; Rezende, R.A.; Sousa, P.; DASilva-Junior, N.; Abreu, A.; Silva, G.; Mion-Junior, D.; Halliwill, J.R.; et al. Morning versus Evening Aerobic Training Effects on Blood Pressure in Treated Hypertension. *Med. Sci. Sports Exerc.* **2019**, *51*, 653–662. [CrossRef]

152. De Brito, L.C.; Rezende, R.A.; da Silva Junior, N.D.; Tinucci, T.; Casarini, D.E.; Cipolla-Neto, J.; Forjaz, C.L. Post-Exercise Hypotension and Its Mechanisms Differ after Morning and Evening Exercise: A Randomized Crossover Study. *PLoS ONE* **2015**, *10*, e0132458. [CrossRef] [PubMed]

153. Walther, C.; Gielen, S.; Hambrecht, R. The effect of exercise training on endothelial function in cardiovascular disease in humans. *Exerc. Sport Sci. Rev.* **2004**, *32*, 129–134. [CrossRef] [PubMed]

154. Green, D.J. Exercise training as vascular medicine: Direct impacts on the vasculature in humans. *Exerc. Sport Sci. Rev.* **2009**, *37*, 196–202. [CrossRef] [PubMed]

155. Thijssen, D.H.; Black, M.A.; Pyke, K.E.; Padilla, J.; Atkinson, G.; Harris, R.A.; Parker, B.; Widlansky, M.E.; Tschakovsky, M.E.; Green, D.J. Assessment of flow-mediated dilation in humans: A methodological and physiological guideline. *Am. J. Physiol. Heart Circ. Physiol.* **2011**, *300*, H2–H12. [CrossRef] [PubMed]

156. Jones, H.; Green, D.J.; George, K.; Atkinson, G. Intermittent exercise abolishes the diurnal variation in endothelial-dependent flow-mediated dilation in humans. *Am. J. Physiol. Regul. Integr. Comp. Physiol.* **2010**, *298*, R427–R432. [CrossRef]

157. Ballard, K.D.; Timsina, R.; Timmerman, K.L. Influence of time of day and intermittent aerobic exercise on vascular endothelial function and plasma endothelin-1 in healthy adults. *Chronobiol. Int.* **2021**, *38*, 1064–1071. [CrossRef]

158. Hood, D.A. Invited Review: Contractile activity-induced mitochondrial biogenesis in skeletal muscle. *J. Appl. Physiol.* **2001**, *90*, 1137–1157. [CrossRef]

159. Andrews, J.L.; Zhang, X.; McCarthy, J.J.; McDearmon, E.L.; Hornberger, T.A.; Russell, B.; Campbell, K.S.; Arbogast, S.; Reid, M.B.; Walker, J.R.; et al. CLOCK and BMAL1 regulate MyoD and are necessary for maintenance of skeletal muscle phenotype and function. *Proc. Natl. Acad. Sci. USA* **2010**, *107*, 19090–19095. [CrossRef]

160. Wu, R.; Dang, F.; Li, P.; Wang, P.; Xu, Q.; Liu, Z.; Li, Y.; Wu, Y.; Chen, Y.; Liu, Y. The Circadian Protein Period2 Suppresses mTORC1 Activity via Recruiting Tsc1 to mTORC1 Complex. *Cell Metab.* **2019**, *29*, 653–667.e6. [CrossRef]

161. Reid, K.J. Assessment of Circadian Rhythms. *Neurol. Clin.* **2019**, *37*, 505–526. [CrossRef]

162. Huang, C.J.; McAllister, M.J.; Slusher, A.L.; Webb, H.E.; Mock, J.T.; Acevedo, E.O. Obesity-Related Oxidative Stress: The Impact of Physical Activity and Diet Manipulation. *Sports Med. Open* **2015**, *1*, 32. [CrossRef] [PubMed]

MDPI

St. Alban-Anlage 66

4052 Basel

Switzerland

Tel. +41 61 683 77 34

Fax +41 61 302 89 18

www.mdpi.com

Antioxidants Editorial Office

E-mail: antioxidants@mdpi.com

www.mdpi.com/journal/antioxidants